D0093502

AN EPIDEMIC OF ABSENCE

A New Way of Understanding Allergies
and Autoimmune Diseases

MOISES VELASQUEZ-MANOFF

SCRIBNER

New York   London   Toronto   Sydney   New Delhi

SCRIBNER
A Division of Simon & Schuster, Inc.
1230 Avenue of the Americas
New York, NY 10020

Copyright © 2012 by Moises Velasquez-Manoff

All rights reserved, including the right to reproduce this book or portions thereof
in any form whatsoever. For information address Scribner Subsidiary Rights Department,
1230 Avenue of the Americas, New York, NY 10020.

First Scribner hardcover edition September 2012

SCRIBNER and design are registered trademarks of The Gale Group, Inc.,
used under license by Simon & Schuster, Inc., the publisher of this work.

For information about special discounts for bulk purchases,
please contact Simon & Schuster Special Sales at 1-866-506-1949
or business@simonandschuster.com.

The Simon & Schuster Speakers Bureau can bring authors to your live event.
For more information or to book an event contact the Simon & Schuster Speakers Bureau
at 1-866-248-3049 or visit our website at www.simonspeakers.com.

Manufactured in the United States of America

1   3   5   7   9   10   8   6   4   2

ISBN 978-1-4391-9938-1
ISBN 978-1-4391-9940-4 (ebook)

The graph on page 7 is reprinted courtesy of the *New England Journal of Medicine*
and appeared in "The Effect of Infections on Susceptibility to Autoimmune
and Allergic Diseases," by J.-F. Bach in the 347(12) issue.

For my mother,
Carmen Socorro Velasquez

Contents

An Epidemic
of Absence

CHAPTER 1

Meet Your Parasites

Mother, it is no gain, thy bondage of finery, if it keep one
shut off from the healthful dust of the earth, if it rob one of
the right of entrance to the great fair of common human life.
—Rabindranath Tagore,
Bengali poet and Nobel laureate

One chilly November morning, I head south from San Diego in a bottom-tier rental car. The standard journalistic paraphernalia—a digital recorder, camera, notepad, and pencils—accompany me in the passenger seat, as well as directions to my meeting point: the last exit before Mexico. I also have a printout of my recent blood work, proof that I'm not anemic, not infected with hepatitis or HIV—that I'm healthy enough for the coming experiment.

As I drive, the radio announcer conducts a gruesome tally of the most recent violence in Tijuana, where I'm headed: two bodies hung from a bridge, a third decapitated, a fourth shot. More than this terrible, ongoing brutality, however, parasites occupy my mind—worms that migrate through flesh, burst into lungs, crawl down throats, and latch on to tender insides. Any traveler might fret over acquiring such hangers-on while abroad. But I'm heading to Mexico precisely to obtain not just one, but a colony. Today in Tijuana I'll deliberately introduce the hookworm *Necator americanus*—the American murderer—into my body.

And for this dubious honor, I'll pay handsomely—a onetime fee of $2,300. If I receive twenty of the microscopic larvae, that's $115 apiece for a parasite that, in the early decades of the twentieth century, was considered a scourge on the American south. Some worried—without condescension, I should add—that hookworm was making southerners dim-witted and lazy, that it was socially and economically retarding half the country. And photos of poor, worm-ridden country folk from the time—followed by their robust health after deworming—clearly show the dire costs of necatoriasis, or hookworm disease: jutting collarbones, dull eyes, and listless expressions on wan faces. They appear as if consumed from the inside.

Hookworm has mostly disappeared from the U.S., the result of protracted eradication efforts in the early twentieth century. But in the usually poor, tropical countries where it's still endemic, it can cause anemia, stunt growth, halt menstruation, and even retard mental development in growing children. Between 576 million and 740 million people carry the parasite. And for all the aforementioned reasons, public-health types consider worm infections a "neglected tropical disease." Helminths, as they're called, are not as obviously fatal as malaria, say, but their constant drag on vitality is insidious. The parasites keep children from learning in school. They prevent parents from working. Some argue that they contribute to the self-reinforcing cycles of poor health and poverty that plague entire nations.

So why am I considering acquiring this terrible creature? Scientists have two minds about parasites these days. Some consider them evil incarnate, but others note that while the above-mentioned horrors are sometimes true, the majority of humans infected with parasites today—upward of 1.2 billion people, or somewhere between one-fifth and one-sixth of humanity—host worms with few apparent symptoms. This camp has begun to suspect that worms may, in fact, confer some benefits on their human hosts.

As early as the 1960s, by which time hookworm had been largely eradicated in the U.S., scientists puzzled over the lack of symptoms in some. "Well-nourished persons often harbor helminths without apparent damage," remarked one physician in 1969. "One may question the wisdom of treating such infections, especially with chemotherapeutic agents with toxic qualities."

Decades of plumbing the mechanisms that allow one creature to persist within another, a clear violation of the self-versus-nonself rules thought to govern immune functioning, has taught immunologists much not only about how wily worms really are, but also about how the human immune system actually works. Parasites like hookworm were ubiquitous during our evolution. Might our bodies anticipate their presence in some respects, require it even? And might some of the more curious ailments of modernity result partly from their absence?

That brings me to my motive: A large and growing body of science indicates that parasites may prevent allergic and autoimmune diseases. And I've got both.

When I was eleven, my hair began falling out. My grandmother first noticed it. I was visiting my grandparents at their beach house that summer when, one afternoon, she called me over, examined the back of my head, and proclaimed that I had a nickel-sized bald spot. Then we all promptly forgot about it. With the sand, waves, and sun beckoning, it just didn't seem that important.

But by the time school started a few months later, the bald patch had grown. A dermatologist diagnosed alopecia areata, an autoimmune disorder. My immune system, normally tasked with protecting against invaders, had inexplicably mistaken friend for foe, and attacked my hair follicles. Scientists didn't know what, exactly, triggered alopecia, but stress was thought to play a role. And at first glance, that made sense. My parents were in the middle of a messy, drawn-out divorce. I was also beginning at a new junior high school that fall; I had, it seemed, much to worry about.

I also had other, better-known immune-mediated problems. I suffered from fairly severe asthma as a child, and food allergies to peanuts, sesame, and eggs. (Only the egg allergy eventually disappeared.) At least once yearly, usually during seasons of high pollen count, my wheezing became so severe that my lips and fingernails turned blue, and my parents had to rush me to the emergency room. There, doctors misted me with bronchodilators, or, during severe attacks, pumped me full of immune-suppressing steroids.

"Aha!" said the dermatologist when he learned of these other conditions. There was a correlation among allergies, asthma, and alopecia, he explained. No one was sure why or what it meant, but having an allergic disease like asthma increased one's chances of developing alopecia.

Years later, I would learn that the co-occurrence of these two disorders was likely evidence of a single, root malfunction. But at age eleven, I accepted on faith that where one problem arose, so, probably, would others. So what to do? Given my age and the relatively small size of the bald spot, the doctor recommended watching and waiting. Alopecia usually corrected itself in time, he said. So we waited.

In a month, another bald spot appeared, on the right side of my head. Then one on the left. Seemingly overnight, a large one opened up just above the middle of my forehead. As more hairless patches appeared, the pace at which new ones emerged accelerated. Every morning, my mother combed and gelled my hair into place to hide the growing expanse of denuded skin; but soon, concealing my bare scalp became nearly impossible. The spots began to converge. I was going bald.

We returned to the dermatologist. This time, he had a less upbeat assessment. The more the disease progressed, he noted, the less likely recovery. The odds worked like this: Only 1 to 2 percent of the population got alopecia areata at all, a bald spot or two that, after a time, usually filled in again. But for a significant minority, maybe 7 percent of those with alopecia areata, the hair loss became chronic. Some progressed to alopecia totalis, total loss of hair on the head. At that point, the chances of a full recovery diminished substantially. Whatever mistake the immune system had made, it became permanent. And of this totalis subset, some went on to develop alopecia universalis—loss of hair on the entire body. For them, recovery was nearly impossible.

None of this sounded good, especially as I was speeding toward totalis and—who knows?—universalis after that. Two treatment options existed, neither of which worked without fail: immune suppression or irritation. Steroids suppressed the immune response and, basically, called off the attack dogs, allowing hair to grow again. Immune stimulation, on the other hand, worked in slightly more mysterious ways. Inflammation induced by an irritant distracted the immune system from less pressing projects, such as attacking hair follicles. Irritation would earn my hair follicles a reprieve. As neither approach was a sure bet, the dermatologist recommended that I try both.

I did, and neither worked—although I developed an oozing blister where I applied the irritant. My alopecia advanced until, by age sixteen, not a single hair remained on my body. I had joined the elite ranks, somewhere around 0.1 percent of the population, of those with alopecia universalis. I put on a hat, which I'd wear more or less nonstop until my early twenties, and tried to get on with my adolescence.

Not until my thirties did I look into what scientists had discovered in the roughly twenty years since that first bald spot appeared on my head. I wasn't too hopeful; surely, I would have heard had a cure been developed. As I contemplated having children, I'd begun fretting about what lay hidden in my genes. The first genome-wide association study of alopecia, published in 2010, showed that the disorder, the most common autoimmune disease in the U.S., shared gene variants with several much worse autoimmune diseases, such as rheumatoid arthritis, type-1 diabetes, and celiac disease. Soon thereafter, my first child, a girl, arrived. Now the results of my investigation had concrete applications. If alopecia suggested a tendency toward immune malfunction, and if that tendency was modifiable, I wanted to know how to better play the cards. I wanted to ensure that my progeny remained free of both allergic and autoimmune disease.

I was right about one thing. Treatments for alopecia hadn't advanced much since my childhood. They still consisted mainly of irritants and immune suppressants, and as neither approach corrected the underlying malfunction, both would require indefinite use. Prolonged exposure raised a host of secondary concerns. Repeated steroid shots, for example, were not only exceptionally painful, they thinned and discolored the skin. Irritants induced swelling, redness, and skin flaking. One powerful immune suppressant called cyclosporine increased the risk of skin cancer. No thanks.

However, the patterns of immune-mediated disease in general caught my attention. The incidence of both autoimmune and allergic diseases had recently increased, and to the degree that scientific literature conveys feeling, in this case it evinced alarm. Scientists threw around the word *epidemic* to describe the rising prevalence of asthma especially, a descriptor usually reserved for infectious diseases, like the prayer-inducing, body-wasting, dead-in-a-day cholera epidemics that terrified the world during the nineteenth century. Generally speaking, however, there was no asthma bacterium, no autoimmune virus. No new plagues were driving this particular pandemic. Instead, we seemed newly vulnerable to immune dysfunction.

If I possessed glasses that afforded me the power to see otherwise non-apparent allergic and autoimmune diseases, I'd be struck by the sheer abundance of people with these problems. Walking down Broadway in New York City, for instance, one of every ten children passing by would have asthma; one in six would have an itchy rash and sometimes blisters—eczema. One of every five passersby would have hay fever. If I could see allergic antibodies directly—immunoglobulin-E—I'd note that half the crowd around me was sensitized to dust mites, tree pollen, and peanuts, among other basically harmless proteins. I'd see pockets full of inhalers, and bags stuffed with allergy medicines. In the satchels of the most severely afflicted, I'd see pills of powerful immune suppressants, such as prednisone. I'd even see a few soon-to-be corpses; about 3,500 people die yearly from asthma attacks.

Americans spend perhaps $10 billion yearly on asthma-related drugs and doctor visits. Direct and indirect costs of asthma combined reach about $56 billion. I'd see these funds flowing from allergic and asthmatic wallets to doctors and drug companies. And I'd observe money *not* flowing from days missed at work, diminished overall productivity, and opportunities lost over a lifetime.

If I took the same walk with glasses that allowed me to see autoimmune diseases, I'd note that one in twenty passersby had one of eighty of these often debilitating conditions. One of every 250 people—it would take about a minute standing in a place like Times Square for such a person to pass by—would suffer from debilitating pain in his or her intestines, what's called inflammatory bowel disease. I'd see scarring and constriction. And in the most severe cases, I'd observe removed lengths of intestine, colostomies (surgically created exits for intestinal contents), and colostomy bags (containers for the effluence) hidden under clothes.

Of every thousand passersby, I'd note one struggling to move legs or arms. These people have multiple sclerosis, a progressive autoimmune disease of the central nervous system. Their vision might blur when they read signs. Their legs might fail to cooperate when crossing the street. The worst cases, of course, wouldn't be out at all. They'd remain at home, perhaps in electric wheelchairs, maybe bedridden.

I'd note glucose monitors on one of every three hundred children frolicking in Central Park's playgrounds, children afflicted with autoimmune diabetes, which is usually childhood-onset. Their skin would bear needle marks from the daily insulin injections required to avoid coma and death.

If my glasses came with headphones, I'd hear a cacophony of worry and desperation: asthmatic teenagers wondering if they'll be able to join friends in a game of baseball; more severe cases focused on walking slowly, so as not to lose breath; eczematics reminding themselves ceaselessly not to scratch, or if they've already scratched, berating themselves for the raw mess left behind.

Those with inflammatory bowel disease might be preoccupied with the pain, sometimes dull, sometimes sharp, that has characterized life since diagnosis. If it's not racking cramps on their minds, they'll likely be strategizing around bowel movements, which arrive all too frequently and with a painful urgency, and which sometimes contain blood. Those with MS might be wondering: How much longer before I can't walk? And everyone will regularly ask: Why can't doctors fix this? Where did this come from? Why me?

The National Institutes of Health estimate that between 14.7 and 23.5 million Americans have an autoimmune disease, or 5 to 8 percent of the population. The American Autoimmune Related Diseases Association puts the number at more than double that—50 million Americans. In the U.S., autoimmune disease ranks among the top ten killers of women. And that speaks to an omission I made for simplicity's sake in the above scenario. Roughly three-quarters of those afflicted with autoimmune disease are female. When I had my autoimmune glasses on, in other words, I'd be seeing mostly women.

Anthony Fauci, director of the National Institutes of Allergy and Infectious Diseases, once estimated that the direct and indirect costs of autoimmune diseases reached a staggering $100 billion yearly. (By comparison, we spend $57 billion on cancer and $200 billion on cardiovascular disease.) That may seem high, but bear in mind that autoimmune diseases, which are chronic in nature, generally strike in the prime of life, and require decades of costly symptom management.

These statistics apply to the richest countries in the early twenty-first century. But immune-mediated diseases weren't always this prevalent. Early hints of immune dysfunction during the late nineteenth century notwithstanding, the allergy and asthma epidemics gained steam during the 1960s, accelerated through the 1980s, and then plateaued by the early 2000s. In that period, depending on the study and the population, you'll find somewhere between a doubling and a tripling of asthma and allergies in the developed world.

Some autoimmune diseases show even more dramatic increases during the late twentieth century. A 2009 study found that the prevalence of undiagnosed celiac disease, a type of inflammatory bowel disease incited by proteins in grains, had increased more than fourfold since the mid-twentieth century. The incidence of multiple sclerosis has nearly tripled. And for some of these diseases, there's no end in sight. The incidence of type-1 diabetes, which more than tripled during the late twentieth century, is estimated to double again by 2020.

What has happened? In 2002, the French scientist Jean-François Bach published a seminal paper for anyone asking that question. The study, which appeared in the *New England Journal of Medicine*, had two graphs side by side, one showing the gradual decline since 1950 of once-common infectious diseases—hepatitis A, measles, mumps, and tuberculosis—next to another showing, over the same period, an increase of autoimmune and allergic disease in the developed world. Nearly everyone contracted mumps and measles in 1950. By 1980, almost no one did. Vaccines had almost eliminated both viruses. In an even shorter period—since 1970— new cases of hepatitis A infection fell to one-fifth their former level. And all the while, new cases of asthma, multiple sclerosis, and Crohn's disease doubled, tripled, and quadrupled, respectively.

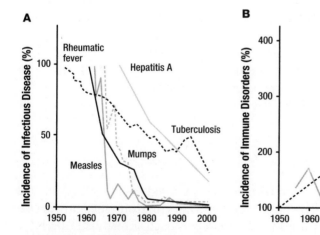

Source: Bach, *New England Journal of Medicine* (2002).

The relationship that Bach so clearly demonstrates, that as infections decline over time, immune dysfunction increases, is evident between contemporaneous regions and populations. The incidence of allergic disease varies by a factor of 20 between the most allergic countries and the least. Vanishingly few children in Albania, for example, have allergy, but one-quarter of Australian children do. The incidence of type-1 diabetes varies even more markedly—350-fold between the most afflicted country, Finland, and the least, China. Are some ethnicities more vulnerable to these disorders than others? Maybe. However, when migrants move from low-risk to high-risk countries, the children born to them in their adopted homelands almost invariably suffer from immune-mediated diseases at rates equal to, and sometimes higher than, the local population. So, if not genetics, what explains the great disparity?

Epidemiologists used to assert that, generally speaking, these disorders increased as you moved from the equator toward the poles. In sub-Saharan Africa they were quite rare. In the U.K., they were all too common. And that seemed irrefutably true even thirty years ago. But evidence of a recent surge of asthma in countries like Brazil and Peru—and urban centers in the developing world everywhere—has undermined this once safely made generalization. Nowadays, you're more likely to hear that allergic and autoimmune diseases correlate with gross domestic product. And for now, that's holding true. The richer the country you call home—or in some cases, the higher your social class within a country—the more likely you are to have asthma, inflammatory bowel disease, and multiple sclerosis.

Critics discount these sweeping statistics for their reliance on questionnaires. Surveys are inevitably vulnerable to recall and cultural biases, they point out. But smaller studies that use objective measures such as wheeze and skin-prick tests, or testing for autoimmune antibodies, have repeatedly revealed the same basic pattern: Immune-mediated disorders arise in direct proportion to affluence and Westernization. The more that one's surroundings resemble the environment in which we evolved—rife with infections and lots of what one scientist calls "animals, faeces and mud"—the lower the prevalence of these diseases.

BETWEEN THE STONE AGE
AND THE NEOLITHIC, NO ASTHMA

In preparing for my Mexico trip, I often pondered another I'd taken, to a place where asthma didn't exist: the Bolivian Amazon. The anthropologists Michael Gurven from the University of California, Santa Barbara, and Hillard Kaplan from the University of New Mexico, Albuquerque, study a horticulturalist people living on the western edge of the Amazon basin. They're called the Tsimane, and they subsist, for the most part, directly off the jungle. They hunt monkeys, tapirs, and other animals with bows and arrows. (They happily use rifles, which some possess; but because they don't regularly participate in a cash economy, they often lack shells.) They fish with weirs, poison plants, and special arrows. And although they have plenty of contact with twenty-first-century Bolivians, their lifestyle is as close to Stone Age living as one can reasonably expect to find these days. That's why Gurven and Kaplan are here.

I caught up with Gurven, smiling, scruffy, and wearing a Phillies cap, at his clinic on the outskirts of a bustling, dusty town in the Bolivian lowlands called San Borja. Horses grazed in a nearby soccer field. Handsome, sand-colored cows wandered about. The occasional sow trotted by.

Gurven belongs to a school of anthropology called human behavioral ecology. The tools come from biology; the novelty is their application in anthropology. To hear him tell it, behavioral ecology emerged in reaction not to the cultural anthropology of the early and mid-twentieth century—Margaret Mead and her study *Coming of Age in Samoa,* for example—but to the period of anxious self-examination that followed. Was the very notion of studying humans imperialistic and exploitative? Could an outsider truly understand "the other," or was she doomed to endlessly project herself on her study subjects?

Behavioral ecology, as applied to the study of people, as Gurven and his students explain to me around campfires during the coming nights, originates in a certain weariness, not necessarily with this self-questioning, justified as it may be, but with the retreat from even trying to comprehend those who inhabit different worlds. Yes, we inevitably project, but people who continue to live as we all once lived can teach us many things, and there are objective ways to measure these things. What's more, anyone interested in these lessons had better move fast. Whatever hunter-gatherers and horticulturalists remain in the world won't be at it for much longer.

Among the Tsimane, Gurven first studied human reciprocity and altruism, why people share in a world of limited resources. He asked questions like: How does a sick person get help in a world without health insurance? And why do people help the ailing when it costs them precious time and energy? He also explored how humans age under the more-or-less constant onslaught of infections. Even here, people live decades beyond their capacity to bear children. According to the most severe interpretations of Darwinian theory, that just shouldn't happen. But for *Homo sapiens,* it does. What are those extra decades for?

As part of his arrangement with the tribe, Gurven gives the Tsimane free medical care. He trucks them to his clinic from the remote villages along the tributaries of the Maniqui River. A doctor examines them. Technicians take stool, urine, and blood samples. In one darkened room, an ultrasound machine peers at their hearts and arteries. We'll revisit the specifics of Gurven's findings later, but, almost incidentally, he's found that the immune system of a horticulturalist living in the Amazon works differently than your average Londoner's or New Yorker's.

Over the past decade, Gurven's clinic has examined more than 12,000 people, almost the entire Tsimane population. In the 37,000 examinations conducted by his staff (they've seen many patients multiple times), no doctor has logged a single case of asthma. If rates approximated those in the U.S. and the U.K., you'd expect at least 1,000 asthmatics. As for autoimmune disease, he's seen fifteen cases—including eleven of vitiligo, a condition in which the immune system turns on pigment-producing cells in the skin, one of lupus, and one of rheumatoid arthritis. If autoimmune disease occurred with the same frequency here as in the developed world, he should have seen roughly six hundred cases. In Tsimanía, in other words, the prevalence of autoimmune disease is one-fortieth what it is in New York City.

What he does see are plenty of infections, which cause half of all deaths among the Tsimane. (Accidents and violence contribute an additional 14 percent.) And parasites are so universal as to be nearly unremarkable. There's lots of giardia and amoebiasis. A few have tuberculosis. Fewer still have a chronic flesh-eating parasite called leishmaniasis. And nearly everyone has hookworm.

He also sees plenty of the wear and tear that comes from an active life: prolapsed uteruses, the result of having many children (the average Tsimane woman has nine), and hernias from heavy lifting. But the diseases of civilization, including cancers of the breast, prostate, ovary, colon, and testicle, are absent. And so is cardiovascular disease.

Are the Tsimane special, genetically immune perhaps? Others studying unacculturated Amerindians in the Amazon have explicitly noted the same absence of allergic disorders, and the suite of diseases so common in modernity. Maybe Amerindians as a group are genetically invulnerable to these diseases. Perhaps, but not likely. Scientists have made similar observations among peoples in Europe, Africa, and Asia. The repeated observation is that people living in "dirtier" surroundings have less allergy and autoimmunity. The reverse holds true as well: Anyone seems able to develop asthma if exposed to the right conditions. And these conditions prevail in places like New York City, London, and Sydney.

WHAT DOES A PLACE
WITHOUT ASTHMA LOOK LIKE?

The day after I find Gurven, we drive an hour through cane fields and pasture to a red-hued river. We pile into a motorized dugout canoe, its sides shored up by planks. The month is August, the Southern Hemisphere winter, and it's chillier than one might anticipate for the jungle. A wind called *el surazo*—the southerly—blows off the vast pampas to the south. (Later I'll learn that this particular winter was so cold that fish and pink river dolphins washed up dead throughout Amazonia.)

After more than an hour of motoring past snowy white egrets, the same species that steps gingerly through the marshland of New York City's Jamaica Bay, we arrive at a Tsimane settlement called Chacal. "Gringolandia," Gurven says softly as several Coleman tents—Gurven's base camp—come into view. "The Tsimane don't live in tents."

There's no central village per se, just a freshly painted yellow schoolhouse next to a field where the men play soccer nightly. The Tsimane live scattered along the river, each family or group of families tending fields of rice, corn, and manioc. Some credit their decentralized way of life with helping them resist Spanish influence. The would-be colonizers found no central authority to usurp, no priests or kings to co-opt. And the Tsimane simply retreated deeper into the jungle before the Spanish advance, which began in the seventeenth century.

Soon enough, we're walking along a narrow path running parallel to the river. As a clearing becomes visible through the underbrush, a Tsimane guide with a boyish face and solemn demeanor named Arnulfo makes a soft hooting sound. Gurven takes up the call as well. High-pitched and elongated like the last syllable of an owl's hoot, the cry serves as a kind of jungle courtesy, notifying those up ahead that we're approaching.

As we pass into the clearing, Gurven and Arnulfo announce their greetings in Tsimane. A group of young boys plays with tops carved from tree nuts. Hammered-in nails serve as points. The children stare at the newcomers expressionless at first, their brows in furrows, but they've seen outsiders before, and they quickly resume their game, winding string around their tops, and then setting their toys spinning with practiced yanks. Two women seated on a large woven mat return the greetings. A little girl lies prone in the lap of one woman, who searches patiently through her hair, extracting lice and nits, and crushing them between her teeth. The men are all gone for the day, we learn, on a hunting trip. We say our goodbyes—it's not good form to visit the women without men present, Gurven explains later—and continue walking.

We see fields of corn, lots of dogs, canoes, exquisitely woven mats, waist-high mortar-and-pestles, and everywhere tools made from jungle materials. It's this mastery of the jungle that strikes me, a twenty-first-century New Yorker with a computer-addled, Internet-spoiled brain, as most impressive. The Tsimane carve slim dugout canoes from tree trunks, and push them through the rivers with long poles. Mats are woven from palm fronds, as are the roofs on their huts. Useful trees and plants surround their jungle homesteads—papaya, banana, and a *tutuma* tree that bears large gourdlike fruits that they then fashion into bowls. They use ginger root to treat insect bites. They sleep on elevated platforms. As Gurven explains, here, one's worth doesn't derive from one's possessions, but instead from one's skills at extracting resources from the jungle. "You could lose everything, yes, but then you just build a new house, get fish, go hunting. Lots of individuals have that ability," he says. "There's a kind of freedom in that."

I could go on about how extraordinary Tsimane adaptations are, but really, I'm here to observe what I can't see directly: the hidden microbial and parasitological landscape. I want to know what that place where the immune system doesn't malfunction looks like. And so how does it look? The answer is, alive.

To Gurven's chagrin, the Tsimane often draw drinking water directly from the muddy river. It's likely teeming with bacteria. Pigs, chickens, dogs, and the occasional pet spider monkey wander about freely. They each bring their unique blend of microbes. Tsimane women make an alcoholic drink by chewing and spitting boiled manioc and letting it ferment. In other words, they regularly imbibe what your average New York health food store touts as "live cultures." And of course, a majority has hookworms embedded in his or her gut.

In short, the Tsimane live in what scientists call "a living environment." Who cares? Much evidence suggests that surroundings like this protect against autoimmune and allergic disease, and for a simple reason: This is the type of environment the immune system has evolved to expect. And when it doesn't encounter the abundant stimulation contained herein, it falls into disarray.

Life here is not easy, of course. Infant mortality, which has improved since vaccinations arrived during the 1990s, remains high. One in five children dies before his or her fifth birthday. By age fifteen, an additional 5 percent have succumbed to disease. Essentially, one-quarter of all children born don't survive to adolescence, and that's an improvement over the early twentieth century. (On the other hand, two of every five Tsimane live to age sixty, one of Gurven's central and somewhat counterintuitive findings.) Despite the ubiquity of infectious and parasitic disease, how-ever, the Tsimane do not appear sickly or starving. They're often missing several front teeth, a result of their fondness for sugarcane and citrus fruit, says Gurven, but otherwise, they seem robust and healthy.

On our return trip, we'll motor down the river, and drive through cane fields on muddy dirt roads. To return home, I'll take a small plane from San Borja over the imposing wall of the Andes to the west, spend a lay-over in the nation's capital, 12,000-foot-high La Paz, and then head back to New York City via Miami in a jet.

That trip passes through a well-defined gradient of allergic disease. I'll have traveled from an area of nonexistent allergies (subsistence living in the jungle) to one of slightly higher (the no-frills Bolivian town) to one of even higher (a large city in a developing country) to a place with the high-est allergy prevalence of all (a large city in the developed world).

The gradient I just described in space also exists in time. If you retrace your own lineage back a few generations, you'll probably find hay fever and asthma lessening with each one. You (like me) may have life-long asthma and food allergies, for example. Your parents, meanwhile, maybe had seasonal hay fever. But relatively few of your grandparents' generation—or great-grandparents, as the case may be—suffered from sneezing or wheezing of any sort. This pattern likely relates not to new exposures, but to the removal of old ones—exposures of the sort still prev-alent in Tsimanía.

Repeated observations like these, backed by piles of experimental evi-dence indicating that the immune system responds differently depending on its history of exposures, have prompted some immunologists to ques-tion the basic assumptions underlying their field. Our understanding of the immune system rests on work mostly carried out during the twentieth century, but by that time, we were living in evolutionarily novel circum-stances. In other words, we may have made a mistake equivalent to study-ing and cataloging an exotic-seeming ecosystem, only to discover that we weren't in the jungle at all; we were actually at the Bronx Zoo.

Or as the Duke University scientist William Parker puts it, "We as immu-nologists are now faced with the unsettling realization that the immune system we have spent all of our effort and energy studying over . . . the past fifty years has turned out to be dramatically different than the system derived by natural selection."

And that brings us to the heart of the matter.

UNDERSTANDING THE IMMUNE SYSTEM
AND ITS DISCONTENTS

You've probably heard peripherally about the many allergens, such as dust mites, peanuts, and tree pollen, which *cause* allergies. Maybe you've heard reference to the infections and toxic pollutants that *provoke* autoimmune disease. Without suggesting that these ideas are totally unfounded, here's an alternative and much simpler model for engendering immune dysfunction. To produce these disorders, you don't need to add something new to your body. All that's necessary, in fact, is the removal of a single critical component of the immune system, and the human organism will collapse in a firestorm of autoimmune and allergic disease.

Immunologists learned this lesson from real-life case studies. In 1982, scientists at Oregon Health Sciences University in Portland described the case of an infant who'd died from multiorgan autoimmune disease—type-1 diabetes, thyroiditis, eczema, diarrhea, and a self-destructive immune response to viral infection. Seventeen other male infants from the boy's extended family had perished the same way, but no girls. The scientists suspected they had a genetic mutation in the X chromosome on their hands.

Boys have only one X chromosome, from Mom. So while girls, who have an X chromosome from each parent, can always refer to workable instructions in their second X chromosome, boys are stuck with whatever defective genes their single X chromosome contains. These boys had apparently inherited a gene that precipitated an immune-system meltdown.

Two more decades passed before geneticists identified the culprit. The gene was named FOXP3 (forkhead box P3 in its full ungainliness). When switched on, FOXP3 changed how white blood cells operated, turning them from aggressors into peacekeepers. In the case of those boys, a spontaneous mutation had disabled the gene. As a result, they couldn't restrain immune aggression. They went thermonuclear on invaders, causing severe collateral damage. And they couldn't tolerate even their own tissues. Mystery solved. Case closed. Except that the finding upended the current understanding of the immune system.

For decades, immunologists had envisioned a system that avoided attacking the self by deleting self-reactive immune cells, and by employing the molecular equivalent of a hall pass system. Cells that belonged—"your" cells—displayed a unique badge (called the major histocompatibility complex, or MHC). Invaders didn't have this badge, and patrols picked them off handily. But here we had cells that possessed the mark of belonging, and were attacked anyway. What's more, healthy individuals tolerated a teeming community of microbes in the gut, organisms that didn't display the requisite hall pass but nonetheless escaped notice. Clearly, the old ideas needed revising.

Scientists, meanwhile, experimentally produced a range of autoimmune disorders by doing exactly what the FOXP3 mutation had done—disabling or hindering peacekeeping cells. Self-directed white blood cells obviously existed in healthy animals; they were a natural part of a functioning immune system. Order was maintained not by destroying these cells, but by restraining them. Disease arose not because lunatic lymphocytes escaped extermination (the old thinking), but because ineffective or absent suppressor cells failed to rein them in. The allergic and autoimmune diseases bedeviling us in modernity stemmed from a failure to police the police.

By the late 2000s, a revised model had emerged. Soon after birth, a wave of autoimmune cells populated the organism. They helped in defense, anticancer immunity, and tissue repair. A wave of peacekeeping cells quickly followed these initial pioneers, restraining them and establishing equilibrium. But keeping the peace in the long run required more suppressor cells. This secondary squadron emerged only after contact with the outside world—with certain parasites and microbes. This dependence was truly weird. It meant that our ability to self-regulate, to maintain homeostasis, was oddly reliant on external stimuli. What a design flaw—unless you considered the human organism in its proper context.

By all measures save sheer size and weight, you're mostly not you at all. The commensal bacteria in your gut, maybe 3 pounds worth, outnumber your cells by ten to one. The collective genome of this microbial community is a hundred times larger than yours, a hefty novel to your trifold pamphlet. That community harbors representatives from the three major branches of life on earth: bacteria (prokaryotes), yeasts (eukaryotes), and archaea (microorganisms that inhabit, among other extreme niches, deepsea hydrothermal vents). You are really an ecosystem, a mutually dependent aggregation of life-forms, what scientists call a superorganism.

Now the reliance on "external" inputs makes a little more sense. How could your genetic self—the You that began when Dad's sperm fertilized Mom's egg—possibly ignore the voice of the majority? The seemingly absurd mistake that prompts immune-mediated disease makes a little more sense as well. Remove or change those stimuli, and of course you'd expect the immune system to lose its bearings. Those signals both guide and stabilize your immune function.

And that, unfortunately, is the story of the past century—the reason some think that the human immune system now malfunctions so spectacularly. We routinely fail to tolerate everything—innocuous proteins (allergies), our own tissues (autoimmune disease), and our commensal flora (inflammatory bowel disease)—because we've done environmentally what that FOXP3 mutation did genetically. By changing our inner ecology, we've hobbled the critical suppressor arm of our immune system.

So here's the question: Can we replace these stimuli? Can I take what's protective about the Tsimane environment and reintroduce it to mine? And can I do it without killing myself in the process, without losing the unprecedented improvement in both quality and length of life that characterizes the developed world?

INFESTED WITH WORMS IN MEXICO

And that brings us back to my impending experiment. I pull off the highway into a eucalyptus-lined parking lot where I'll meet my hookworm donor, a medical school dropout named Garin Aglietti. Warehouse-sized outlets of major American brands—Marshalls, Nike, Levi's, McDonald's—surround us. I join a group of forlorn-looking elderly people waiting under a tent. A bus passes by here to ferry them across the border. They belong, I presume, to the daily migration of Americans who travel to Mexico to buy cheap drugs.

Aglietti arrives in a tan Jeep Cherokee with Nevada plates. He's wearing baggy jeans, a blue shirt, and silver-rimmed wraparound sunglasses. He removes them to reveal blue eyes in a round, open face. In brief, Aglietti's story goes like this: In the 1990s, he developed psoriasis, an autoimmune disorder of the skin. He'd also suffered from asthma for most of his life. Mostly he fretted over the conditions known to accompany psoriasis, such as cardiovascular disease and autoimmune arthritis. All-too-frequent chest pains incited a cascade of worry. "I felt like it was killing me," he tells me. "I was way too young to be getting chest wall pain."

Allopathic medicine—also known as modern medicine—didn't offer much by way of treatments. Then in the early 2000s, Aglietti heard about a Japanese scientist named Koichiro Fujita. Working in Borneo in the 1990s, a time when Japanese children seemed increasingly prone to developing eczema, Fujita had noticed that Bornean children had exquisite skin and no allergies. They also harbored plenty of parasites. Was there a link?

Back in Tokyo, Fujita took the extraordinary step of self-infecting with tapeworm. His hay fever cleared up. His skin became clearer and less muddled. He started preaching that the modern world was too clean for our own good. Corporate funders began withdrawing support from his lab.

Aglietti decided to follow Fujita's lead. Tapeworms have an intermediate and definitive host. In the former, they form a cyst; in the latter, they live as an intestinal worm. In 2005, Aglietti traveled to Kenya, toured cattle slaughterhouses searching for tapeworm cysts, found two, and swallowed them. Soon thereafter, Aglietti's psoriasis plaques softened. A few months later, they'd almost entirely disappeared. But once a tapeworm matures, it begins releasing rather large, semi-self-propelled egg-filled sacks called proglottids. They slither out one's rear and down one's leg in search of new intermediate hosts.

When they began passing, Aglietti felt as if sweat were dripping down his leg in the absence of any perceivable heat. "It's just a very unclean feeling psychologically," says Aglietti. "I just couldn't deal with it." He terminated the experiment with antiworm drugs. After passing a three-foot-long tapeworm, he set off in search of another, less psychologically disturbing parasite. This time, he settled on hookworm. Now he sells hookworm to others in Tijuana.

As we walk along the highway toward Mexico, Aglietti asks me almost gingerly why, with my apparently extensive knowledge of parasites, I didn't travel to some corner of the developing world, as he did, and acquire parasites naturally. I don't have the time, I say. But as we pass through turnstiles into a walled corridor, the no-man's-land that separates the two countries, I'm wondering the same thing.

No doctor or scientist I've yet met would recommend traveling to Tijuana to acquire hookworm. Not only is this approach completely outside the realm of what's proven to work scientifically, those like Aglietti who offer the service—at least two operations exist as of this writing—do so outside the scientific and medical establishment. No standards of quality or care exist save those that are self-imposed. And there's just as little accountability if anything goes wrong.

The cons of what I'm about to do are therefore significant. Illness and death are the most obvious. But I'm most worried about encouraging Aglietti, who seems perfectly nice, and his ilk. I'm not sure they deserve more attention than they've already received. On the other hand, self-infecting with hookworm has become an underground phenomenon of sorts, an unconventional treatment for often desperately ill people. I want to see what these individuals go through, how the process works.

And that brings me to the pros: I've heard fantastic tales of remission from people who've come this way before. Some I can confirm. Many more I cannot. There's nothing like seeing with your own eyes to settle questions like these. The potential benefits are also considerable—not worrying about peanuts, not wheezing, no more hay fever, no red, swollen eyes when cats jump in my lap. Sprouting a full head of hair would really be icing on the cake. Most important, success might point the way toward the Holy Grail of prevention—not for me, but for my children.

We pass through another revolving gate that's strangely reminiscent of the unjumpable, floor-to-ceiling turnstiles in the New York City subway, and we're suddenly at a small plaza with a fountain in Mexico. No more American chain stores. Small shops with colorful signs dominate. A friendly young man with thick black eyebrows and hair gelled into spikes pulls up. He drives us to a neighborhood near the ocean. We park in front of building with a Mexican flag waving from the second-floor balcony. A sign says UNIDAD DE MEDICINA HOLÍSTICA—Office of Holistic Medicine.

While Aglietti confers with the doctor upstairs, our driver, Andrés, the doctor's son, tells me he's twenty years old, and was just admitted to medical school. He adds that his lifelong asthma forced him to stop playing sports years ago. Some months ago he infected himself with hookworm, and now it's much better. He began playing soccer again.

Aglietti returns and tells me the doctor is ready. I follow him to a clean, spare office on the second floor. A T-shirt with SAY HELLO TO MY LITTLE FRIENDS over an image of a gaping hookworm maw—four flat fangs lining the top, and vague indents where the eyes should be—is pulled over the back of a chair. The four "teeth" suggest *Ancylostoma duodenale,* generally considered more pathogenic than the *Necator americanus* I'll receive today. *N. americanus* has just two teeth that are boxier and, somehow, less sinister-looking.

Dr. Jorge Llamas enters, dressed in black slacks and blazer, and worn black loafers. He has a paunch, jowls, and a robust head ringed with trimmed white hair. Others who've passed this way have expressed great affection for him, and I can see why. He projects an easygoing, friendly manner that's reassuring and soothing.

"We're divorced from nature," he tells me. "And it's hurting us." He relates a story about an American woman who, after having lived in Acapulco for a time, returned to the U.S. to find she'd acquired parasites. She had them removed. Suddenly she was diagnosed with Crohn's disease. He mentions that as a child in Guadalajara, his father took him to the beach often, where swarms of mosquitoes fed on him. "It made my immune system strong," he says. He's never had allergies. He rails against the modern obsession with cleanliness. Everyone is mindlessly following the U.S.'s lead, he says. And everyone is getting U.S. diseases. "We need to stop and think."

He ends his holistic doctor spiel to take my medical history. Do I wake up at night? (Yes.) How many times, and what happens when I do? (I go back to sleep.) Just go back to sleep? (Yes.) How often do I exercise? (Three times a week.) What's my religion? (None.)

"That must be a lonely existence," he says, and notes something on my chart. He begins explaining the "known" emotional states associated with asthma and alopecia—stress and depression, respectively. "We create our own realities," he says at one point. "We're even creating this reality right now."

As we continue drifting into what I'm fairly sure is pseudoscience, I grow disconcerted. I'm here to acquire parasites, which is among the stupider things I've done. But the experiment is meant to probe what I imagine are universal principles of systems biology—relationships forged over mind-numbingly long periods of coevolution. None of which is hocus-pocus. I attempt to correct course. I ask to see Aglietti's blood work. I've found no evidence that hookworms can transmit viruses between people, but they're born as eggs in one human's bowel movement and, after hatching into larvae and piercing the skin, they pass directly into another human's bloodstream. The precautionary principle applies.

Soon we're shuffling through a year or two's worth of tests. I confirm that Aglietti is clear of the major viruses—HIV, cytomegalovirus, hepatitis—as well as *Strongyloides stercoralis,* a nasty worm that, uniquely among soil-transmitted helminths, can reproduce in the host. I'm as satisfied as I'm going to be.

"Are you nervous?" asks Llamas.

"Do I look nervous?"

He shrugs. "A little."

We move to a room at the back of the building. Aglietti has donned a light blue doctor's overcoat with *Worm Therapy* embroidered in script over his right pectoral. He's smiling and seems excited. With a pipette, Llamas removes what I'm assuming is larvae-laden water from a beaker, and squirts it onto an absorbent bandage. Given my apparent health, Aglietti and Llamas have recommended thirty worms, not the twenty or twenty-five I was assuming.

The bandage goes on. Within a minute, I feel a tickling, itching, nearly burning sensation—rather like a mild case of stinging nettles. That's the microscopic larvae burrowing through my skin. Before anyone knew a parasite caused it, the distinctive itch had gained notoriety around the world, earning monikers like "ground itch," "miners' itch," "water pox," or the more poetic "dew poison." Now scientists understand that hookworm larvae leave their outer cuticle, discarded inside-out like a sock, embedded in your skin. Your immune system responds savagely. But the now-naked larvae are already long gone.

Each larva will find its way into a capillary, and hitch a ride on my venous blood flow, like rafters on a river. They'll pass through the thunderous pump of my heart, which causes me no small degree of anxiety. And once they've arrived at the capillaries of my lung, they'll burrow out of the circulatory system, into the bunch-of-grapes-like sacs called alveoli. They'll then follow the coordinated sweeping motion of millions of hairlike cilia up- and outward—the so-called mucociliary escalator—over the pharynx, where windpipe and food pipe branch, and plunge down into the esophagus.

They'll miraculously survive the hydrochloric acid bath of my stomach and finally—after an odyssey through my body lasting several weeks—arrive at my small intestine, the final destination. They'll latch on to my intestinal wall. They'll mate. Large individuals will reach a centimeter in length. The females will lay perhaps 10,000 microscopic eggs daily, all the while grazing on intestinal tissue to the tune of 0.04 milliliter of blood per day. Assuming they all survive, that's eight drops for every ten worms, or twenty-four drops daily paid to host a thirty-strong colony—not much, but not nothing either. And they can live for five years, maybe longer. The eggs, which require a week or two in tropical conditions to become infective larvae, will pass out with my stool—which, in New York City, means they end up in a wastewater treatment plant.

I might get a mild cough in a week or so, Aglietti explains. Flulike symptoms are common. Then "epigastric pain" once the worms attach. If I start coughing, I shouldn't spit out the discharge.

"Swallow it," he says. "That's your medicine."

Then Aglietti, who's periodically glanced at his wristwatch since the bandage went on, says, "Okay, we're past the possibility of anaphylaxis." He's referring to a potentially fatal allergic reaction usually associated with bee stings or, these days, peanuts. Anaphylaxis is treated with a shot of Adrenalin, which he has handy. Llamas hands me a box containing three pills of mebendazole, a deworming drug. "This is your out," he says. "Here in Mexico we take two. But in the U.S., being the U.S., they take three."

By now I have a headache. I'm filled with feelings of disgust, hope, and wonder—disgust with myself for agreeing (with myself) to this experiment; hope that the experiment may do some good; and wonder at the parasite's biology, its ability to pierce skin, navigate circulatory systems, and, in the coming weeks, arrive at my small intestine. Underlying these sentiments is a recently acquired, quasi-religious faith in evolution—confidence that the organism knows what it's doing, and won't kill me in the process. For an obligate parasite, a dead host is, after all, a useless host. For better or worse, we're now in this together.

CHAPTER 2

Homo Squalidus:
The Filthy Ape

[O]ne can properly think of most human lives as caught in a precarious equilibrium between the microparasitism of disease organisms and the macroparasitism of large-bodied predators, chief among which have been other human beings.

—William H. McNeill in *Plagues and Peoples*

Judging by our natural parasite load, *Homo sapiens* ranks among the filthiest of primates. This observation may be an accident of self-interest: We know more about our parasites than those of other species because they're important to us, and that knowledge gives a false impression of abundance. But there are several reasons to think that human parasite load is, in fact, unusually high.

First is our restlessness as a species. By 15,000 years ago, when paleo-Indians crossed the Bering Land Bridge connecting Siberia and North America, human beings had learned to live in almost every habitat on earth, from tropical jungle and Australian desert to temperate Eurasian woodland and northern tundra. Our omnivorous nature and adaptability, enabled by technology, allowed our rapid radiation around the globe.

A single species sprawled across so many niches—not to mention our close contact with animals after domestication began in earnest roughly 12,000 years ago—is also one exposed to many parasites. By one count, 80 percent of the roughly four hundred parasites that call the human body home are zoonotic, meaning they jumped from other species at some point in the past and adapted to their new home. "Homo sapiens ranks among the most parasitised of all animals," write the parasitologists R. W. Ashford and W. Crewe in *The Parasites of Homo Sapiens*. "There must be few parasitic species which have never had the opportunity to infect a human."

Ashford and Crewe include only eukaryotes—organisms whose cells have a defined nucleus—in their assessment. I'm using the term much more broadly: any organism, single-celled, multicellular, or viral that requires the human body to complete its own life cycle, and that can cause disease.

Our extreme sociality as a species has also amplified our parasite load. Some anthropologists now argue that, our comparatively enormous brain aside, the trait that distinguishes humans from other great apes is the ability to cooperate. We can and do work as a team; teamwork makes us more effective. But cooperation also presupposes close living, and since the birth of agriculture 12,000 years ago, and even before, we have aggregated in ever larger communities. With each increase in size, human communities became more complex, more structured, and in some ways better able to harness and direct human ingenuity and energy. They also grew more pestilential, more miserable, and more unfriendly to that very same promise.

One view of the arc of human history since the late Paleolithic is as a constant push toward larger human networks, an inexorable movement toward globalization checked by the amplification of disease that same trend occasioned. The filth reached an apex in the West with the rapid urbanization of the Industrial Revolution in the late eighteenth and early nineteenth centuries. Some rightly worried that the emerging mechanized civilization would drown in its own putrescence. That anxiety ultimately sparked sanitary reforms whose benefits we continue to reap today. These improvements ushered humanity through its second great epidemiological transition. The first such transition occurred when hunter-gatherers settled down to farm. And the third transition is ongoing: Old bugs that have evolved resistance to antibiotics are resurgent; and most important for the purposes of this book, chronic degenerative diseases with no obvious infectious cause characterize the modern diseasescape.

So what about our outsized parasite load? For our purposes, the parasites of the Paleolithic hold special interest, the hangers-on we began losing during the second epidemiological transition. We spent a long time with those organisms. And long periods of coevolution produce entangled relationships.

Paleolithic means, more or less, "ancient stone" in Greek. The word refers to our toolmaking, which has improved dramatically in the 3.2 million years since our ancestor Lucy the Australopithecine lived in East Africa. During the Paleolithic, scattered groups of thirty to seventy people predominated. The scorched-earth approach to parasitism—replicate like mad, the host be damned!—would have been self-defeating in these circumstances. Any parasite that killed its host quickly, and for which humans were the only host, would rapidly drive itself extinct.

As a result, parasites from the Paleolithic usually establish long-term residence. They tend to have a "softer" touch, at least compared with the plagues of later times. To suggest that their constant presence for millions of years permanently affected our immune function is to misconstrue the depth of our entanglement. They altered our immune function the way that atmospheric oxygen modified our lungs, or dry land our limbs. Which is to say, much of our immune system evolved precisely to manage the problem of parasites. They constituted a dominant feature of the landscape in which we evolved.

HUMAN EVOLUTION AS TOLD BY OUR PARASITES

How parasitized were we? Modern-day hunter-gatherer groups, such as the Pygmies of central Africa, the Xavante of Brazil, and the San of southern Africa, are almost universally parasitized, but the loads are light. Most have a few worms, but no one has too many. Using modern hunter-gatherers as our guide, however, may lead us astray. They live in a much more crowded world compared with even a hundred years ago, let alone sixty thousand. And they may have acquired parasites from settled peoples, who in turn acquired them from animals.

Our primate relatives may better illuminate our own primeval parasite load. And wild chimpanzees host a veritable ecosystem of intestinal worms, blood flukes, and unicellular protozoa. Again, no individual seems heavily infected. As it turns out, the native human repertoire of parasites more resembles that of baboons than of chimpanzees. That's likely due to the long time we spent living on the savanna. Indeed, our parasites tell us much about where we've been, and whom we've met along the way.

Take the tapeworm. Human-adapted tapeworms, which can grow twenty-five feet long, require two hosts to complete their life cycle: the intermediate, in whose tissues they encyst, and the definitive, where they reproduce. Three species of tapeworm generally infect humans, one that uses cows as an intermediate host, one that uses pigs, and one that uses both.

Scientists have generally blamed the domestication of pigs and cows for our having acquired these rather large worms, but that was before Eric Hoberg, a scientist with the U.S. Department of Agriculture, took a closer look. He found that human tapeworms were most closely related to those of the large felids, canids, and hyenas of Africa, not Eurasia, where we domesticated most animals. Our tapeworms diverged from these African relatives between 1 million and 2.5 million years ago, roughly about when our tool-using, fire-taming *Homo erectus* forebears began scavenging, and maybe hunting regularly on the savanna. We ascended a rung in the food chain and, rather like an ecological rite of passage, we inherited the top predators' parasites.

And what of our hominid relatives encountered along the way? There are more than three thousand species of lice infesting birds, rodents, ungulates, and probably most living things with fur or feathers. (Lice-free creatures include the egg-laying duck-billed platypus, the scaly anteater, and hairless dolphins and whales.) These tiny biting pests have inhabited the primate pelage for at least 25 million years. Gorillas and chimpanzees each have their own, unique species, but humans mysteriously have two: one that lives on the head, and the other in the pubic area. Did the two species diverge from a single human-dwelling ancestor? Not exactly.

In 2007, David Reed at the Florida Museum of Natural History announced that human head lice were most closely related to chimpanzee lice. We shared a common ancestor with chimps some 6 million years ago, which matches the divergence of our respective lice. However, our pubic lice, colloquially known as crabs, descended from gorilla lice. The most recent common ancestor of gorillas and humans lived about 7 million years ago, but Reed found the pubic louse to have diverged from the gorilla louse much later, about 3.5 million years ago. How was that possible? "We'll never know if it was sex or something more tame," he told the *New York Times*. But at least the lousy acquisition shed light on another long-standing mystery: when our lineage lost its body hair. For the gorilla louse to colonize the pubic niche, native lice must have already disappeared, the thinking went. By that time, the crotch area must have already been an island of coarse hair in a sea of bare skin.

The naked ape, as we know, eventually began covering up with clothing. Again, lice tell us when that occurred. A subspecies of the head louse inhabits our clothing. (This louse carries the dreaded epidemic disease typhus.) And the head louse evolved into its new niche, woven cloth, roughly 107,000 years ago.

When modern *Homo sapiens* left Africa some 60,000 years ago, the descendants from previous hominid outmigrations still inhabited Eurasia: the Neanderthals in the West, with whom we shared a common ancestor some 350,000 years ago; *Homo erectus* in the East, who left Africa perhaps 1.8 million years earlier; and also in the East, the recently identified Denisovans, close relatives of the Neanderthals. We interbred a little. Between 1 and 4 percent of all human DNA from people outside of Africa comes from the Neanderthals. Melanesians and some Southeast Asians carry a slightly larger quantity of Denisovan DNA.

But while we interbred only fleetingly, we permanently adopted one of these other hominids' parasites. A "race" of human head lice found only in the Americas differs dramatically from the two Old World races. According to DNA analysis, this louse diverged from the human head louse some 1.18 million years ago, long before *Homo sapiens* departed from Africa. David Reed thinks the insect hopped aboard from archaic hominids living in Asia. *Homo erectus* and the Denisovans disappeared, but their lice survived in the hair of those modern human pioneers who eventually pushed all the way to North America.

About 30,000 years ago, an unknown artist drew graceful pictures of buffalo, horses, lions, and hyenas on a wall deep in a cave in southern France. These remarkable sketches in the Chauvet-Pont-d'Arc cave not only madden present-day artists with their unschooled mastery, but provide a glimpse into a world teeming with large game.

A lesser-known cave in northern France, the Grottes d'Arcy-sur-Cure, also contains paintings, although of inferior quality. But this cave tells us something else about life in those times. At some point roughly 30,000 years ago, someone defecated toward the back of one grotto. The stool this person left behind fossilized and, thirty millennia later, scientists found that it contained eggs from what's since become the most common of all worms infecting humans: *Ascaris lumbricoides*. The giant roundworm now inhabits an estimated 1.2 billion people—one-sixth of humanity—mostly in the developing world. But not long ago, everyone, including Europeans and Americans, was rife with this worm.

Again, scientists have generally blamed our domestic animals—in this case our pigs, which host a similar species—for the giant roundworm. But if these fecal remains are any indication, we had the parasite some 20,000 years before domesticating the wild boar. Some suspect that the Grottes d'Arcy-sur-Cure feces may actually have come from a bear, which would render the idea moot. Other evidence corroborates a Paleolithic acquisition, however. Namely, Amerindians had ascaris worms nearly 4,000 years before the Spanish introduced pigs to the Americas. Their progenitors migrated across Beringia before the advent of agriculture. We can absolve our pigs, it seems. When we domesticated boars, we gave them our roundworm, not vice versa.

HEALTH IN THE LATE PALEOLITHIC

After leaving Africa, modern humans arrived in Australia by 45,000 years ago—and probably earlier—and in Europe by 40,000 years ago. There they encountered tree-dotted grasslands teeming with mastodons, woolly rhinoceroses, horses, cave bears, mammoths, saber-toothed lions, and bison. At the millennial scale, the climate was schizophrenic, with rapid retreats and advances of massive ice sheets from the north. The overall trend, how-ever, was toward a great chill that would culminate 20,000 years ago with much of the British Isles and Scandinavia completely entombed in ice, and what's now the English Channel an arctic steppe.

For tens of millennia before this cold climax, early Europeans enjoyed robust health. Judging from the skeletons they left behind, they were tall and big-boned. They had plenty of food and exercise. Compared with later periods, they suffered from little infectious disease. By one count, men averaged 5'8.5", and women 5'4" tall, about modern proportions. The oval cross-sections of their upper leg bones—a sign of thigh and hamstring muscles pulling back and forth for a lifetime—suggest that they walked a lot. More sedentary populations, like farmers and especially modern office workers, have femurs with a more circular cross-section.

As climatic conditions changed, the bones of these Europeans changed as well. By 20,000 years ago, the glaciers had pushed people southward. And late-Paleolithic Europeans lost nearly four inches in height—men to about 5'5" and women to just over 5'. Leg bones became less robust, the cross-section less oval and more circular. Toe bones began to atrophy about 40,000 years ago in East Asia, and 26,000 years ago in Europe, indi-cating that for the first time shoes became widespread.

At least since the 1980s, anthropologists have faulted agriculture, its less protein-rich and diverse diet, and the amplification of disease brought by settled living, for the decline in health seen when people first began farm-ing. These arguments partly rested on studies conducted in the Americas, where some communities that began to cultivate maize did, indeed, suffer poorer health compared with hunter-gatherer precursors. But additional research has complicated the picture. In some cases, farmers were health-ier than their immediate forebears. More important for our purposes, in western Eurasia, human health began to decline thousands of years before the advent of agriculture.

By the late Paleolithic, the big, easy-to-hunt game had also become scarcer, and communities resorted to less glamorous fare like shellfish and, in the Middle East, hares. They'd become less mobile, and a pitting of the bone called porotic hyperostosis started to appear more frequently. The condition, which results from anemia, could signify an iron-poor diet, more disease, an increasing parasite load—or all of the above.

The anthropologist Brigitte Holt attributes these changes to increased crowding and a more sedentary lifestyle, the result of glaciers having swal-lowed up land, and a growing human population. For perhaps the first time in our evolution, we began to experience the negative consequences of success: crowds and scarcity.

Geneticists find evidence of population expansions, especially in warmer climes, beginning even earlier. Some 41,000 years ago in sub-Saharan Africa, hunter-gatherer groups like the San and Biaka increased thirteenfold. The proto-Yorubans and Mandenka in West Africa septupled 31,000 years ago. And 22,000 years ago, near the glacial maximum, northwestern African populations tripled. Six millennia later as Europe thawed, the population increased elevenfold.

Why does this matter? As crowding enhanced pestilence, pestilence began leaving its mark on our immune-system genes, and that had consequences for our susceptibility to inflammatory diseases. At some point between 100,000 and 500,000 years ago, for example, a spontaneous mutation inactivated a gene called caspase-12 that aids in recognizing bacterial invaders. Having the original, nonmutated version meant a quick and decisive response to bacterial pathogens. The mutated version, however, translated to a slower, more lethargic response. When the inactive version of the gene arose, for perhaps hundreds of thousands of years, natural selection neither favored nor disfavored it. However, between 100,000 and 60,000 years ago, something changed in the diseasescape. Carriers of the inactivated version began having more offspring than noncarriers. They suddenly had an advantage.

Why would a nonfunctional gene be advantageous? The inactivated gene, it turns out, protects against sepsis. The severity of sepsis is partly determined by the invading bacterium, and partly by one's own immune response. An overwhelming counterattack can spur blood clotting, organ failure, and even death. Today one-third of people struck with sepsis die, but people with two copies of that mutant gene are almost eight times less likely to succumb to sepsis than those without. So there's our answer: The gene spread because people began encountering more sepsis-causing pathogens. Those with the active ancestral version tended to melt down more often.

Other genes responded to the shifting diseasescape, although with different end results. The nonfunctional variant of a gene called CARD8, for example, also began spreading. The gene inhibits an inflammatory cascade. The nonfunctioning version is, therefore, like a broken off-switch; the lightbulb stays on indefinitely, and inflammatory processes keep chugging along. So in contrast with the caspase-12 mutation, the "null" CARD8 gene variant *improved* one's germ-fighting vigor. The downside of a broken off-switch, however, is a tendency toward inflammatory diseases, such as rheumatoid arthritis.

Like humans, other animals that aggregate in large groups and are exposed to many pathogens tend to lose functionality in this gene (meaning they have a prolonged inflammatory response). Mice, cows, and horses have the nonfunctional version. Cats and dogs retain the working gene, however. Chimpanzees, gorillas, and orangutans—primates that live in relatively small troops—also retain the functional version. By contrast, rhesus monkeys, which aggregate in groups that number in the hundreds, appear to be losing it.

In our case, the nonfunctioning version has generally become more common in direct proportion to how long one's ancestors have farmed. Very few hunter-gatherers—10 percent of the San and just 4 percent of Pima Indians—have the "broken-off-switch" version. People who started farming within the past 4,000 years, relatively recently, have it at a greater frequency.

These two gene variants, one that turns down the immune response and the other that turns it up, epitomize the immune dilemma: Overwhelming force (the ancestral caspase-12 gene) seems the obvious first choice. But if you're responding with the nuclear option every day, you'll inevitably blow yourself to smithereens. On the other hand, if you're under regular assault, you need some sort of constant response (the broken off-switch of the nonfunctional CARD8 gene)—but then you risk inflammatory disease.

The immune system has always had to navigate these pitfalls—on the one hand, potentially destroying the self, and on the other, being destroyed by opportunists. Recognizing the dangers inherent in this balancing act is important for understanding our genetic proclivity to develop autoimmune disease in modernity. In all likelihood, gene variants now associated with autoimmune and allergic diseases helped resist pathogens in the past. And they almost certainly didn't cause as many problems in the process.

THE NEOLITHIC:
FROM EDENIC FILTH TO PANDEMICS

Roughly 12,000 years ago—and likely earlier—someone in the Levant deliberately planted a seed, maybe cared for the seedling, and then harvested the adult plant. Agriculture was born. It arose independently at least seven times: in Mesopotamia (wheat and barley), sub-Saharan Africa (millet and sorghum), Southeast Asia (rice and bananas), China (millet and rice again), the Papua New Guinea highlands (taro root), Mesoamerica (corn, beans, and tomatoes), and South America (potatoes).

Another revolution co-occurred. Someone in eastern Anatolia had second thoughts about immediately killing a lamb or sheep, instead adopted the animal, and eventually raised a herd. Humans have a long history of interaction with animals. Chauvet cave in France contains 26,000-year-old footprints of a child accompanied by a very wolflike dog. Westerners traveling among hunter-gatherers in the twentieth century regularly tell of women nursing and chewing food for adopted wild animals. But raising entire tribes of animals represented an escalation, a new symbiosis that was part mutualism and part parasitism. Animals gave milk, hide, flesh, and muscle power. In exchange, humans fed, nurtured, and protected them from predators.

By 8,000 years ago, and probably earlier, humans had domesticated cattle from wild aurochs in the Near East and India. We'd tamed the pig from the wild boar maybe as early as 13,000 years ago in Mesopotamia, and again in eastern Eurasia. Chickens came from jungle fowl in Southeast Asia. And horses arrived from the grasslands of present-day Kazakhstan by 5,500 years ago.

The new closeness of hominid, bird, ungulate, and pig allowed an unprecedented exchange of parasites and pathogens. Settled living also created new ecological niches for animals not directly domesticated. Wolves, some think, began scavenging around human camps somewhere in Eurasia perhaps 14,000 years ago, initiating the taming process that eventually produced the dog. Agriculture also implied stored grain. Rodents came knocking. In the Middle East, a small desert feline followed 10,000 years ago, the forebear of all domestic cats. Every new arrival brought its parasites. Pigs increased exposure to *Trichinella spiralis,* the dreaded muscle- and brain-burrowing worm, and the reason you should always thoroughly cook pork. Cats brought *Toxoplasma gondii,* which cycles between felids and their rodent prey. Dogs perhaps brought the hookworm species *Ancylostoma duodenale.* Rodents contributed their own worms—*Hymenolepis diminuta* and *H. nana.* So far, these parasites were relatively benign, especially compared with the killers to come, but they had a cost, and early town-dwellers show signs of a heavier parasite load.

Residents of Catalhöyük, a 9,000-year-old settlement in central Turkey often billed as the world's first town, suffered from chronic anemia and porotic hyperostosis, a pitting of the bones. Around the same time, the first verifiable case of tuberculosis appeared. A twenty-five-year-old woman and a child buried in a now-submerged village off the coast of present-day Israel both have telltale signs of active tuberculosis.

In the Levant, however, early agriculturalists were generally healthier than their hunter-gatherer precursors. Their teeth were better. Male farmers especially lived longer. (They also suffered from head trauma nearly six times less often than their nomadic ancestors.) However, one aspect of life worsened: Farmers had more bone lesions suggestive of inflammatory disease. To anthropologists, this indicates not necessarily an infection, but a beefed-up immune response. As we saw earlier, as encounters with pathogens become more common, pro-inflammatory tendencies become more advantageous. The downside: a tendency to develop chronic inflammation. Here was evidence of that trade-off written in bone.

The broader diseasescape was shifting. Parasites from the Paleolithic, when humans lived in small groups, adopted a marathoner approach: long-term persistence in the host while inflicting as little damage as possible. But with larger settled groups, another method became viable: the microbial blitzkrieg.

THE AGE OF PLAGUES

By five millennia ago, human settlements throughout the Middle East had grown large enough to sustain epidemic disease. How many people was that? The estimated population required to maintain a virus such as measles, to which you develop lifelong immunity once infected, has been repeatedly revised, from 1 million people to 500,000, and now to 200,000.

Around 5,000 years ago, plagues appear, first in stories, and then in human remains. The 4,000-year-old epic of Gilgamesh features Erra, god of war and pestilence. Egyptian papyri from the same period describe a poxlike affliction. By 3,500 years ago, mummies display skin lesions that resemble pox. One of these preserved bodies belonged to King Ramses V, who, while in his early thirties, expired suddenly in 1157 B.C. From the sores on his desiccated remains scientists have extracted a pox-looking virus—direct evidence that the affliction had arrived in ancient Egypt.

These new plagues came from our animals. Smallpox is closely related to a virus that infects gerbils and camels in the Levant and North Africa. Camels were domesticated by 5,000 years ago in southern Arabia. Measles diverged from rinderpest, a virus that afflicts cattle, sometime within the past 2,000 years, and probably more than once. The modern strain may be just 200 years old.

The pandemics repeatedly changed the course of history. In 430 B.C., a plague struck Athens while it was under siege by its on-again-off-again rival Sparta. A four-year-long epidemic ensued, killing one of every four people within the city's walls, including Pericles, the ruling despot, himself. The Spartans, meanwhile, appeared immune. Some think they brought the affliction. Whatever its source, historians credit this, the first documented plague, as precipitating the decline of Athenian, and more generally Greek, influence in the eastern Mediterranean.

The Romans soon took up the slack, only to suffer plagues of their own. In A.D. 166, Roman troops returning from the east brought back to Rome a scourge that would eviscerate the empire. At its apex, the pandemic killed five thousand Romans daily. Ultimately, it felled one in every ten people in the Roman Empire.

And then there was malaria. Agricultural activity around the Mediterranean—felling trees, building roads, irrigating—created endless habitat for the human-adapted anopheles mosquito, which carried the most fearsome of malaria parasites, *Plasmodium falciparum*. By A.D. 100, malarial fevers had emptied the densely populated Pontine Marshes just south of Rome on the Tyrrhenian Sea. And the city itself wasn't spared.

At the height of Roman power, when Rome was a cosmopolitan city of 1 million people with a sewer system, the cloaca maxima, and aqueducts supplying clean drinking water, Romans suffered mightily from malaria and other diseases. You can see it in their stature. When they settled in central Europe, they were on average 4 centimeters (1.6 inches) shorter than the surrounding population. The height differential had nothing to do with genes. After the collapse of the Roman Empire in the fifth century A.D., Romans grew taller as they dispersed from their megalopolis to eke out a living from the land.

On more than one occasion, however, Rome's very pestilence saved the city. In A.D. 69, Emperor Vitellius occupied Rome, but his soldiers, mostly from Gaul and Germania, died in masses after camping in a marshy area near the Tiber. They had no immunity to malaria. And when Attila the Hun invaded the Italian peninsula in the fifth century A.D., word of a vicious malaria season may have kept him from sacking Rome itself.

In the east, however, disease foiled the restoration of empire. In the sixth century A.D., after the Western Roman Empire had collapsed, the Byzantine emperor Justinian set out to reconquer former Roman territory in North Africa. He succeeded, but then a plague struck his capital, Constantinople. On the worst days, it killed ten thousand people. The illness, which some suspect was smallpox and others the bubonic plague, took another two hundred years to burn out. By then, 100 million people are estimated to have died. At the time, world population hovered around 190 million. If the Justinian Plague hadn't struck, we might all be speaking derivations of Greek. And when an army of nomads rode out of the Arabian Peninsula in the seventh century under the banner of a newly minted monotheism, they might not have encountered old, moldering civilizations ravaged by disease. Within a hundred years, the Islamic Caliphate stretched from Iberia in the west across North Africa to India's doorstep in the east.

The more globalized and interconnected human experience became, the more disquieting the plagues. During the thirteenth century, the Mongols established the largest empire the world had ever seen. These famous horsemen governed a swath of the Eurasian continent that stretched from the Pacific Ocean in the east to the Danube River in the west. And just about smack in the middle of the landmass, they awoke a terrible affliction.

Yersinia pestis, the bacterium responsible for the bubonic plague, was native to burrowing rodents called marmots that lived in the Central Asian steppe. Fleas transmitted the bug, and, presumably having learned the hard way, people who lived near the rodents had strict taboos on hunting them. Outsiders, however, didn't know any better, and the new empire brought strangers from all corners of Eurasia.

Records outside of Europe are fuzzy on when and how the bubonic plague spread throughout Central Asia. But it must have. And in 1347, twelve Genoese sailing ships newly arrived from Caffa, a trading port on the Black Sea, brought it to Sicily. Early observers describe walnut-sized "burn blisters" in armpits, necks, and groins that oozed blood and swelled to the size of a goose egg. The agony usually lasted three days. Most infected people died. The horrific affliction traveled along Mediterranean ports, and then inland. Burghers fled to the countryside, but the illness followed them. Surgeons tried to protect themselves with beaked, spice-infused masks, but died anyway. By 1353, that first—and worst—wave of the Black Death finally subsided. With the casualties of subsequent waves tallied in, the malady killed one in three Europeans, maybe more. Traveling along the trade routes that now spanned the Eurasian landmass, it was a truly global pandemic.

By some estimates, throughout human history, the bubonic plague and smallpox killed more people than all other infectious diseases combined. The two checked human population growth for millennia. Of course, as Jared Diamond argues at length in *Guns, Germs and Steel,* this very pestilence became an asset in its own right, an accidental agent of biological warfare that Europeans unleashed in the Americas beginning in the late fifteenth century. Following first contact, Amerindian populations collapsed to one-tenth their former abundance in the space of a few decades. The incomparable filth of Eurasian civilizations guaranteed, in a sense, their eventual triumph.

THE BACKGROUND INFECTIONS: AN EVER WORMIER WORLD

So let's say you're a worm surveying this great human drama. You, like everyone else, just want to live and procreate. That means attaching yourself to some host, mating, shedding fertilized eggs, and, to the degree that it's possible, improving the odds that those eggs find their way to a new host. If you're a human-adapted parasite, about 70,000 years ago, you were kicking yourself over your decision to specialize in this particular hominid. That's roughly when, for some mystifying reason—a massive volcanic eruption in Indonesia perhaps, or climate change—*Homo sapiens* almost went extinct. Judging by the loss of genetic diversity around that time, we dwindled to as few as two thousand individuals. Humans appeared doomed, and so did our parasites. But ever since that near miss, our hangers-on have almost certainly congratulated themselves.

During the late Paleolithic, we spread around the globe, carrying our parasites to all continents save Antarctica. And then, during the Neolithic, we settled down to farm. For organisms that depended on their hosts predictably encountering their own waste, this lifestyle change presented the ultimate boon.

All evidence suggests an intensification of worm infection during the Neolithic. To begin with, farming and irrigation in the tropics and subtropics created new habitat not just for mosquitoes, but for the skin-piercing schistosomes. Indeed, the field of paleo-parasitology begins in the early twentieth century with the discovery and rehydration of schistosome eggs extracted from the preserved bladders of 3,200-year-old Egyptian mummies. And Chinese mummies from the same time also harbored the parasite.

The salient symptom of schistosomiasis, blood-tinged urine, caused anxiety in ancient Egypt. Desperate to protect themselves, but unaware of how the disease was transmitted—did it enter through the anus or the penis?—Egyptian hunters donned penis sheaths while stalking prey in marshlands. (Like hookworms, schistosomes infect their victims through the skin.) Millennia later, urinating blood had become a rite of passage for pubescent Egyptian boys, a form of male menstruation that signaled imminent manhood.

More temperate climes offered little reprieve. In 1991, a glacier in the alpine region along the Italian-Austrian border yielded the mummified body of a Neolithic man. Scientists named him Ötzi after the nearby Ötz Valley. He was 5,300 years old and, along with an arrowhead lodged in his shoulder, he had whipworm.

Six thousand years ago, lakeside communities in Switzerland and Germany had fish tapeworm. And then, as their diet changed, they acquired more cow tapeworm. In both Gallo-Roman and medieval times, tapeworms began making class distinctions. The rich, who evidently preferred their fish and beef undercooked, got tapeworms more often than the poor. (In the twentieth century, the fish tapeworm earned a reputation as one that favored Jewish grandmothers, who acquired it while taste-testing raw gefilte fish, a delicacy that contains carp.)

By medieval times, the ascaris-trichuris combination had become ubiquitous in Europe, showing up in latrines across the Continent. "The European historical period seems to be written on an *Ascaris* and *Trichuris* parchment," observes the French parasitologist Françoise Bouchet.

In the Americas, the world was just as wormy, and long before the advent of agriculture. Chileans had fish tapeworm 6,100 years ago. The widespread prevalence of hookworm in pre-Columbian times has prompted some to propose an alternate migration route for the peopling of the Americas. Why? Hookworm requires time in warm soil. That's where fertilized embryos molt into infective larvae. These parasites couldn't have survived crossing the frigid Bering Land Bridge 15,000 years ago, the thinking goes. Some pioneers must have come via a warmer route, or perhaps by sea in a voyage that lasted less time than the life span of these parasites. A third possibility: The University of Nebraska paleoparasitologist Karl Reinhard points out that humans re-create the tropics with fire, housing, and clothing wherever we go. The greater climate may be frigid, but the microclimate we produce tends to be perfectly humid and toasty.

When Europeans arrived in the Americas, they also imported their particular parasites. The whipworm-roundworm duo showed up in sediments from the colonial period in Williamsburg, Virginia, and Philadelphia. Fish-loving Norwegian immigrants brought their fish tapeworm to Minneapolis. Residents of the Five Points neighborhood in New York City, a notorious slum in downtown Manhattan around a pond that's now paved over, had loads of whipworm and ascaris. And Chinese laborers brought exotic liver flukes to California around the turn of the nineteenth century.

Sediments laid down in colonial Albany tell a story of parasite load increasing with population expansion. The Dutch West India Company founded the city as a fur-trading outpost in 1614, but by the mid-eighteenth century, the city had become an important military outpost for the British. Barracks and stockade walls went up. The population grew. And, judging by the quantity of worm eggs left in sediments, Albany residents stewed in their own filth.

Cows and pigs wandered freely. Residents emptied chamber pots into drainage ditches, or they used the contents to fertilize vegetable gardens. Scientists find the parasite eggs aggregated in rings around houses and vegetable beds. The town's denizens probably constantly imbibed infective eggs with their veggies. A teaspoon of privy soil from the time contained more than 150,000 parasite eggs. Affluence didn't help: Parasites afflicted rich and poor alike. And conditions only worsened after the Revolutionary War. The town grew from 3,500 people in 1790 to 50,000 in 1850—and then to more than 90,000 in 1880.

For a time, hygiene improved. In the early nineteenth century, the city prohibited open-air dumping, and an innovation—stone-lined cesspools—arrived. Both developments lessened the spread of parasites. Then a public sewer system began operating in the 1880s—a blessing, you'd think—but it drained right into the Hudson River, which supplied the town's drinking water. Albany residents were now drinking from their own sewer.

The town's experience spoke to larger changes afoot: namely, energy from fossil fuels; an abundance of consumable products, which translated into previously unimaginable quantities of waste; and the feverish growth of cities. The Industrial Revolution, which would forever change human experience, was in full swing.

THE APEX OF SQUALOR— THE INDUSTRIAL REVOLUTION

Over a century earlier, in England's Derbyshire countryside, a new sort of building had gone up. Erected in 1771, Cromford Mill spun cotton thread not with weavers seated at spinning wheels, but with flowing water from the River Derwent. With an array of cogs and pulleys—a clockmaker helped design it—the mill could perform the work of a hundred cottagers. At that point, crude coal-fired engines had existed for decades, but as coal began to power mills like this, the Industrial Revolution began in earnest.

Towns and cities had been filthy places for centuries. Chamber pots and other refuse often went right into the street. Roving bands of pigs served as edible garbage-disposal units, a practice that stretched back to the Neolithic Levant. But the rapid urbanization that occurred during the Industrial Revolution, and the concurrent population explosion, brought this storied filth to a new intensity. "Industrialism, the main creative force of the nineteenth century, produced the most degraded urban environment the world had yet seen; for even the quarters of the ruling classes were befouled and overcrowded," writes the historian Lewis Mumford.

In 1801, London had about 100,000 people—the only city in Britain with a population that large. Fifty years later, it had grown to 2.5 million, and ten other English towns had passed the 100,000-person mark. In 1701, 5.06 million people lived in England. A century later, 8.66 million did. In 1851, there were 16.74 million Britons. And the crowding affected health.

In the early nineteenth century, life expectancy in England and Wales was forty-one. In cities, however, it was considerably less—thirty-six in 1840s London, and just twenty-six in Liverpool and Manchester. Extremely high infant mortality accounted for these frightful statistics.

In many towns and cities, nearly half of all children died by age five from typhoid, dysentery, and, later, cholera. Cities survived as entities only by feeding off the constant influx of rural immigrants.

And the crowded, dirty conditions allowed old parasites to acquire new virulence. Modern analysis suggests that we carried *Mycobacterium tuberculosis,* the bacterium that causes tuberculosis, out of Africa with us—that it wasn't, as has long been suggested, acquired from cows. But a wave of tuberculosis swept Europe at the very beginning of the Industrial Revolution. The "white plague" accounted for two of every five deaths among the urban working classes. The better-off weren't spared. The poet John Keats, the novelists Anne and Emily Brontë, and Charles Darwin's daughter, among other notables, succumbed to consumption. The pale, ethereal look of the consumptive even gained a desirable mystique. The poet Lord Byron once remarked, "I should like to die of a consumption . . . because the ladies would all say, 'Look at that poor Byron, how interesting he looks in dying.'"

Britons responded to the high infectious load by shrinking. After increasing in the late eighteenth century, the height of military recruits decreased during the early nineteenth. Rural men remained taller than urban, and Scots and northerners taller than Londoners and others from the urbanized southeast—a pattern now reversed. Britons would regain their lost height during the late nineteenth century after sanitary reforms, but first they had to manage the sheer quantity of sewage produced by their cities.

In times past, "rakers" or "gong farmers" collected excrement from cesspools and sold it as fertilizer to farmers near London. During the wars with Spain in the 1600s, nitrogen extracted from sewage went into gunpowder. But as London grew, and farms moved farther afield, these recycling practices became impractical. And, ironically, a new type of toilet compounded the problem. Rather than store excrement in a cesspool— the old way—the newly invented "water closet" swept it away with a pulse of water. The city wasn't equipped to handle the outflow of sewage, however, and the waste flowed right into the city's major waterway.

"The Thames is now made a great cesspool instead of each person having one of his own," lamented one observer in 1840. And worse, the river was tidal. Depending on the moon's position in the heavens, it flowed backward, forward, or simply stood still. London was continually bathed in its own effluence.

The situation came to a head in July 1858 in what was later branded "the great stink of London." The Thames had grown so putrescent—"a fermenting sewer," in the words of the sanitary reformer Michael Faraday—that Parliament, recently rebuilt on the riverbank, couldn't convene.

"Parliament was all but compelled to legislate upon the great London nuisance by the force of sheer stench. . . . We are heartily glad of it," wrote the *Times* that June of 1858. And finally, the ruling classes moved to address what today we'd call a public-health catastrophe. They weren't necessarily motivated by altruism, but by a well-founded worry that the Industrial Revolution would sputter out in its own filth. After all, how could laborers work if they were sick and dying?

Sanitary reforms began in earnest. London hired an engineer, Joseph Bazalgette, to design a sewer system to carry the city's effluence a safe distance downstream. Three decades later, the Thames had become a different waterway—"the cleanest metropolitan river in the world, which it remains," in the words of the historian Stephen Halliday.

THE UNITED STATES:
LAND OF AN INNOCENT GRIME

At the dawn of the nineteenth century, as industrialization convulsed Britain, the newly independent United States remained predominantly a country of farmers, homesteaders, and frontiersmen. In the first U.S. census, conducted in 1790, just one of twenty Americans lived in cities. Most Americans lived in villages or homesteads.

Perhaps because of their mostly rural existence, Americans seemed especially dirty to European eyes—"filthy, bordering on the beastly," in the words of the English visitor William Faux. "Dirty hands, heads, and faces everywhere," he noted. The U.S. was the developing world. Shoes were prohibitively expensive. Many went barefoot, except during winter. Mosquitoes, ticks, and ants attacked in hordes. Flies swarmed over food. In 1818, another English visitor described "a sort of out-of-doors slovenliness . . . bits of wood, timber, boards, chips, lying about, here and there, and pigs and cattle trampling about in . . . confusion."

Given Charles Dickens's descriptions of London around the same time—"dogs, undistinguishable in mire. Horses, scarcely better; splashed to their very blinkers . . . crust upon crust of mud, sticking at those points tenaciously to the pavement"—some of this antipathy must stem from inborn English contempt toward former subjects. And superficial grime aside, skeletal remains from the turn of the eighteenth century indicate that Americans were better fed, taller, and more robust than their English contemporaries.

American unkemptness was not, apparently, unhealthy, but rather characteristic of a mostly rural people who worked the land. Some Americans may have even venerated grunge. "In the eyes of hard working New

England or Midwestern farm families, dirt was seen as something positive, even healthy," writes Suellen Hoy in her book *Chasing Dirt*. "Above all, it gave life and livelihood in the form of crops."

Then the revolution transforming Britain arrived at American shores. American cities began to grow. Between 1820 and 1850, the number of people living on an average block of lower Manhattan grew from 157.5 to 272.5. Irish immigration, sparked by famine, helped fuel the growth. Like London before it, New York became pestilential. Farm animals lived in basements. People slept on boards laid above muck. Refuse of all sorts—including dead animals and many tons of manure—clogged city streets. (An estimated 130,000 horses lived in New York City in the late nineteenth century. Each produced, on average, 22 pounds of manure and a quart of urine daily. That's roughly the equivalent of forty-five dump trucks full of equine waste per day.)

In 1865, the sanitary reformer Stephen Smith described a city drowning in rotting vegetables and fruit, dead animals, ashes, and human excrement. "It is a melancholy fact that fifty per cent of the mortality of cities is estimated to be due to such causes, and is hence unnecessary," he later said. The *Medical Times* concurred. "The country is horrified when a thousand fall victims in an ill-fought battle"—probably a reference to the Civil War—"but in this city 10,000 die annually of diseases which the city authorities have the power to remove, and no one is shocked." That indifference was about to change.

CHOLERA EPIDEMICS SCARE UP REFORM

In 1817, the world learned of a new disease—cholera, a diarrheal affliction that could kill in a single day. The *Vibrio cholerae* bacterium is native to the Indian subcontinent around the Bay of Bengal, where it lives in brackish water, scooting about with a single flagellum. Vibrio bacteria are not, as a rule, pathogenic to humans. But scientists now know that *V. cholerae* is really a hybrid organism—a vibrio bacterium that, upon infection with a virus, acquired new powers of virulence.

This virus-bacterium fusion likely occurred repeatedly throughout history in and around the Bay of Bengal—or, more accurately, in the bowels of someone who lived in that area—but it took the faster ships of the Industrial Revolution, first sleek sailing ships, and then metal steamers, to spread cholera beyond southern Asia.

In the early 1820s, the disease passed through Afghanistan into Persia, reaching the Caspian Sea. When it receded, Europe and the Americas breathed a sigh of relief. However, in 1829, it clawed its way outward again, passing through Russia and Hungary, and arriving in western Europe by 1831. The following year, it crossed the Atlantic, striking Montreal first, and then New York City. By 1834, the disease had traveled to the Pacific coast.

The death toll was dramatic. Half of those afflicted died. In the second pandemic, that translated to 100,000 Hungarians, and slightly more Frenchmen killed. In American cities, cholera dispatched 5 to 10 percent of the population. By one estimate, 150,000 Americans perished.

Cholera continued to sweep the world in waves roughly once every decade—1839, 1863, and 1881—but it was hardly the only epidemic disease. Yellow fever, a mosquito-borne virus that also periodically swept through the U.S., claimed just as many lives, and probably more. As late as 1878, it killed nearly 10 percent of the populations of Memphis and Vicksburg. It would halt the initial efforts at constructing the Panama Canal.

But for reasons that historians still struggle to explain, cholera inspired a new level of dread. "More than any other infectious disease, cholera brought the world together," writes Christopher Hamlin in *Cholera: The Biography*. "The fate of all might be in the bowels of any." Maybe the indignity with which cholera killed—endless vomiting, relentless "rice water stool," a pale blue pallor—terrified American hearts in a way that other plagues hadn't. You could be healthy in the morning and dead by evening. But again, yellow fever, called *el vómito negro* in Spanish—the black vomit—hardly treated its victims any better. ("Yellow" refers to the color of the flag raised when a ship carried stricken crew members.)

Perhaps cholera's foreignness—it came from the Orient—played to latent and growing fears of "the other," of African Americans, of the newly arrived Catholic Irish and Jewish immigrants. Maybe it reawakened a cultural memory of the Black Death, also from the east, centuries earlier. Or perchance the powers that be worried that cholera would spark social unrest. In some European cities, mobs rioted when cholera appeared.

Whatever the reason, the disease spurred sanitary reforms. No one yet understood how illness spread, but hygienic practices had gained credibility on the battlefield. Florence Nightingale's experience first in the Crimean War, and later running hospitals in Britain, helped substantiate notions that cleanliness—clean wound dressings, waste removal, scrubbed floors—could stem disease. American nurses brought home similarly instructive experiences from the Civil War.

New York City, meanwhile, found that quick action by authorities could quash epidemics before they gained momentum. In 1866, with cholera poised to ravage the city yet again, the newly formed Metropolitan Health Board quarantined the steamship *Virginia,* which had sick passengers on board. Doctors moved house-to-house, removing the indisposed to dispensaries. Squads operating day and night reported new cases. Although they didn't understand how diarrheal disease spread, the intervention made a difference. Just six hundred people died. Where not so many years prior, the American president Zachary Taylor had, as cholera gripped the nation, called on citizens to fast and pray in atonement, now authorities understood that decisive action could halt a disease's spread. And so began a new era of public-health initiatives.

The Old Croton Aqueduct began supplying freshwater to New York City beginning in 1842. Wastewater treatment plants went up in the 1890s. And George Waring Jr.—a dandyish mustached fellow who trotted around on a horse—established a two-thousand-person army of white-clad sanitation workers to keep the city clean. When he died in 1898 from yellow fever contracted in Cuba, mourners christened him the "fever slayer" and the "apostle of cleanliness."

Sanitary reforms worked almost better than anyone could have hoped. And to this day, they're held in the highest regard. A 2007 survey in the *British Medical Journal* ranked the "sanitary revolution" as the most important medical development of fifteen options during the preceding 170-odd years—more important than antibiotics (no. 2), vaccines (no. 4), or germ theory (no. 6). But somewhat paradoxically, sanitary reforms operated on incorrect notions of how disease spread. They looked to the "miasma theory" of disease, which had roots in ancient Greece. Illness came from *miasmata,* sanitarians thought, foul smells emanating from swamps, graveyards, sewers, and even disturbed earth. Even George Waring Jr., who'd designed an advanced sewer system for Memphis, went to his grave thinking that the germ theory of disease was completely wrongheaded.

By 1900, however, the once-fringe germ theory of disease had gone mainstream. An Englishman named Edward Jenner had, in the 1790s, developed the first smallpox vaccines from a related virus that infected cows. In the mid-nineteenth century, the Frenchman Louis Pasteur had disproved the theory of spontaneous generation, and in the process invented protocols for food sterilization and preservation. Moreover, he developed vaccines for anthrax and rabies. The German Robert Koch, meanwhile, isolated the bacteria that cause tuberculosis and cholera in the 1880s, further proving that certain bugs cause certain diseases. Suspecting germs, not *miasma,* physicians such as Ignaz Semmelweis in Vienna and Joseph Lister in Edinburgh dramatically reduced mortality in hospital settings with simple disinfection protocols.

And in 1928, the scientist Alexander Fleming returned from vacation to his lab in London to find mold growing in petri dishes meant to cultivate bacteria. The clear, bacteria-free margin around the bluish spots caught his attention. This mold eventually yielded the world's first antibiotic—penicillin. First used on Allied soldiers during the latter years of World War II, penicillin became publicly available thereafter. Other antibiotics derived from soil microorganisms followed, such as strepto-mycin, which, unlike penicillin, could treat tuberculosis.

Sanitary reforms and the new understanding of disease afforded by germ theory, vaccines, and antibiotics transformed human existence. Someone born in western Europe in the mid-nineteenth century could expect to reach age forty-five. A hundred years later, life expectancy had pushed into the seventies. By the year 2000, women, who generally live longer than men, routinely passed the eighty-year mark in developed countries everywhere.

Average life spans, however, belie the true accomplishment. Death had always disproportionately struck the very young. Sanitary improvements, vaccines, and antibiotics had their greatest impact on infant mortality. If you had four children in the early nineteenth century, you reasonably expected one to die in the first year of life, a victim of dysentery, measles, typhus, or a number of maladies that today are easily treated or vacci-nated against. Mortality was higher in urban slums and among the poor, of course, edging toward half of all children born in some places.

But by the mid-twentieth century, child mortality in the U.S. and U.K. had declined by almost an order of magnitude. Just 3 of every 100 chil-dren born died in the first year. Today, roughly 5 of every 1,000 children in the U.K. die before age one, a fiftyfold improvement. (The U.S. lags at 7 per 1,000 births.)

Infant mortality among modern-day hunter-gatherers, like the !Kung in southern Africa or the Aché in South America, approximates that of pre-modern Europe. About one-quarter of children die young, often from infectious disease. This pattern was likely a constant throughout human evolution. Which makes the "mortality revolution" of the past 160 years all the more extraordinary. We'd finally escaped this eternal and terrible fact of existence—the premature death of our progeny.

This triumph over infectious disease represented the culmination of a trend that began some 60,000 years earlier when humans first dispersed from Africa, a progression toward greater dominion, for better or worse, over life on earth. The landscapes we first encountered in Eurasia, the Americas, and Australasia teemed with a richness and variety of ani-mals—mammoths, cave bears, woolly rhinoceroses, giant wolves, huge sloths, and hippo-sized wombats—that seems fantastical today. But by

10,000 years ago, these animals, as well as our hominid cousins the Neanderthals in western Eurasia and other relatives in the east, were gone.

For millennia, *Homo sapiens* has left a trail of extinction in its wake. Even as our own collection of parasites grew ever larger, we simplified external ecosystems. With the advent of germ theory and the sanitary movement in the nineteenth century, this unarticulated habit of liquidation entered a new phase. And with the same zeal that characterized our pursuit of animals we liked eating, we went after the microbes and parasites that had, since time immemorial, eaten us. In the process, we may have created a biological novelty: an immune system evolved to deal with a bustling collection of hangers-on that, in the space of a generation or two, found itself suddenly alone.

THE FIRST IMMUNE MALFUNCTION:
HAY FEVER

In March 1819, a physician named John Bostock presented a case report—his own, it turned out—to the Medical and Chirurgical Society of London. He described "a periodical affection of the eyes and chest" that began in mid-June each year. He thought the sun caused the malady. But Bostock was describing hay fever.

The affliction was new in Britain, and apparently quite rare. In 1828, Bostock described another twenty-eight cases of "catarrhus aestivus," or summer mucus discharge. "One of the most remarkable circumstances respecting this complaint is its not having been noticed as a specific affection, until within the last ten or twelve years," he wrote. More impressively, it appeared only among the upper classes. "I have not heard of a single unequivocal case occurring among the poor," he observed.

Almost fifty years later, the Manchester physician Charles Blackley, himself a hay fever sufferer, inhaled a collection of pollens, and correctly concluded that pollen, not sun or heat, caused the ailment. He made several other telling observations: Hay fever had been less frequent even thirty years prior, and mostly unknown in earlier times; whereas it was once a disease of nobility, now it afflicted the educated class as well; and somehow farmers, who inhaled pollen on a regular basis, never developed the disease. "[T]he persons who are most subjected to the action of pollen belong to a class which furnishes the fewest cases of the disorder, namely, the farming class," he wrote.

Blackley offered two explanations: Either education made one more vulnerable to hay fever, or farmers' continual exposure to pollen protected against hay fever. If the latter explanation held true, he predicted that continued urbanization would greatly increase the prevalence of the disorder. How prescient he turned out to be.

By then, money and status had inserted themselves into the story. Precisely because of hay fever's consistent association with affluence, it became, like gout (a well-fed, rich man's affliction) and consumption (the sensitive romantic's ailment) before it, fashionable. For the London physician Morrell Mackenzie, the English "proclivity to hay fever" was "proof of our superiority to other races." He noted its preference for higher classes as evidence of that eminence. "One of the most singular features of this complaint is, that it is almost exclusively confined to persons of some *education,* and generally to those of fair social position," he wrote.

Not so fast. In 1911, the American physician and nose-dripper William Hard countered that hay fever was now an "American speciality . . . the English compete with us no longer."

"In no other country is the Hay Fever travel toward certain regions so thick that railways serving those regions might well enter Hay Fever with the Interstate Commerce Commission as the basis for part of their capitalization," he bragged. "In no other country does Hay Fever give so much employment or cause so much prosperity. It has come to deserve to be a plank in the national platform of the Republican party."

He was referring to what had become a lucrative business in the U.S.: retreats for the gilded class during hay fever season. They sprang up in New Hampshire's White Mountains, New York's Adirondacks, and along the shores of the Great Lakes. By the sounds of it, everyone who was anyone clamored for admission into this sneezing elite.

"Only individuals of the highest intellectual grasp, and the strongest moral fibre have the disease," said a resorter named George Scott. "If it were not for the hay fever, I might have lived all my life among those who are not classed among the intellectual giants of America." For some, the disease typified everything that had gone wrong in the newly mechanized civilization. "Contemporary civilisation alone had produced the peculiar combination of causative agents so deleterious to nerve force," wrote the American physician George Beard, who suffered from hay fever.

The apparent absence of hay fever among African Americans was claimed as another proof of the superiority of the white race in the U.S. Its nonexistence in Africa and Asia showed the eminence of the English colonizers. (Only they got it abroad.) And the apparent absence of hay fever in Scandinavia, France, Italy, Spain, and Russia indicated the primacy of the English race even among Europeans. "The fact of exemption from hay fever of savages and practically all of the working classes in civilised countries, as well as other considerations, suggests that we must look upon hay fever as one of the consequences of higher civilisation," wrote one physician working in Britain.

The science behind these observations was certainly not very rigorous. And given the status attached to having hay fever, the epidemiology must be taken with several grains of salt. And yet the pattern described is suggestively specific. The U.K. and the U.S., the two nations that first noted the curious affliction, were also among the first nations to urbanize and industrialize. They were the first to experience the disaster of the modern city, and among the first to institute major sanitary reforms. They had the first populations—the newly moneyed merchant and professional classes—with both the desire and means to clean up. Scandinavia, Italy, Spain, Russia, and to a lesser degree France, on the other hand, remained largely agrarian countries with a mostly rural populace until later.

Something without biological precedent had occurred in these populations: the removal, perhaps for the first time in human evolution, of certain microbes and parasites from the human organism. Our bodies would never work quite the same way again.

CHAPTER 3

Island of Autoimmunity

Nothing makes sense in biology except in light of evolution.
—Theodosius Dobzhansky

One blustery spring day in Sardinia, I find myself wandering around a crumbling Bronze Age tower made of black volcanic stone. A neurologist named Stefano Sotgiu accompanies me. He expounds on the mysterious structure looming over us. It is called a *nuraghe*; *nur* translates roughly to "hollow" or "pile of stones" in the language spoken before the Romans brought Latin to Sardinia. More than seven thousand of the edifices dot the island, and they were already ancient—1,500 years old—when the Romans arrived more than two millennia ago. They serve as an important cultural reference point: While the rest of Christendom begins its calendar with the birth of Jesus, Sardinians, Sotgiu explains, divide history into pre- and post-*nuraghe*—a nearly 2,000-year head start.

No one knows why they were built, although some have theorized that the towers provided escape from the constant biting of mosquitoes. And the blood-feeding insect has certainly weighed heavily on the Sardinian psyche; in Cagliari, the island's capital to the south, the local Virgin Mary is Our Lady of *Bonaria*—or "good air," the opposite of *malaria,* which derives from the Latin for "bad air." These days, however, archaeologists are apt to characterize the towers as status symbols, Bronze Age versions of modern-day skyscrapers.

For Sotgiu and me, the *nuraghi* furnish a ready-made metaphor for our conversation. Sardinians inexplicably have one of the highest prevalences of autoimmune disease in the world—the reason I've come to the island. They're between two and three times as likely to develop multiple sclerosis compared with mainland Italians, or people from nearby islands, such as Corsica to the north or Sicily to the southeast. And they're second only to the Finns in their vulnerability to autoimmune (type-1) diabetes.

Why?

The fallback explanation imputes genes, which is probably roughly accurate. But Sotgiu has taken the "bad genes" hypothesis a step further than usual. He's asked what these "autoimmune" gene variants might be for— what's their purpose?—and why are they so common among Sardinians?

He didn't have to delve far into the past for an answer. Malaria was endemic on the island until just sixty years ago. His parents, uncles, and aunts all survived it, and their parents and grandparents before them. Until the mid-twentieth century, malarial fevers were a rite of passage on the island, and had been for millennia. Those who were resistant to the parasite survived; those who weren't died. Gene variants that helped defend against it became enriched among Sardinians. Much like the mysterious *nuraghi* that characterize the island's landscape, these protective genes now define the Sardinian genome. These same variants, Sotgiu thinks, also increase the chances of developing autoimmune disease. The price Sardinians pay for excellent antimalarial defense is a tendency to fall ill with multiple sclerosis.

I follow Sotgiu up a winding staircase of rough-hewn stone. We emerge on a dais about sixty feet high with a commanding view. A wind-whipped grassy plain stretches before us, framed by flat hills in the distance. Trenches dug on Benito Mussolini's orders in the 1930s cut through the meadow below. They were meant to drain the malarial marshes once and for all. Cone-shaped, thatch-roof huts, a design common to old Europe, rise here and there from the tall grass. They're mostly used to house sheep these days, says Sotgiu. But people lived in them once. Birds hover at our eye level, held aloft by the northwesterly wind, which in Sardinia and greater Italy has its own name—*il vento maestrale.*

Sotgiu has light brown hair, a trimmed goatee, and a relaxed, convivial manner. As we survey the plain, in carefully spoken English he ticks off a list of the extremes that define his homeland. Sheep outnumber people on the island by nearly two to one. The island is the least populated area of Europe. It has the lowest birthrate in Italy, and one of the lowest in Europe. It suffers from high unemployment. The younger generation often leaves to join a Sardinian diaspora in Spain and mainland Italy, among other places. Sardinia also produces an usually large number of centenarians, people who reach age one hundred.

Traditionally, outsiders have regarded Sardinia as a backwater, he explains, an island of rough peasants and shepherds who prefer to be left alone. (Later I learn that when the animated American show *The Simpsons* is dubbed into Italian, groundskeeper Willie, who's Scottish in the U.S., has a Sardinian accent.) The feeling of distrust is mutual. "We have suspicion with regards to outsiders," says Sotgiu. For three thousand years, wave upon wave of would-be rulers arrived from the sea. Sardinians responded by retreating to the island's interior. "That's probably why we weren't changed by all the invasions," says Sotgiu. This insularity, both figurative and literal, partly explains the Sardinian genome's uniqueness, and its vulnerability to autoimmune disease.

Today, 1 in 430 Sardinians has multiple sclerosis, a degenerative disease of the central nervous system that, as it progresses, steals one's ability to move limbs, to see, and eventually to breathe. (That's the official number, but Sotgiu confides that unpublished data put it higher still.) One in every 270 Sardinians has type-1 diabetes, an autoimmune condition in which the immune system attacks the body's insulin-producing organ, the pancreas.

The stats weren't always like this here. In Sardinia, there's a distinct Year Zero for autoimmune disease. Just after the eradication of malaria in the 1950s, immune-mediated diseases began increasing precipitously. Sotgiu thinks the timing isn't coincidental. Malaria may have selected for autoimmunity-prone genes. But infection with the malaria parasite *Plasmodium falciparum* likely protected against the dark side of the very genes it helped shape. In this aspect, Sotgiu's hypothesis departs from more run-of-the-mill invocations of genetics. He suspects that the highly specialized Sardinian immune system functions properly only in the context of the invader it evolved to thwart. Sardinians need to engage with their old foe, in essence, to avoid the demons lurking within.

A BLIGHT OCCASIONED
BY DEFORESTATION AND IRRIGATION

Four plasmodium parasites cause malaria in humans, *P. vivax, ovale, malariae,* and *falciparum.* (A fifth, *P. knowlesi,* also infects humans, but it's considered native to macaque monkeys.) Of the four, *P. falciparum* is by far the most deadly. Female mosquitoes carry the parasite between people. When a mosquito that has just fed on an infected person bites a second victim, it injects a little plasmodia-laden saliva. The parasite migrates to the liver and replicates. Second-generation clones then disperse to the bloodstream, seeking red blood cells, which they invade. Once established in erythrocytes, they replicate again—doubling four times until they reach sixteen total—and then burst forth, leaving the husk of the red blood cell like discarded chaff. Symptoms include cyclical fevers and chills. Sometimes a dry cough sets in. At its worst—and especially in children—*P. falciparum* causes cerebral malaria. Parasitized red blood cells "stick" in capillaries around the body, prompting convulsions, coma, brain damage, and death.

For perhaps 10,000 years, *P. falciparum* has chipped and hammered ceaselessly at the human genome in the malaria belt, a swath of land stretching from tropical Africa, around the subtropical Mediterranean rim, throughout the river valleys of the Levant, and across South Asia all the way to Papua New Guinea. Trade has helped spread the parasite far and wide. And by the time Geoffrey Chaucer wrote *The Canterbury Tales* in the fourteenth century, the "ague," as the English referred to malarial fevers, was known and feared as far north as the British Isles.

At the turn of the recent millennium, *P. falciparum* infected between 350 million and 500 million people yearly. One million, roughly the population of Dallas, died from the infection. (Both the incidence and mortality have since declined by about one-third.) Most of the victims were children five years old or younger living in sub-Saharan Africa. The high mortality has probably always been thus. And the death caused by *P. falciparum,* especially among children who haven't yet passed on their genes, has probably constituted the greatest selective pressure on human genes of any single pathogen. At least, that's the argument invoked by scientists seeking to explain some of our bizarre and, on the face of it, counterintuitive-seeming defenses against *P. falciparum*.

In 1949, the British scientist John Haldane tried to explain the apparent clumsiness of our adaptations to the parasite. He argued that the necessity of surviving malaria had prompted peoples around the Mediterranean to evolve apparently self-injurious traits. Exhibit A was a peculiar anemia called thalassemia. (*Thalassa* means "sea" in Greek. Populations most often afflicted hailed from coastal areas.) The anemia, Haldane argued, resulted from an advantageous adaptation to *P. falciparum*.

Red blood cells use a complex molecule called hemoglobin to transport oxygen around the body. When *P. falciparum* invades the cells, it devours that hemoglobin. Genes had arisen that made the hemoglobin less palatable to the parasite. Altering the design of this all-important molecule, however, had inevitable consequences. Children receive two copies of every gene (except those on the X or Y chromosomes, which are sex specific), one from each parent. So while possessing one version of the thalassemia variant protected against cerebral malaria, two copies caused anemia and premature death.

Parents who each had one thalassemia gene, therefore, would have two children like themselves—children who were genetically protected from cerebral malaria at no cost. Terrific. They'd also have one child without any protective genes. Less terrific, but acceptable. And they'd have one child with two copies of the protective gene. That child suffered from congenital anemia, and probably expired young. Terrible. But in the calculus of natural selection, two children with inborn resistance to a ubiquitous pathogen, a third left to fend for herself, and a fourth born sickly still translated to a net advantage. And so the thalassemia trait spread.

In 1954, A. C. Allison, a scientist studying malaria in East Africa, arrived at the same conclusion regarding a different trait, one he dubbed "sickle cell." Carriers had crescent-shaped red blood cells. As with the thalassemia trait, having one copy of the gene protected from cerebral malaria. Having two copies, however, condemned one to anemia and early death.

In the grander scheme of things, these defenses seemed inelegant. But that was Haldane's point: When it came to evolution, elegance had no bearing on survival. "[T]he struggle against diseases, and especially infectious diseases, has been a very important evolutionary agent," he wrote. "[S]ome of its results have been rather unlike those of the struggle for life in its common meaning."

The seminal work essentially proving Haldane's malaria hypothesis occurred in Sardinia. In the 1950s and '60s, scientists found that the frequency of thalassemia variants among Sardinians related directly to altitude.

The higher one went in Sardinia—there's less malaria at higher altitudes—the less frequent these genes became. In high-altitude Tonara, for example, 5 percent of the population carried the thalassemia-beta gene. But in Sassari, which sits on a low-lying plain in the island's northwestern quadrant, 25 percent had the trait. That matched the historical prevalence of malaria. (Sotgiu himself has one version of the thalassemia gene, as does his wife, which meant genetic testing during her pregnancy to ensure that their children wouldn't suffer from congenital anemia. They don't.)

High altitude couldn't protect many Sardinians, however. Compared with neighboring Sicily and Corsica, Sardinia is relatively flat, a feature I noted when I landed in late May. From the air, Corsica's snowcapped peaks contrasted mightily with the rolling greenery of Sardinia as it loomed into view. Much of Sardinia falls between sea level and 600 meters (2,000 feet) in altitude. In other words, *P. falciparum* probably infected most of the island's population for thousands of years. And the ubiquity of malarial infection is important for Sotgiu's hypothesis.

Whereas thalassemia emerges from defined gene variants, autoimmune diseases tend to be much more genetically complex. They involve a constellation of genes. And while scientists have identified some variants that increase the odds of developing autoimmune disease, many with these genes never develop disease, and plenty without do. Unlike thalassemia or sickle-cell anemia, in other words, in autoimmunity, genes don't absolutely predict destiny. Environment plays a critical role. What about the Sardinian environment has changed in the past half century?

That brings us to the heart of Sotgiu's idea. Malaria disappeared from the island roughly sixty years ago. He thinks that genes that protected against *P. falciparum* in Sardinia's past are only now, in malaria-free modernity, causing autoimmune disease.

HOW LONG ON THE ISLAND, AND FOR HOW LONG MALARIAL?

Beginning 2.6 million years ago—roughly when our *Homo habilis* ancestors first began flaking stone tools in Africa—the earth entered a period of glacial cycles. Geologists refer to them collectively as the Pleistocene. Each time glaciers advanced from the poles, sea levels dropped worldwide by up to 400 feet. During those periods, Sardinia and Corsica formed a single landmass. And Corsica was likely separated from the Italian mainland by no more than six miles of water.

So it's not surprising that hominids, probably Neanderthals, inhabited Sardinia long before modern humans left Africa, maybe as early as 170,000 years ago, which is two ice ages back, and about the time *Homo sapiens's* so-called mitochondrial Eve—the African woman from whom every human alive today has inherited his or her mitochondrial DNA—had her fateful brood. By the time modern humans arrived in Sardinia, maybe by 20,000 years ago, and definitely by 13,000 years ago, the pygmy hippos, elephants, and pigs that once roamed the island were long gone, probably from overhunting, as were the hominids who had pursued them. The new arrivals nonetheless found a land rich in shellfish, deer, and a now-extinct giant hare. Sea levels gradually rose, submerging the return route to Corsica and the mainland, and isolating these settlers in the middle of the Mediterranean. These Stone Age pioneers would provide the raw genetic material that the later malarial plague would sculpt. But first, they most likely enjoyed many millennia blessedly free of malaria.

Plasmodium falciparum is itself a recent arrival to the human organism. Wherever the parasite is endemic, humans have evolved different ways of resisting it. The sheer variety of defenses suggests that *P. falciparum* first infected human beings after we dispersed from Africa. If we'd had the infection before leaving our homeland, the thinking goes, we'd all have inherited similar defenses.

Other circumstantial evidence supports a recent acquisition of *P. falciparum*. After the last Ice Age, human-adapted mosquitoes, which carry *P. falciparum*, boomed in sub-Saharan Africa, evidence perhaps of a more mosquito-friendly postglacial climate, and of people living in more crowded conditions.

And then there are clues carried in the DNA of the parasite itself.

Decades ago, in line with our long-standing habit of blaming domestic animals for our parasites, scientists faulted chickens as the source of *P. falciparum*. But genetic analysis revealed that chimpanzees harbored a more closely related plasmodium species. A subsequent study, however, identified an even closer relative living in gorillas. So, for now, the story of the human acquisition of *P. falciparum* goes like this: At some point within the past 10,000 years, a mosquito that had just fed on an infected gorilla somewhere in central Africa bit someone nearby. At that moment, a malarial Eve colonized *Homo sapiens*. And in no time at all, the parasite had burned through the combustible forest of human bodies out of Africa, through the Middle East and beyond.

Precisely when the *P. falciparum* parasite arrived in Sardinia remains unclear. It probably colonized the eastern Mediterranean before arriving in the western. Egyptian texts began referring to an annual epidemic linked to the Nile's floods 5,000 years ago. The earliest direct evidence, in the form of plasmodium DNA, comes from a 4,000-year-old Egyptian mummy. But the anemia caused by adaptations to malaria, such as thalassemia, was already well known in ancient Greece and Anatolia, suggesting a much longer coexistence with the parasite. Deformed bones unearthed in a now-submerged 8,100-year-old village off the coast of Israel display telltale signs of thalassemia. And seafaring Sardinians, who during the Bronze Age traded and raided regularly in the eastern Mediterranean, would have encountered the parasite and presumably brought it home.

Despite the ample evidence of sustained early contact with malarial regions, however, historians usually blame the Carthaginians for bringing *P. falciparum* to Sardinia from North Africa around 2,600 years ago. And when it did arrive, the parasite found the island extremely hospitable. A marsh-friendly geography provided plenty of stagnant water, ideal habitat for mosquitoes. Like other peoples around the Mediterranean rim and beyond, Sardinians also inadvertently made the landscape more mosquito-friendly. They felled trees for timber, or to create pasture, increasing the number of sunlit puddles. And at least since Roman times, the island has had a reputation for pestilence.

In the centuries since, as foreign powers kept arriving by sea, settling coastal areas, ruling, cursing the malarial fevers, and pining for home, the Sardinians themselves retreated to the island's interior. Unlike in mountainous and relatively malaria-free Sicily, which assimilated wave after wave of settlers—Phoenicians, Greeks, Romans, Vandals, Arabs, and Norman crusaders—the Sardinian genome never really absorbed fresh blood.

Comparative genetic studies find that the Sardinian branch of the larger regional family tree separated early, and remained apart. Mainland Europeans are more closely related to Iranians than Sardinians. Although in some coastal towns, like the old Catalan capital Alghero, scientists can discern traces of outside genetic influx, by and large, whatever genes the constant, relentless whittling of malaria selected for in Sardinia, they were never diluted.

And then modernity came knocking.

WELCOME TO THE TWENTIETH CENTURY: SARDINIA CLEANS UP

Unlike much of Western Europe, which experienced a gradual decline in infectious disease beginning in the nineteenth century, Sardinia's epidemiological transition occurred relatively recently, and all at once. Beginning in the late 1940s, the Rockefeller Foundation, which had spearheaded efforts to eradicate hookworm from the U.S. decades earlier, sent people across the island on bikes and mules. Armed with tanks of DDT, they sprayed marshes, puddles, and any other standing water they encountered.

The effort proved remarkably successful. In 1947, authorities reported nearly 40,000 cases of malaria on an island of roughly 1.2 million people. Three years later, new cases had fallen to zero. After causing untold illness and death for millennia, in the space of a few years, the plasmodium parasite vanished. And about a decade later, the incidence of multiple sclerosis began inching upward.

Generally speaking, the prevalence of multiple sclerosis around the world runs on a north–south gradient, increasing toward the poles and decreasing toward the equator. And Sardinia lies at what should be a low-prevalence latitude compared with, say, Scandinavia or Scotland, two areas of notoriously high MS prevalence. So when Sardinian scientists first documented the upward trend of MS in the 1970s and '80s, noting that Sardinians born in the 1960s developed the condition more often than their parents or grandparents, they doubted that the perceived increase was even real. The trend was probably an artifact of improved diagnosis, they thought, or improved survival of MS patients.

But comparative studies quashed these doubts. In the mainland Italian province of Ferrara, the incidence held over time at 2 cases per 100,000 people yearly. In Sardinia, the incidence accelerated. Between the late 1970s and the early 1990s, it more than doubled from 2 to 5 new cases per 100,000 people yearly. And by the late 1990s, it had increased again to 6.8.

Then a paper in the mid-1990s directed Sotgiu's attention to malaria's paradoxical impact on the human genome. The study identified variants in Africans that influenced the outcome of malarial infection. One lessened the risk of cerebral malaria but increased the risk of anemia. A second increased the risk of anemia but lessened the risk of cerebral malaria. How did these genes work? They up- or down-regulated production of an inflammatory signaling molecule called tumor necrosis factor alpha, or TNF alpha. Excess TNF alpha allowed you to quickly beat back the parasite, but also increased your risk of complications. On the other hand, less TNF alpha spread over a longer period of time—the second variant— avoided cerebral malaria, but increased your chances of becoming anemic. You just couldn't win.

High TNF alpha was also, Sotgiu knew, a salient feature of autoimmune disease. Indeed, drugs commonly used to treat inflammatory disorders, such as Remicade and Humira, blocked this signaling molecule. The idea that the constant grinding pressure of malaria in Sardinia might have selected for genes that naturally elevated TNF alpha, and thus increased the risk of autoimmune disease, took hold in Sotgiu's mind. He began seeking evidence in the Sardinian genome. And it didn't take long to find.

White blood cells express a receptor called the human leukocyte antigen. They use the receptor, which you can envision as a molecular grasping claw, to present pieces of invaders to other immune cells, to instruct them that something unsavory is afoot, and to show them what the interloper looks like.

In 2001, geneticists found that HLAs known to predispose to multiple sclerosis occurred with greater frequency in Sardinia than almost anywhere else in the world. Most important for Sotgiu's hypothesis, these variants occurred with greatest frequency in those areas of Sardinia most ravaged by malaria in the past. If malaria were to leave a footprint on the Sardinian genome, you'd expect just this sort of pattern.

Sotgiu did not find, however, that gene variants that increased TNF alpha were similarly distributed around Sardinia. Instead, the entire Sardinian population had TNF-alpha-enhancing variants regardless of altitude or malaria-infection intensity in former times. The variants were enriched in the Sardinian population as a whole—ten times more common than in nearby Sicily, for example. Sardinians, it seemed, were hardwired for a relatively strong inflammatory response.

Now the question was, did all this innate inflammatory potential do Sardinians any good when faced with the plasmodium parasite? Sotgiu tested the idea directly. He mixed P. falciparum with white blood cells culled from Sardinian MS patients—patients who had both the HLA gene variant that increased the risk of autoimmune disease, and the gene variants that elevated TNF alpha. He used healthy Sardinians and MS patients from mainland Italy as controls. All groups mounted a similar inflammatory response to bacterial products. (Plasmodia are not bacteria, but protozoa—distant relatives of animals.) But compared with the two control groups, white blood cells from Sardinian MS patients had a knack for destroying the parasite. They were about one-third more effective at controlling the *P. falciparum* compared with controls. "It's immunological memory that's genetic," says Sotgiu.

Other observations also hinted that Sardinians had naturally amped-up immune defenses. Compared with Sicilians, Sardinians had about twice as much of an enzyme called chitotriosidase circulating in their blood. Elevated levels of the enzyme meant greater defensive potential, but also translated to an increased risk of both MS and stroke. The key observation was that high quantities of circulating chitotriosidase didn't increase the risk of MS everywhere. Whereas Sardinians in general produced about twice as much of the enzyme as Sicilians, sub-Saharan Africans, where MS was exceedingly rare, churned out about 40 percent more than even Sardinians, and three and a half times as much as Sicilians. What was the difference?

In Africa, these immune defenses mobilized in the context of infection. In Sardinia, they remained elevated in the absence of infection, a genetic consequence of having battled malaria for millennia. Without a real firefight, these once-advantageous adaptations had become detrimental.

As it happened, both the observation that malaria prevented autoimmune disease and its logical conclusion—that deliberate infection might halt autoimmune disease—had some history.

LEANING ON PATHOGENS, AND FALLING WHEN THEY DEPART

In the mid-1960s, the British scientist Brian Greenwood arrived in Ibadan, Nigeria. He hoped to gain experience working in the type of infectious environment that had become rare in western Europe. He'd previously worked with patients afflicted with rheumatoid arthritis, an autoimmune disease of the joints, in the U.K. And so he immediately noticed that the disorder was relatively absent in Nigeria. He formally investigated the apparent dearth. Of the 100,000 patients seen at the local hospital over the previous decade, he could find only 104 diagnoses of autoimmune disease. That was about one-sixth the frequency seen in hospitals in England and Wales. To rule out the possibility that poor diagnostic abilities accounted for the apparent disparity, he surveyed nearly six hundred

Nigerian villagers. The study turned up only two mild cases of autoimmune arthritis.

He knew that genetics couldn't explain the discrepancy. African Americans, many of whose ancestors hailed from this very part of West Africa, suffered more from systemic lupus erythematosus, another autoimmune disease, than both whites in the U.S. and Africans in Africa. So what distinguished the West African environment from the North American? Most obviously, parasitic infections, including *P. falciparum,* were relatively abundant.

Greenwood then made an observation that presaged Sotgiu's by three decades: Africans with malaria had antibodies directed at the plasmodium parasite; and they also had lots of an antibody called rheumatoid factor, which bound to the body's own tissues. In the U.K., elevated rheumatoid factor was associated with autoimmune diseases such as rheumatoid arthritis and lupus. In Africa, however, rheumatoid factor was strictly associated with malarial infection. It helped defend against the parasite. Greenwood's insight was that immune tendencies that helped repel invaders in one context caused autoimmune disease in another. So what about reintroducing the infection to prevent the autoimmune disease?

Back in Britain, Greenwood tested the idea in rodents. He infected rats prone to developing rheumatoid arthritis with the rodent-adapted *Plasmodium berghei.* They had much milder rheumatoid arthritis. The same was true of mice bred to spontaneously develop lupus, a devastating systemic meltdown in which the rogue immune system targets many organs, from skin to lungs to kidneys. Infection with *P. berghei* protected these mice as well.

These splendid results presented only one problem: How was any of this happening? What mechanisms explained these outcomes? Scientists knew that malaria suppressed the immune system, but these experiments occurred decades before the genomic era and its more nuanced understanding of immune function. There was no easy way to explain the results—to harness them. And so Greenwood moved on to other projects, becoming a respected researcher in the malaria field.

In the intervening decades, however, scientists continued to note the relative absence of autoimmune disease in sub-Saharan Africa, along with the ubiquity of parasitic infection, especially malaria. Eventually, the British researcher Geoff Butcher again directed attention to the relationship. He argued that malaria had selected for genes that predisposed to lupus in the developed world, and these genes explained African Americans' greater vulnerability to this particular autoimmune disease in the U.S. Yet those very same gene variants helped protect against malaria in Africa without causing autoimmune disease.

In the 2000s, scientists embarked on a new round of animal experiments testing the relationship. Mice with a gene known to increase the odds of developing lupus, called Sle3, did indeed display a remarkable capacity to fight off experimentally induced pneumonia and to resist sepsis. The "autoimmune" gene helped defend against microbial incursions.

What about people? Some "autoimmune" genes were indeed enriched in populations historically plagued by malaria. A variant that predisposed to lupus appeared relatively frequently in sub-Saharan African and Southeast Asian populations—peoples from the so-called malaria belt. East Africans with two copies of this gene had half as much cerebral malaria as those without. In relatively clean Hong Kong, meanwhile, those carrying two copies had a 70 percent elevated risk of developing lupus. How did the gene work? A Cambridge University group found that the variant, which made some white blood cells more aggressive by, essentially, incapacitating an off-switch, also made mice very good at clearing plasmodium parasites.

Here's the point: The tendencies underlying autoimmune disease have an evolved purpose other than causing misery by autoimmunity, and this purpose relates to defense. More important, in the context of the infections these variants evolved to handle, scientists repeatedly observe that autoimmunity materializes much less often. And that observation helps explain another mystery: why genes that predispose to autoimmune disease have become more prevalent in our recent evolutionary past.

WHY RISK AUTOIMMUNITY, UNLESS T'WAS NO RISK AT ALL

With age-related diseases, such as heart disease or dementia, one can always argue that, in the past, no one lived long enough to develop them. One can assert that, because they generally strike after childbearing age, natural selection was blind to the degenerative diseases of old age.

That argument is harder to make, however, for autoimmune diseases. They tend to strike in the prime of life, one's twenties and thirties, or earlier. And even if they don't kill outright, they generally impose a high cost on fitness. Even minor-seeming symptoms can have a significant impact. I've often contemplated how the loss of eyelashes from alopecia universalis, for example, would have affected my ability to survive 20,000 years ago. There's the deer in my sights and—*zzzttt!*—a gnat lands in my eye. Prey lost. Clan starved. Fecundity diminished. Genes deselected for. (And this scenario totally disregards my exercise-induced asthma. I'd better have perfect aim, because I would have been unable to pursue a wounded deer very far.)

Everything matters, even eyelashes. And with enough time, we can assume that natural selection would have pruned genes that prompted costly immune malfunction from the human genome. That precisely the opposite has occurred—that genes that predispose to type-1 diabetes, Crohn's disease, psoriasis, lupus, and celiac disease have become more frequent in the past 30,000 years—is therefore significant.

The population geneticists Luis Barreiro and Lluís Quintana-Murci argue that increased crowding in the late Paleolithic and Neolithic enriched autoimmune-associated gene variants. More recently, intensified exposure to new diseases from animals also cranked up the selective pressure. The more crowded and pestilential conditions became, in other words, the more advantageous these genes. Perhaps the benefits from possessing amped-up defenses always outweighed the costs incurred by autoimmune disease. Or maybe these genes, as the above studies suggest, didn't cause as much autoimmune disease in past environments.

Celiac disease, a kind of inflammatory bowel disease triggered by a protein in wheat, oats, and other grains, presents a thought experiment to test the premise. Of nine gene variants associated with celiac, natural selection has positively selected for four since the beginning of agriculture. That's paradoxical: Why would genes that predispose to a disease triggered by the very diet that's becoming more common—cultivated grains—spread? Probably because, in the past, these genes didn't cause said disease to the same degree. Indeed, even sixty years ago, far fewer people were diagnosed with celiac disease.

Type-1 diabetes presents an even starker case study. In one analysis, of eighty gene variants associated with the disorder, fifty-eight have become more frequent during recent millennia. Autoimmune diabetes usually has a childhood onset, and before scientists developed insulin injections in the 1920s, it was probably universally fatal. There's no stronger negative selective pressure than death before procreation. So if these genes always caused as much type-1 diabetes as they're causing today, they would have very quickly erased themselves from the human genome. But they didn't. They grew more common. They spread. Again, we can infer that in past environments these genes didn't produce the same vulnerability to auto-immune disease.

As we've seen, scientists have a pretty good idea how these gene variants enhance immunity. Lupus patients handle malaria with unusual grace. So do, apparently, Sardinians. What about type-1 diabetes? Finnish children with immune-system genes associated with the disease also have a talent—not for fighting malaria, but for resisting microbes that invade through the gut, like the Coxsackie and polio viruses. And mice bred to develop type-1 diabetes adeptly manage infections by bacterial parasites such as *Mycobacterium tuberculosis*, which causes TB. (As with malaria infection in lupus-prone mice, mycobacterial infections prevent auto-immune disease in these diabetes-prone mice.)

The same genetic proclivities that lead to type-1 diabetes in modernity probably equipped people to handily repel intracellular invaders in the past. More generally, these observations suggest that the genes underlying autoimmune disease aren't random mutations. They aren't unfortunate acquisitions. They belong to a very particular genetic tool kit that has evolved to help us survive in an ever-dirtier world.

So why didn't we ratchet up the immune response even further? There's a ceiling, it turns out, to how far a mammal can turn up the heat. Consider the wild sheep of St. Kilda, an archipelago off the northwest coast of Scotland. For decades, scientists have closely monitored a five-hundred-animal herd on one of the islands. Recently, they examined a decade's worth of the sheeps' blood samples, looking especially at self-reactive antibodies—a predictor of autoimmune disease in modern "domesticated" humans. Then they compared these levels to the animals' reproductive success.

The researchers found that self-reactive antibodies conferred an immediately apparent advantage in the wild. During the worst winters, more than half of the herd died off. But animals with more self-directed antibodies better survived the harsh conditions and the high parasite loads. Here was direct evidence of the advantage of autoimmune propensities. Yet, although overt autoimmune disease was absent, another drawback was evident: Sheep with this "autoimmune" tendency didn't have as many lambs. The highly functioning immune system had a cost: fewer offspring.

Rudi Westendorp, a scientist at the University of Leiden who studies aging, explained this dynamic to me. As a general rule for mammals, and especially for those like *Homo sapiens* with long gestation periods, the jumpier and more inflammation-prone the immune system, the greater the chances of interfering with reproduction. The fetus is essentially a foreign organism lodged in the mother. And Mom's immune system must maintain a delicate balance of tolerating the developing progeny while retaining enough firepower to zap pathogens. There is therefore an upper limit to how savage the mammalian immune system can become, a threshold beyond which immune acuity begins to interfere with reproductive success.

"When you go into this pro-inflammatory mode, it has a fitness cost," says Westendorp. "And it's selected against."

Now imagine moving those sheep into a parasite-free apartment in London—into clean, well-fed, couch-potatoey modernity. Without the immune activation of real infections, those long-lived sheep would, you can bet, develop actual autoimmune disease. Why? Because real infections engage the immune system in a very particular way. They induce suppressor cells of the sort I mentioned in chapter 1.

TRAITORS OR SAVIORS? THE CELLS THAT TOLERATE PARASITES

The immunologist Eleanor Riley at the London School of Hygiene and Tropical Medicine studies the immune response to malaria in sub-Saharan Africa. And she's found that, as is the case with other chronic parasitic infections, long-term infection with *Plasmodium falciparum* elicits a strong regulatory T-cell response. Where malaria is endemic in rural Africa, people tend to have more circulating regulatory T cells. In urban centers where it's less prevalent, however, people have fewer.

Scientists argue about whether this is good or bad. Those in the con camp point out that more T-regs correlate with higher parasite loads. You're giving them a free pass, they contend. And real-world evidence suggests that limiting your own T-regs helps fight off malaria. Consider the Fulani of Burkina Faso, West Africa. Italian scientists have found them to be unusually resistant to malaria. Sickle-cell genes aren't behind their resistance, however. Instead, they have a high frequency of those HLA gene variants associated with autoimmune disease. And they have a genetic defect in their regulatory T cells. Their suppressor cells are hobbled. As a result, they have an unusually amped-up inflammatory response compared even with nearby ethnic groups, such as the Mossi. And they eliminate invading plasmodia parasites with great aplomb. The price paid by the Fulani for incapacitating their own suppressor cells, however, is a greater vulnerability to autoimmune diseases, like type-1 diabetes.

The Fulani case highlights both the costs and benefits of T-regs, and suggests that at least in the Fulani environment, the advantage gained by disabling your own off-switch may outweigh the evident disadvantage.

But the Fulani approach to surviving malaria is just one. There are others. For example, Riley and her colleagues find that where *P. falciparum* is endemic, *more* regulatory T cells—the opposite approach—can improve chances of avoiding deadly complications, like cerebral malaria. Unlike the Fulani, these people don't beat off the parasite outright. Instead, they employ what you might call the "tolerate thine enemy" strategy: Let the parasite persist, allow it to steal more resources than is perhaps ideal, but avoid demolishing your own house in the process.

Sardinians offer further evidence that balance, not unrestrained aggression, is key to surviving malaria. In this case, the evidence comes in the form of one autoimmune disease that's thankfully absent in Sardinia, ankylosing spondylitis. A certain human leukocyte antigen variant (called HLA-B*2705) predisposes to this painful inflammatory condition of the spine, whose symptoms include fused vertebrae and a stooped posture. What's this HLA gene variant good for? The variant enhances defenses against retroviruses, such as hepatitis C and HIV. The gene is oddly distributed around the world. It occurs only in populations that were relatively free of malaria, like those at high latitudes. In populations like Sardinia's, however, the HLA variant is almost entirely absent.

Alessandro Mathieu and Rosa Sorrentino at the University of Cagliari think that possessing this variant in malarial regions like Sardinia was simply too dangerous. Unrestrained ferocity may be advantageous when dealing with viruses, but the same savagery when facing *P. falciparum* is likely to explode the boiler.

Indeed, the Fulani case notwithstanding, during malaria season in rural areas of Africa, Riley finds that both regulatory and attack cells increase in concert, not just one or the other. "The balance of inflammatory to anti-inflammatory immune responses is in my view crucial," Riley says. Dealing with malaria requires finesse, not brute force. Too many T-regs, and the parasite will suck you dry. But too many attack cells, and you run the risk of malignant malaria.

Why the obsession with suppressor cells? Generally speaking, people with autoimmune and allergic diseases have a dearth of, or dysfunction in, these regulatory T cells. Extrapolating from Riley's work in Africa, in the malarial times of yore, Sardinians likely had many more T-regs in circulation, not only from *P. falciparum* but also from other chronic infections, such as tuberculosis and worms. (As recently as the 1970s, half of the schoolchildren in the capital, Cagliari, had helminths.)

Would the T-regs induced by these parasitic infections counterbalance Sardinians' hardwired autoimmune tendencies? In that environment, would the Sardinian knack for managing plasmodium infection confer all benefit and no cost? The evidence from sub-Saharan Africa indicates that the answer is yes. Conversely, the emergence of autoimmune disease in Sardinia suggests that when formerly malarial regions clean up elsewhere, they may be in for an unpleasant surprise.

The larger issue is one of evolutionary context. Genes that predispose to autoimmune disease evolved in a radically different environment than today's, one characterized by multiple chronic infections—worms, malaria, tuberculosis—and a comparative surfeit of T-regs. When you remove those environmental stimuli, the regulatory aspects they induced subside. But the idiosyncrasies encoded in our genes remain fixed. They're permanently written into our genetic code.

Rick Maizels, whom we'll hear more about later, uses the term *allelic rheostats* to describe these gene variants—alleles that tweak aspects of immune function, but that, in the absence of infection, are "likely to overshoot," and cause autoimmune and allergic disease. They have a Janus-faced nature; they produce opposite outcomes, enhanced defense or degenerative disease, depending on the greater context. And while Sardinians, with their unique history and exposure to malaria, are an extreme example, the lesson is generalizable.

No matter who we are, we evolved with many more parasites and commensals, both large and small, than we generally encounter today. The implication—and let's face it, the hope—is that reestablishing contact with some of these organisms can rebalance the immune system. For his part, Sotgiu dreams of using a stand-in for *P. falciparum* to help Sardinian MS patients. And the idea has precedent. A decade ago, the neurologist Giovanni Ristori and colleagues at La Sapienza University in Rome injected twelve MS patients with bacillus Calmette-Guérin (BCG), a weakened mycobacterium used to immunize against tuberculosis. The progression of MS, as measured by magnetic resonance imaging, lessened dramatically, and remained markedly diminished two years later. The *M. bovis* bacteria in the BCG vaccine could, for reasons that weren't entirely clear, correct the malfunction underlying this particular autoimmune disease, and stop it in its tracks.

And scientists elsewhere are now developing this idea using not bacteria, but multicellular parasites—worms—to address autoimmune disease.

CHAPTER 4

Parasites to Heal the Gut

Discovery consists of looking at the same thing as everyone
else and thinking something different.
—Albert Szent-Györgyi, Nobel laureate

In the summer of 1995, Joel Weinstock found himself stuck on a plane
going nowhere. A lightning storm had rumbled into the Chicago area,
delaying his flight back to Iowa City indefinitely. It was an annoying development, and yet, sitting there without distractions—no phone calls or
papers in need of completion—had provided a welcome respite.

Weinstock, a gastroenterologist, had just attended a meeting in New
York City on inflammatory bowel disease, his specialty and a central pre-
occupation of the field. He also happened to be writing a chapter about
IBD for a forthcoming book on autoimmune diseases, and his mind
wandered to a long-standing question: Why had the prevalence of IBD
increased so dramatically in the past fifty years? In some populations, it
had ballooned from 1 case per 10,000 people, to 1 in 250—a fortyfold
increase in the space of two or three generations.

The disorder came in two versions: Crohn's disease, which often affects
the small intestine but can appear anywhere from "gum to bum"; or ulcerative
colitis, painful lesions in the large intestine, the broad loop of gut that culmi-
nates with the anus. Mysteriously, the incidence of ulcerative colitis tended
to increase first in a given population, followed by a spike in the incidence
of Crohn's. Symptoms, including bloody diarrhea, weight loss, anemia, mal-
nutrition, and sometimes death, mimicked those of gastrointestinal infection.
But scientists had yet to isolate an IBD-causing bacterium or virus. Indeed,
chronic inflammation in the absence of an obvious, legitimate target was the
defining characteristic of IBD—a flamethrower switched "on," but with no
enemy in sight. And the damage could be extensive: scar tissue buildup that
interrupted the flow of digested material; blood loss and anemia; even a gradual liquefaction of one's own tissues that produced an opening, a fistula, to
other areas of the abdominal cavity.

The gut, of course, wasn't sterile to begin with. It harbored a teeming community of microbes, some of which synthesized important molecules for the host, such as folate and vitamin K. These days, scientists compare the diversity of the gut ecosystem to that of a rainforest's, but in the 1990s, they waxed less poetic. They accepted, however, that the human body was, at its core, full of bacteria. One line of thought held that, for whatever reason, IBD resulted from an immune system that mistook friendly commensal bacteria for mortal enemies. This disease of progressive self-destruction stemmed from a case of mistaken identity.

Geneticists were valiantly trying to pinpoint the genes that predisposed people to IBD. But the genetic explanation never sat right with Weinstock. To his mind, the great variation of IBD in place and time strongly suggested an environmental factor. The increase had occurred in just half a century, a nanosecond in evolutionary time. "Bad" genes couldn't spread that quickly. Studies of identical twins also found significant disparity. If one twin had Crohn's, for example, the chances that the second would develop it were only 50 percent. For ulcerative colitis, concordance was lower still—19 percent. Clearly, when it came to IBD, genes did not determine destiny.

Sitting on the plane, Weinstock rewound progress in the field to its very beginning. In the 1930s, Burrill Bernard Crohn, a physician at Mount Sinai Hospital in New York City, described what would eventually be called Crohn's disease. All fourteen original patients were Jewish. History subsequently showed that IBD could afflict any ethnicity; but what if, Weinstock wondered, the high prevalence among Jews in New York City sixty years earlier was a clue to its cause?

Others had explicitly linked IBD, which generally struck in the third or fourth decade of life, with socioeconomic status while young. The cleaner one's circumstances during childhood, scientists found, the greater one's chances of developing IBD in adulthood. Hot, running water and a flush toilet while growing up elevated one's risk later. Drinking from a well or stream, and defecating in an outhouse, or the bushes, lowered it. Clearly, these were markers of affluence. But what about affluence predisposed to the disease?

As an hour's delay turned to two, and two turned to four, Weinstock engaged in a mental exercise. He imagined holding a mirror up to a map of the IBD prevalence in the U.S. Generally speaking, the prevalence increased as one moved from south to north. The reflection he now beheld in his mind's eye was reversed, something that increased as one moved from north to south. He went a step further: Rather than ask what new exposures might cause IBD—pollutants, drugs, and diet, for example—he asked what might prevent it. Is it possible, he wondered, that something that protected people from developing IBD had disappeared?

At the time, in addition to contributing to a book on autoimmune disease, Weinstock was also editing a tome on parasites. He'd spent more than a decade studying the blood fluke *Schistosoma mansoni*. The parasite was a marvel of evolution. Microscopic, torpedo-shaped, and fork-tailed, *S. mansoni* larvae awaited their prey in freshwater. After penetrating the skin of swimmers or waders, they navigated the bloodstream, sojourned in the lungs and liver, and settled down to mate, with the female installing itself in a groove along the male's body, in the veins draining the bladder and colon. They then shed fertilized eggs with the host's feces and urine.

These eggs, which sometimes ended up lodged in the liver and elsewhere in the body, could cause chronically inflamed, spherical lesions called granulomas. The lesions, which greatly resembled the inflamed ulcers of Crohn's disease, gave scientists a nifty model for studying IBD. But by most other measures, *S. mansoni* infection differed substantially from IBD. For one, the flukes could survive for decades in their human hosts, and the occasional egg-induced granuloma aside, the persistence of one rather large organism—about one centimeter long—inside another incited much less inflammation than you'd expect.

Indeed, as he edited the book on parasites, Weinstock had struggled and failed to concoct terrible things to say about these, and other, worms. One-third of humanity still carried helminths, the great majority with no symptoms. More had likely been exposed at some point in their lives, probably as children. Provided you were well nourished and didn't have too many, parasites, he thought, were relatively benign.

At that moment, these three seemingly separate puzzles—the weirdly benign parasites, the suggestive prevalence patterns of IBD, and the disease's mysterious increase during the twentieth century—melded into a single phenomenon. The inverted map he'd imagined, a protective factor becoming more pronounced as you moved from north to south, mimicked the historical prevalence of worms in the U.S. The historical increase of IBD followed their eradication. Worm infections, he thought, had protected against IBD in the past.

Back in Iowa, Weinstock talked over the idea with two colleagues, the gastroenterologists David Elliott and Robert Summers. Immunologically speaking, that worm infection might prevent IBD made immediate sense. In those days, immune responses were broadly categorized as belonging to one of two types: T-helper-1 or T-helper-2. The first, Th1, marshaled immune defenses against unicellular bacteria and viruses, such as salmonella or smallpox, that sought to invade your cells. The second response type, abbreviated as Th2, pursued much larger, multicellular invaders, such as worms and blood-feeding insects. The painful, red swelling around an infected cut or pimple exemplified a Th1 response. The itchy red bump of a mosquito bite typified Th2. Immunologists thought that the two responses were mutually exclusive. If you turned one on, the other would shut off, and vice versa. Gastroenterologists viewed IBD as resulting from excessive Th1. Introducing worms to this situation could presumably ramp up the Th2 response, and shut down the chronic Th1-type inflammation driving the malady.

In that moment was born the kernel of an idea that, in many ways, ran counter to the trajectory of Western medicine since the triumph of germ theory a century earlier. Germ theory held that infectious agents caused disease. Remove the offending microbe, or prepare immune systems with vaccinations, and you'd cure or prevent the associated malady.

But Weinstock was proposing both a more complicated model for the origin of inflammatory bowel disease, and a different kind of cure. Parasites, Weinstock knew, exerted a powerful influence on the host immune system. Until very recently, they'd been a constant presence over the course of human, and likely all mammalian, evolution. He was arguing that over millions of years of coexistence, the human immune system had adjusted to that presence, and even come to rely on it. So the sudden disappearance of parasites in the twentieth century left the immune system off balance. One consequence of that imbalance was a greater predisposition to IBD. No infectious agent caused this disease, in other words; it was prompted by a conspicuous absence. And no vaccination or antimicrobial would fix it. Addressing the problem would require an ecosystem restoration of sorts.

"Hygiene has made our lives better," Weinstock says. "But in the process of eliminating the ten or twenty things that made us sick, we've gotten rid of exposure to things that made us well."

WHENCE CAME INFLAMMATORY BOWEL DISEASE?

In 1859, a judge sentenced the London physician Thomas Smethurst to death by hanging for the murder of his mistress, forty-three-year-old Isabella Bankes. She'd succumbed to a case of diarrhea and fever that suspiciously resembled poisoning, and doctors claimed to have found traces of arsenic in her stool.

A second test, however, failed to find any poison. And a subsequent autopsy revealed widespread ulceration and scarring of her intestines, evidence not of foul play but of a long-standing condition.

"Such an acute inflammation of the large intestine, however produced, is sufficient in itself to cause death," wrote Samuel Wilks, the physician who conducted the autopsy. Smethurst received a queen's pardon,
although he was later reconvicted, of bigamy. (He was already married
when he wedded Bankes.)

Wilks, meanwhile, dubbed Bankes's condition "simple ulcerative
colitis." And although at least two descriptions of a similar affliction—
inflammation and scarring with no apparent cause—appeared earlier,
historians generally refer to Wilks's detailed case study as the earliest
verifiable case of inflammatory bowel disease.

During the late nineteenth century, there was a marked acceleration of
new cases, especially in London and Dublin. Between 1883 and 1908, more
than three hundred patients checked in to London hospitals with debilitating inflammation of the gut. Treatment options in those days included
lots of sour milk, opium, and enemas of boric acid or silver nitrate, two
antiseptics. One hundred forty-one, nearly half the total cases, died from
the ailment.

During the early decades of the twentieth century, more reports of unexplained intestinal inflammation emerged in France, Germany, and Italy. In
1909, by which time germ theory had gained currency, scientists convened a
meeting at the Royal Society in London to discuss this odd and often deadly
illness without obvious cause. They were accustomed to seeing infectious
dysentery, intestinal tuberculosis, and various cancers of the intestinal tract,
but whence came this inflammation in the absence of infection?

From the earliest descriptions, the disease disproportionately struck the
upper classes—"well-to-do, well-nourished persons in excellent health," as
William Allchin, a London doctor, put it. This pattern was, of course, the
inverse of infectious diseases such as cholera and typhoid that flourished in
undernourished populations living in crowded, filthy conditions.

And then, in May 1932, Burrill Bernard Crohn presented a paper at the
American Medical Association in New Orleans. He described "chronic
necrotizing and cicatrizing inflammation"—inflammation that liquefied,
scarred, and constricted the intestines. The fourteen patients, all Jewish,
were mostly young and in the prime of life. He called the affliction "ter-
minal ileitis," then "regional ileitis." (The ileum is the final, ten-foot-long
segment of the small intestine.) Eventually, gastroenterologists christened
the condition Crohn's disease.

To Weinstock, the acceleration of IBD in the upper classes after the
Industrial Revolution—the very same population that, I should point
out, first saw a spike in hay fever—was consistent with a loss of parasites.
The preponderance of Jews in New York City afflicted with Crohn's also

pointed to helminths—or a lack thereof. In the U.S., New York was the first large city to clean up—to provide potable water, organize garbage collection, pursue sewage management, and pave streets. Weinstock surmised that dietary restrictions (no pork, salting meats) limited Jews' contact with parasites before everyone else. And newfound affluence—even simple materials like shoes could make a difference—would have further curbed infestation with helminths.

Elsewhere in the country, worm infection remained quite widespread well into the twentieth century. Just over one-sixth of autopsies conducted during the late 1930s revealed *Trichinella spiralis* cysts, the dreaded muscle- and brain-burrowing worm acquired from eating raw or undercooked pork. By the 1960s, however, trichinella appeared in just 4.2 percent of autopsies. (Today, the Centers for Disease Control and Prevention logs fewer than twenty-five cases of trichinosis yearly, mostly from hunters eating undercooked puma and bear meat.)

Of course, trichinella was far from the most prevalent worm in the U.S. In 1909, when the newly established Rockefeller Foundation set out to eradicate hookworm disease from the Southeast, some 40 percent of children examined harbored parasites. Workers funded by the foundation dispersed throughout the South to instruct people on the proper construction and use of privies, or outhouses, at county fairs. Dig a hole away from streams, wells, or other water supplies. Place an inverted box with a cutout hole over the pit. Keep it covered when not in use. Make sure the box is banked and sealed with dirt. Occasionally fill this pit, dig a new one, and move the privy.

The antihookworm effort, based largely on education, was remarkably effective. In the 1910s, 61 percent of Floridians carried hookworm. By the 1930s, that had fallen by almost half, to 34 percent. By the 1950s, infection rates halved again, to 18 percent. Other states saw similar improvement. Hookworm prevalence in South Carolina declined from 37 percent in the 1910s to 24 percent two decades later. (By the late 1980s, a mere 2 percent of southerners showed any sign of worm infection.)

And yet it would be some time before Americans were, as a whole, worm-free. At the end of World War II, an estimated one-third of North Americans and Europeans—members of industrialized countries putatively on the vanguard of civilization—still had worms. Between 40 and 60 percent of children on both continents had pinworm, a helminth whose main symptom, an itchy bottom, seems so comparatively benign that it's often overlooked by public-health campaigns.

"One cannot have experienced the war without having been impressed anew, and depressed, by the amount of parasitism in the world," said the parasitologist Norman Stoll in a 1947 speech that's now widely regarded as a classic. For effect, Stoll calculated the combined weight of the microscopic eggs shed annually by giant roundworms in the 335 million infected inhabitants of China. They would reach 18,000 tons, he estimated. That's roughly the equivalent of one hundred very large blue whales.

By the mid-twentieth century, we'd invented planes and cars, mastered the secrets of atomic energy, and stood poised at the cusp of a space age. Despite these remarkable advances, parasites continued infesting humans here and elsewhere. Progress toward their elimination was uneven. They lingered on for decades in some communities after they were gone from others. Indeed, it was this very uneven pace of eradication that, to Weinstock's eye, went a long way toward explaining the equally irregular prevalence of IBD.

African Americans remained infested with worms longer than whites, a legacy of the country's tortured race relations. So did other disadvantaged minority groups, such as Cherokees living on a reservation in North Carolina. Caucasian ethnicity didn't necessarily prevent worm infection. In 1965, more than two-thirds of those surveyed in poor Clay County, Kentucky, had helminths. Poverty, which in the early twentieth century protected against IBD, was clearly the great worm enabler.

This pattern was roughly the inverse of IBD's advance around the U.S. After Dr. Crohn first described the disease in New York City, IBD spread outward. First, it struck white northeasterners. (President John F. Kennedy, born into a wealthy Massachusetts family, suffered terribly from colitis for much of his adult life.) A decade or so later, white southerners saw an uptick. By the 1970s, black communities north and south were getting more IBD. And by the 1990s, Native Americans moving off their reservations had it more often. Those who remained on the reservations, however, remained relatively IBD-free. Weinstock estimated that about a decade after a given community lost its worms, the incidence of IBD increased.

The gaping hole in Weinstock's hypothesis concerned the so-called global South. Worm infestation was common in Africa, South America, and South Asia, he knew, but little data existed on the prevalence of inflammatory bowel disease. Did this apparent dearth signify a true absence, or poor diagnostic ability? Weinstock and Elliott asked around at conferences. African and Asian doctors had often trained in Europe or North America, they found. They knew the IBD symptom set, and they simply didn't observe it in populations rife not only with worms, but also malaria, dysentery, and other infectious diseases.

Indeed, in 1988, 130-odd years after Samuel Wilks described Isabella Bankes's ulcerative colitis in London, a doctor in Soweto, South Africa, documented the "first 46 patients" ever treated for ulcerative colitis there. They hailed from nearby Johannesburg, he noted, not rural areas. And they belonged overwhelmingly to "the upper educational group" and "higher educational categories."

A MEDICINE LIKE NO OTHER

Weinstock has thick black hair streaked with gray. When he grins, his face takes on a playful, almost impish air. He's fond of saying that science is about curiosity and discovery. "I'm still twelve years old," he says. He has a slow and deliberate way of speaking. He often begins his talks with a seemingly offhand, but actually quite profound factoid: As judged by weight, human stool is 60 percent living bacteria. He never states the implication: With all our unprecedented brain processing power, our ability to travel to space, our parsing of the laws that govern the universe, we are still, at some level, nothing more than microbe-generating and -dispersing machines.

"We're part of our environment; we're not separate from it," he told me once. "And we *can't* be separate from it."

I'm accustomed to hearing such vaguely pleasant platitudes from conservationists and ecologists. But the chief of gastroenterology at Tufts Medical Center? Coming from him, it has a certain gravitas, not least because, in the late 1990s, Weinstock began looking for a way to restore, in a manner of speaking, the ancestral parasite fauna of the human gut. He started searching for a worm suitable for human experimentation.

At that point, experiments on animals had shown that worms could prevent not just IBD, but other inflammatory and autoimmune diseases. Inject them first with schistosome eggs, and mice were invulnerable to experimentally induced colitis. Even mice that were genetically IBD-prone didn't develop the disease after exposure to worms. And parasites' protective effects extended far beyond the gut. They could prevent the mouse version of multiple sclerosis. From the lab of Anne Cooke, a scientist at the University of Cambridge in the U.K., came news that just the extract of schistosome egg prevented autoimmune diabetes in mice bred to develop the disease.

The time had come for human trials, and Weinstock knew he had to proceed carefully. Considerable time and effort had gone into eradicating worms from the U.S. And while modern sewers precluded the spread of parasites between people, he knew that even the possibility of contagion would scuttle the project. He needed a worm that not only prompted negligible symptoms, but that also couldn't propagate on its own. These criteria eliminated most human-adapted species right off the bat. And casting about, his attention fell on pigs.

Iowa, the number-one swine-producing state in the U.S., had about 15.5 million pigs. With a population of just under 3 million people in 2000, that was five pigs for every person. Pig farmers regularly encountered a porcine whipworm species called *Trichuris suis*. (Humans had their own whipworm species called *Trichuris trichiura*.) The worm seemed to cause few, if any, symptoms in people. Unlike hookworms or flukes, *T. suis* didn't migrate through tissue—a plus. And while the worm could temporarily colonize the human digestive tract, human and porcine insides differed sufficiently that, for whatever reason, it never sexually matured. The worms perished after roughly two months in the human gut. There was no chance, Weinstock thought, of spreading infection.

In 1999, Weinstock, Elliott, and Summers commenced a safety trial on seven patients, four with Crohn's and three with ulcerative colitis. The volunteers downed a single glass of Gatorade laced with 2,500 whipworm eggs. The scientists monitored their disease severity for the following twelve weeks. Symptoms improved steadily for four weeks, but then started regressing. By roughly ten weeks, they were back to square one—beleaguered by inflammation.

Weinstock changed the protocol to 2,500 eggs every three weeks ongoing. Constant dosing kept the disease in remission. And it was perhaps testament to the painful and difficult-to-treat nature of IBD that the scientists easily recruited two more groups of volunteers, thirty people each, willing to gulp down thousands of whipworm eggs.

Beginning in 2004, they drank 2,500 microscopic *T. suis* eggs with Gatorade at three-week intervals. After six months, twenty-three of twenty-nine people with Crohn's disease—nearly 80 percent—improved. Twenty-one—nearly three-quarters—went into total remission. This first study wasn't blinded: The researchers knew who'd received worm eggs and who placebo. But a subsequent double-blind study on ulcerative colitis saw thirteen of thirty patients sent into remission. Two of every five patients improved. No major side effects were reported in either trial.

With the 2005 publication of his studies, Weinstock officially departed from the main thoroughfare of modern medicine, one defined by germ theory. He set out from this mostly black-and-white worldview to clear a new path, one that sought to incorporate, and exploit, biological relationships that were gray-hued and nuanced. (Was a parasite just a parasite if it benefited its host?)

He'd taken an idea in the air for some time prior—that human disease should be viewed through the lens of human evolution—and turned it into a practical approach for treating an otherwise inexplicable malady. His work raised the prospect not just of relatively benign treatments for diseases like IBD, but of preventing them altogether. If inflammatory diseases emerged because certain stimuli were missing, then one could theoretically preempt the dysfunction by replacing these stimuli beginning at an early age. With enough foresight and planning, we could—in theory—preclude these terrible disorders in the future.

But we're getting ahead of ourselves.

A YOUNG WOMAN
CURES HER INCURABLE DISEASE

In the summer of 2002, a young woman on vacation in Central America began suffering a curious range of symptoms. The twenty-one-year-old, whom I'll call by her first name, Lisa, had traveled through Costa Rica and Panama when she was struck by severe diarrhea, nausea, and loss of appetite that failed to resolve. Like anyone traveling abroad, she had anticipated some intestinal discomfort, so she visited doctors along the way. They prescribed one antibiotic or another, which improved her symptoms temporarily. But the malaise always returned. She ignored her gut problems as best she could and focused on enjoying her holiday.

But when she returned to her native Switzerland, she became concerned. She'd lost 16 pounds, and her bowel movements continued to arrive frequently and with painful urgency. Neither antibiotics nor antiparasitics had an effect on the malady. Finally, a specialist made the diagnosis: Lisa had Crohn's disease.

Then in her early twenties, the most pressing question for Lisa was how to manage the disorder in the long term. "I'm a young person," she recalls thinking. "I have a couple more decades ahead of me. I need a plan." She started on prednisone, an immune-suppressing steroid, which was immediately helpful. The pain subsided. Inflammation decreased. But the steroid also had side effects: She was constantly hungry; she sweated continuously and profusely; she occasionally retained fluid in her legs; depression set in. And she fretted over the possibility of premature osteoporosis, a loss of bone density that can occur with prolonged steroid use.

Alternative therapies, including Chinese medicine, bioresonance, and avoiding wheat, dairy, and eggs, didn't help. As soon as she stopped taking prednisone, the inflammation invariably returned.

Then, in the winter of 2006, she had a conversation that would change her life. An old friend knew someone who'd tried Weinstock's *Trichuris suis* eggs with great success.

By then, a German company called Ovamed was producing pharmaceutical-grade whipworm eggs based on Weinstock's protocol. The company, which harvested the eggs from miniature Danish pigs raised in hyperclean conditions, had developed a manufacturing process that met the approval of European regulatory authorities. (The University of Iowa owns the patent.)

Lisa ordered ten doses of the *T. suis* ova, or TSO. At that point, she'd been taking prednisone fairly regularly for about five years. Now she began imbibing 2,500 eggs at two-week intervals, a "sip" of salty liquid from small vials stored in the fridge. After two months of taking TSO, she tapered off steroids. Her Crohn's remained controlled. Inflammation never reappeared. "I've never gone back since," says Lisa.

The treatment was expensive—about €300 every two weeks, or €7,800 per year (roughly $9,250 at the time). At this point, insurance companies don't cover it. But for Lisa, the results have justified the expense: remission without the side effects associated with steroids. (She still takes another immune suppressant, called Imurek, however.)

"It's really changed my life," says Lisa, who's now a practicing psychologist in Zurich. "I've been sick for a long time. And since [beginning TSO], I really don't feel sick anymore. I'm able to live my life as I used to before."

IS A PIG-ADAPTED WORM SAFE IN PEOPLE?

Not everyone was convinced that Weinstock's worms were safe, however. There was some evidence from animal experiments that worms, which skewed the immune response, could make other infections—in this case a bacterium called *Campylobacter jejeuni*—worse. The idea made sense: If helminths pushed the immune response toward Th2 when Th1 was warranted, opportunist bugs might more easily gain purchase.

Others voiced concern over the unpredictability of any worm, even those supposedly adapted to a human body. Sometimes pinworms, which were native to humans, where they lived in the intestine, found their way into liver and lung. Parasites that colonized unfamiliar hosts could end up in even stranger places. Dog worms could penetrate human lung and liver. Worms native to deer, in which they cause few problems, sometimes killed moose. Raccoon worms occasionally burrowed into human brains, causing neurological complications and death. And even *T. suis*, largely asymptomatic in the domestic pig, could migrate to the kidneys of wild boars, a closely related species.

"There is no predicting where *T. suis* larvae will go in humans, the abnormal host in this controversy," warned Herbert Van Kruiningen, a scientist at the University of Connecticut. "It may only be a matter of time and numbers of larvae before retinal or [central nervous system] disease occurs in a patient 'treated' with *T. suis*."

Summers and Weinstock disagreed. In all the literature—and in the presumably millions of encounters between man and pig whipworm in Iowa alone—there was no report of anything going wrong, they countered. Moreover, by then, some three thousand patients had taken TSO. None had reported adverse side effects.

But in 2006 came a case report of a sixteen-year-old boy who'd taken TSO for his Crohn's. He'd swallowed five doses, and then his doctors discovered what they deemed an adult whipworm in his large intestine. That wasn't supposed to happen, they thought. What's more, his IBD symptoms had worsened. His physicians blamed the worm.

Weinstock and Elliott again thought differently. The inflammation observed was likely due to the disease itself, they argued, not the therapy. As for the putative adult worm, worms of various sizes appeared in treated patients. But no patient had ever shed eggs in their stool, the gold standard for determining that the worms had reached sexual maturity.

At its heart, the back-and-forth was over acceptable risks. All treatments pose risks and, one hopes, benefits. But absent large trials with TSO—they're in the works—no one yet knows the actual risk-to-benefit ratio of TSO treatment. Weinstock argues that one larger-than-anticipated worm is certainly not reason enough to scrap the entire approach, especially when the more mainstream immune-suppressant treatments like Humira and Remicade carry not insignificant risk, including cancers, severe infections, and death. "There is nothing wrong with having a worm," says Weinstock. "Billions of people have them."

And there's certainly a need for new—some might say any—effective treatments for inflammatory bowel disease. Even ignoring the side effects, current treatments only work about half the time. And three of every four people with IBD end up undergoing bowel surgery.

For one person, however, the potential unpredictability of a worm adapted to pigs was too much to countenance. Only the real thing, the human-adapted whipworm, would suffice.

WHERE TO FIND
A ONCE-UBIQUITOUS PARASITE?

About the time that Weinstock was assembling his first volunteers for *Trichuris suis* testing, in 2003, a twenty-eight-year-old man in New York

City was, after a bout of abdominal pain and loose, bloody stools, diagnosed with ulcerative colitis. I'll call him Rick.

Rick's colitis didn't respond to the usual anti-inflammatory drugs. Only high doses of the steroid hydrocortisone helped. Prolonged use of steroids came with those already enumerated risks, of course—obesity, premature osteoporosis, and increased risk of infection. Within a year of diagnosis, Rick's gastroenterologist urged him to check into a hospital for a round of intravenous cyclosporine, a powerful immune suppressant also used to prevent the rejection of transplanted organs. That would, they hoped, bring the inflammation under control. But the damage might be so extensive they'd have to remove some, or all, of his colon anyway.

Needless to say, Rick wanted to avoid a colectomy and the colostomy bag that came with it. (Colectomy patients evacuate through a surgically created opening in their left side.) Cyclosporine posed its own dangers, including kidney and liver problems, increased risk of some infections, and elevated odds of developing certain cancers.

These might be acceptable contingencies for a disease with no cure—a genetically predetermined disorder—but everything Rick learned about ulcerative colitis suggested that it was environmentally induced. None of Rick's immediate family had the disease. And the prevalence of colitis varied greatly around the world, with poorer countries having a much lower incidence. Given the many signs pointing to some environmental factor, why not address the disease by changing the environment?

Rick researched tirelessly, seeking an alternative approach. The bloody stools continued unabated. Whatever his thinking on the failings of Western medicine, he had to do something, and fast. If not, he risked developing a condition called toxic megacolon: enlargement of the colon from extreme, self-perpetuating inflammation, and then its possible rupture.

He found his way to Weinstock's research. The theoretical framework—that worms were ever-present over the course of human evolution, and that, in their absence, the immune system might malfunction—made immediate sense. If he hoped a parasite would manipulate his immune system, however, then the parasite should be adapted to the human organism, he reasoned. *T. suis* was, of course, native to pigs.

Rick decided to acquire the human whipworm, *Trichuris trichiura*. His gastroenterologist thought the idea misguided, and declined to monitor the experiment. He found a new one. His wife at the time, a surgeon-in-training, thought the scheme crazy, and refused to participate. They later divorced. Even scientists he contacted for help considered the idea bonkers. But Rick saw his options as stark and clear: "Do I go take cyclosporine, which gives you blood cancer, or a worm from the earliest-known coprolites?" he says. "I mean, even Ötzi had this worm." (Ötzi was the 5,300-year-old man discovered in an Italian glacier with whipworm in his gut.)

The human whipworm measures roughly 45 millimeters in length, the span of two quarters placed side by side. It's variously estimated to live between one and three years. These days, nearly 1 billion people carry it in their large intestines. The Centers for Disease Control and Prevention ranks it as the third most prevalent roundworm in the world.

The surprising lack of disease caused by the worm is a common theme in parasitology texts, an almost begrudgingly conceded observation that inevitably undercuts the gathering gravitas. After listing the many potential side effects of whipworm infection—clubbed fingers, delays in cognitive development, a prolapsed rectum—and dubbing it "a major public health problem of global significance," one typical textbook states, with what sounds like disbelief, that "studies that have been performed have shown a remarkable absence of immunopathology despite heavy infections with the parasite."

Parasitologists still argue about what the worm, an obligate parasite, really takes from the host. It embeds itself headfirst in the intestinal wall, which would appear quite invasive. Females lay somewhere between 3,000 and 20,000 eggs daily, definitely a drain on resources. But unlike hookworm, whipworm doesn't suck blood. It may survive off intestinal secretions. What's so bad about a parasite that quaffs mucus?

For Rick, the only question was how and where to find human whipworm. In the U.S., the parasite persisted in scattered pockets of the rural Southeast and Puerto Rico. The occasional patient presented with trichuriasis in New York City hospitals. But as he studied the epidemiology, he realized that the major obstacle wasn't locating the worm; it was ensuring that he found only whipworm.

The parasite tended to co-occur with other helminths, like the giant roundworm, *Ascaris lumbricoides,* the most prevalent human-adapted worm on earth. Ascaris's capacity to protect from inflammatory bowel disease was untested, however, and it was disturbingly large. Rick also worried about tapeworms, which posed another set of potential complications. If tapeworms, which cycle between intermediate and definitive hosts—humans and pigs, for example—mistake you for an intermediate host, they can burrow into important tissues, such as the brain, killing you in the process. Dangerous viruses—hepatitis C and cytomegalovirus, to name two—infections that were lifelong once you acquired them, also tended to co-occur in wormy environments. He needed to minimize exposure to these pathogens. And so, in 2003, he headed to a place he thought relatively safe. He flew to Thailand.

Rick declined to provide too many details—to protect the people who helped him, he says—beyond that he collaborated with NGO types working on worm eradication in rural areas. And eventually these collaborators brought him a whipworm-egg-containing stool sample from an eleven-year-old girl in southern Thailand. "Hopefully one day I can thank her," he says.

His travails were just beginning. You'd think it would be easy to germinate a parasite that, in the wild, effortlessly—some might say tenaciously—infects around one-sixth of humanity. But for months, Rick couldn't get the worm eggs to embryonate and become infective.

Like chicken eggs, which require time warmed by a hen, geohelminth eggs need to incubate in the right soil at a certain temperature and humidity. This embryonation period can last between two weeks and one month. Without proper embryonation, the eggs remain duds: They pass right through without hatching.

Rick tried sterile conditions, closed containers, open containers, washing the eggs first with antibiotics, and bleaching them. Nothing worked. They remained noninfective. "I was very frustrated," he says. Eventually, he simulated conditions like those found under a tree in the tropics: aerated, nonsterile, and humid soil. A year and a half after beginning his search—and after repeated trips to Thailand—late in 2004, he swallowed a batch that took. He acquired a whipworm colony.

Three months later, Rick tapered off all immune suppressants. In mid-2005, two years after receiving a diagnosis of ulcerative colitis, he was in complete, drug-free remission. "I returned to what I thought was one hundred percent normalcy," he says. At this point, Rick took an unusual step. He sought out a scientist willing to study him.

The theory of the coevolution of man and worm was all well and good, he reasoned, but he wanted scientific documentation of what, exactly, the parasites were doing to his intestines. So in 2007, Rick, by then living in San Francisco, called up a young Malaysian parasitologist at the University of California, San Francisco, named P'ng Loke. Loke was inclined to ignore the query, but after hearing Rick's improbable story over lunch, he agreed to study him.

I meet Rick one November evening in the café of a Whole Foods in Torrance, California, just south of Los Angeles proper. He has a slight build, large expressive eyes, and a square jaw covered with a few days of stubble. He talks with the firmness and precision of someone in executive management. He seems tired.

Tomorrow, on December 1, his case study will appear in the prestigious journal *Science Translational Medicine*. At least in some circles, Rick's large intestine and the worms inhabiting it are about to become world-famous.

Some months earlier, I'd seen pictures of Rick's whipworms. P'ng Loke, who recently moved to New York University, had shown them to me—photos from four endoscopies conducted over five years. I was astonished at first, and then enthralled by the scores of small white worms embedded, like curlicues, in Rick's intestinal wall.

"He may have gotten more than he realized," Loke had said, pointing to a red crack. "You can see they're causing some damage."

More remarkable, however, was the damage they were apparently preventing. Over the course of the study, Rick's colitis flared twice, both times preceded by a drop in whipworm egg production. White-tinged ulcers appeared. When the worms aged and died, they lost their curative effect, Rick concluded. Each time, he brought his colitis back under control by redosing. The ulcerous lesions looked minor. But if you considered the excruciating pain associated with even the tiniest cold sore, which they resembled, they seemed agonizingly enormous.

Rick's flare-ups afforded an opportunity to measure the difference between disease and health. And although his was a single case study, this was the great contribution of Loke and Rick's collaboration. As expected, inflamed areas were saturated with pro-inflammatory signaling molecules of the sort associated with autoimmune diseases in general, most notably one called interleukin-17. Less expectedly, however, the worms boosted mucus production.

Scientists have noted that inflammatory bowel disease often presents with a relative paucity of mucus. Mucus maintains the minuscule, but crucial, barrier between the intestinal lining and the resident microbiota. We live intertwined with an ecosystem of microbes, but really, the ecosystem remains at a slight distance. If the bacteria come too close, some think that the immune system responds as if facing invasion—with inflammation. So one line of thinking holds that inflammatory bowel disease results from having lost this protective layer. Rick's case suggested that worms could restore the mucous layer, a "bystander effect," Loke says, of trying to fight off worms.

When I ask Weinstock, in an e-mail, about Loke's study—one largely inspired by his work—he's both cautious and optimistic. "This is data from just one patient," he warns. But the finding on mucus production is new and potentially important. What's more, the study shows for the first time that human whipworm can help IBD. "[T]his is a wonderful case report again showing the potential benefits of helminth exposure," he says. "Undoubtedly, there will be much more to come."

Others are less sanguine. A few days later, as Rick's case study makes the media rounds, one prominent gastroenterologist calls the study "irresponsible."

"It's ridiculous and incredibly inappropriate," Stephen Hanauer, a member of the board of trustees of the American College of Gastroenterology, tells CNN, referring to both the study and the coverage it engendered. "You're driving people to go on the Internet and buy these worms, and these are potentially pathogenic organisms. These eggs can invade the systems of people who are immune-suppressed and cause infections."

Even Rick declines to precisely describe his embryonation process for fear of inspiring copycats. But it's hard to ignore one simple fact: If Rick had heeded the advice of his original gastroenterologist, there's a good chance he'd have lost some or all of his colon—that he'd be wearing a colostomy bag right now.

"It's incredibly encouraging, at least in my personal experience, that something like a live therapy like this can actually have a positive effect," says Rick. It "could open up a very different approach to . . . the treatment of disease."

We're going to need all the help we can get. One remarkable aspect of Weinstock's work is that, in the nearly two decades that he's thought about worms and IBD, the epidemiology of the disease around the world has shifted exactly as one might predict if worms protected against the disease, and formerly wormy populations suddenly lost their parasites.

The gradient of IBD increasing from south to north across North America has narrowed, although it still hasn't disappeared entirely. East Asians, formerly considered to be mostly immune to the disorder, are also seeing an uptick, although absolute numbers are still far below those of the West. Of note, worm-eradication efforts began decades later in Japan and South Korea than in the U.S.—not until well after World War II and the Korean War, respectively.

Worse, when populations from developing tropical countries immigrate to cleaner, temperate ones, they seem to have an even greater propensity to develop IBD than natives. South Asian immigrants to the U.K. have less IBD than native Britons, but their U.K.-born children have two and a half times the risk.

And India, which has developed rapidly in recent decades, has also seen an increase of IBD. The prevalence first spiked in those very regions, such as Kerala, that saw the earliest improvements in public hygiene. "Persistent improvement of sanitation is surely desirable but may have some adverse effects too," observed a Mumbai physician in 2005. Other Indian scientists have seized the opportunity—different segments of the population at different phases in the epidemiological transition—to test Weinstock's ideas in real life. And so far, their finds support his basic premise.

People who develop Crohn's disease tend to come from cities, and judging from their weak reaction to hookworm proteins, have lived a relatively parasite-free life. Southern Indian patients with type-1 diabetes, meanwhile, are fourteen times less likely to have been infected with mosquito-borne filarial worms than the population at large.

And then there's the question of allergic disease. Does the historical appearance of both IBD and hay fever in the same population in the same period—the well-heeled classes of the late nineteenth century—mean anything? Could the loss of worms drive both phenomena? The complex relationship between allergic disease and parasites is the subject of the next chapter.

CHAPTER 5

What Is Asthma For?

We should think of each host and its parasites as a superorganism with the respective genomes yoked into a chimera of sorts.

—Joshua Lederberg, Nobel laureate, in *Science*

In 2004, the parasitologist David Pritchard placed a damp piece of gauze against his left forearm. The bandage held an unknown quantity of hook-worm larvae—10, 25, 50, or 100. Pritchard and nine others, including several colleagues at the University of Nottingham—scientists who'd collectively spent decades studying the parasite around the world—were infecting themselves with the parasite. The immediate goals of the experiment were twofold: to determine how many worms a twenty-first-century Briton could handle, and to prove that said Briton could handle some worms without debilitating symptoms—or worse, dropping dead.

Considerable unease surrounded the project. The ethics board was horrified. And even Pritchard's wife worried he'd spread the parasite around. But Pritchard and his colleagues suspected that *Necator americanus* didn't quite deserve the anemia-causing, growth-retarding, poverty-entrenching reputation it had earned. They thought hookworm might have therapeutic value in the treatment of allergic disease. More generally, the scientists were directly addressing a question at the heart of the modern study of allergies: What, exactly, was the allergic response for?

A single antibody type called immunoglobulin-E, or IgE, drove the nose-running, eye-burning, nonstop-sneezing misery of hay fever season. It also facilitated the hives, spontaneous vomiting, and potentially deadly throat constriction of food allergies. It was responsible for those few people every year who, after a bee or wasp sting, go into face-swelling, blood-pressure-dropping anaphylactic shock and fall dead. The abiding question was, assuming that unleashing a murderous fury on cat hair, birch pollen, or bee stingers was not IgE's evolved "purpose," then what was IgE there to do?

All mammals have IgE, including marsupials, a lineage that separated from our own 110 million years ago. Birds, which went their own way 200 million years earlier, have an antibody with a similar function called IgY. IgE was therefore ancient, dating back to well before that insect-eating, shrew-like creature that, as it scurried about in the dinosaurs' shadows 100 million years ago, was destined to become the progenitor of all placental mammals.

The antibody's conservation across mammalian lineages, and a functional analogue in birds, implies that IgE is important. Nature doesn't generally waste valuable resources on useless attributes. But ever since scientists identified it in the 1960s, the antibody's reason for existence had remained obscure.

In a roundabout way, parasite immunologists first proposed a reason for IgE's existence. They posited that IgE, which was elevated in worm infections, was really a component of our parasite-controlling machinery. Yet while elevated IgE signified allergies in places such as London and New York, in populations harboring worms, high IgE failed to correlate with allergic disease. Someone who harbored parasites in the tropics could have hundreds of times as much circulating IgE as a hay-fever-stricken Londoner or New Yorker. But this person generally failed to sneeze from dust, to develop hives from tree nuts, or to have allergies of any sort. In wormy environments, elevated IgE was curiously uncoupled from allergic disease.

That observation had prompted scientists to venture the following: Maybe these parasite defense mechanisms worked properly only in the context of the worms they had evolved to manage; perhaps the cure to allergic disease lay in artificially stimulating IgE levels; maybe a living parasite could do the trick.

As you might imagine, this last idea prompted contentious debate. Pritchard's experiment stemmed from a certain fatigue over the interminable back-and-forths. To his mind, the most direct way to prove that parasites protected against allergies was to infect an allergic person with a worm and, he hoped, watch the allergies disappear. Pritchard wasn't yet at that step, of course. He first had to prove safety. But he was headed in that direction, toward an unconventional treatment for allergic disease.

THE ABSENCE OF ASTHMA IN RURAL AFRICA

Three decades before Pritchard pressed that larvae-bearing bandage against his arm, a British researcher named Richard Godfrey had traveled to the Gambia, a sliver of a country along the Gambia River in West Africa. Godfrey was interested to know the prevalence of asthma there compared with Britain. What he discovered must have seemed incredible, impossible even: He couldn't locate a single asthmatic in a randomly selected group of 231 children and adults from rural areas. He perused medical records from village clinics. Again, not a single file in 1,200 described symptoms resembling asthma. In rural Gambia, it seemed, no one wheezed.

In Banjul, the Gambian capital of 44,000, however, the story was different. The city was rapidly Westernizing, and the local hospital saw eight asthmatics daily on average, most of them regulars. In all, Godfrey identified forty-four patients with the lung disease. Curiously, they came almost exclusively from the upper classes. In other words, in the Gambia, asthma afflicted the richest segment of the most urbanized population. What could explain the pattern? Rural Gambians had more than two and a half times as much IgE as their asthmatic urban counterparts. They had many more parasites.

"The idea that allergic disorders may represent the continuing activity of an immune system made redundant by man's cleanliness has attraction," Godfrey wrote in 1975. And then he presaged, in a manner of speaking, Pritchard's experiments. "If true, it is perhaps not too fanciful to look forward to the treatment of allergic disorders by a harmless preparation of parasite antigen able to stimulate IgE."

Godfrey tested the idea directly. He alternately exposed lung fragments (removed for other reasons during surgery) to African and British serum, and then to allergens, in this case pollen. If the tissue was bathed in the African before the British serum, it failed to respond to the pollen. Exposure to the African serum prevented immune cells from learning the "hay fever" response. But when he reversed the order—British serum followed by African—the pollen prompted an allergic response. African serum, which had been tempered in the heat of battle with parasites, possessed almost magical antiallergenic properties. British serum, on the other hand, exhibited the opposite tendency; it enabled allergy.

"One theoretical approach to prevention or treatment of allergic diseases would be deliberately to induce high IgE responsiveness—for example, by artificial infection with parasites," wrote the editors of the Lancet in an article accompanying Godfrey's study. The mere mention of the idea sparked a firestorm of revulsion and alarm. Many scientists were certain that precisely because worms triggered IgE, such an experiment would cause the very allergies it ostensibly sought to prevent.

"I was horrified to find your editorial . . . terminate with the but half logical, and certainly wholly dangerous suggestion," wrote one scientist. Another was "astonished" by the proposition. A third researcher warned that Malay children, nearly half of whom carried parasites, had plenty of hay fever and asthma.

But then came an almost breathless letter from a thirty-three-year-old British parasitologist named J. A. Turton. He'd infected himself with 250 hookworm larvae, ostensibly to study IgE levels. He'd suffered severe stomach pain, which eventually subsided, and then an unanticipated miracle: The hay fever that had plagued him since age eight disappeared.

"During the summers of 1975 and 1976 I remained completely free from all symptoms," he told the *Lancet*'s readers. This was, he acknowledged, a single example. But the results suggested that worms did, indeed, squelch the allergic response, and that they had therapeutic potential. At the very least, his observation countered the argument that worms, by elevating the allergic antibody, made allergies worse. Said Turton, "I clearly cannot agree . . . that parasitic infections would be expected to exacerbate existing allergies."

STUDYING WORMS
IN THEIR NATURAL HUMAN HABITAT

David Pritchard knew this backstory when, in the late 1980s, he first arrived on Karkar Island, a still active volcanic islet about ten miles off Papua New Guinea. He'd written his doctoral thesis on the intriguing relationship between worms and the allergic response. But he'd come to the lush island to investigate a more fundamental aspect of the parasite-host relationship: How did one organism take up residence inside another and, despite a fully responsive immune system bent on its murder, live for years in what seemed like peace?

"No one really, really understood how a worm was not rejected by the body," he says. "We set off to Papua New Guinea to study people who were naturally infected."

When Pritchard arrived, many Papuans still lived a mostly rural, village-centered existence. And they were almost universally infected with *Necator americanus*—about 95 percent prevalence—along with other parasites. Provided that infections weren't too heavy, Papuans seemed to live in relative harmony with their "snek bilong bel"—the "snakes in their bellies." Pritchard and his team collected stool, handed out deworming medicine, analyzed parasite loads, and measured antibody levels. Years passed. The time in the jungle provided ample opportunity to think. And before long, he began having doubts about the reigning immunological paradigms in general.

For one, he became less sure that the immune response seen in worm infections was really meant to kill parasites at all. It rather seemed that the response protected the host—from itself. The condition called elephantiasis, a painful and grotesque swelling caused by a threadlike worm, illustrated this dynamic. The worms, transmitted by mosquitoes, resided in the body's lymph system. Most people harbored them with few symptoms. A minority, however, counterattacked savagely. These unfortunates developed the painful, disfiguring disease. Clearly, the all-guns-ablaze approach could backfire. Sometimes your own immune response was your worst enemy. Occasionally tolerance was the best reaction.

Pritchard wasn't the only one asking these questions. Others who studied naturally worm-infested populations began to wonder if the elevated IgE really benefited the host or the parasite. Much of the IgE didn't have an obvious affinity with worm proteins. It seemed more like a smoke screen, a deliberate blurring of the host's defenses by the worm, the immune equivalent of nonsense. Others thought that maybe this so-called polyclonal IgE was a deliberate self-obfuscation by the host, a way to avoid severe responses of the type that sometimes killed extremely allergic people.

For his part, Pritchard found that those people who churned out more IgE than everyone else tended to have wimpier worms. Their parasites were smaller and produced fewer eggs. And this observation brought him to the conclusion that while the ability to churn out IgE by the bucketful in a wormy environment had its evident advantages, those who possessed this innate talent would, if they'd grown up in London, be most prone to develop allergies. "Allergy could be regarded as an evolutionary 'hangover from parasitism,'" he wrote in 1997. And those with the worst hangover would fare best in an environment rife with worms.

More important, we weren't necessarily condemned to endure allergic disease. Allergy remained relatively rare on Karkar Island. As long as these worm-fighting tools remained in their proper context—battling parasites—they caused few problems. And if we required contact with parasites to keep the immune system functioning optimally, then these organisms weren't, strictly speaking, parasites. They were mutualists.

Again, Pritchard wasn't alone in these musings. Others were finding that worms could benefit their hosts in unexpected ways. Researchers in Thailand observed that harboring giant roundworms lessened the risk of cerebral malaria. The worms didn't directly protect against the plasmodium parasite as a vaccine might. Rather, the helminths dampened the host's immune response, preventing the immune firestorm that led to malignant malaria. (Others have replicated this finding, but it remains controversial.)

In Australia, repeated outbreaks in humans of the dog hookworm *Ancylostoma caninum* during the 1990s underscored another important point: Only worms that had coevolved with a given host conferred benefit.

New arrivals, on the other hand, could cause significant disease. In this case, the dog-adapted hookworm successfully colonized humans, suggesting a new species in the making. But the parasite didn't establish itself very gracefully. Unlike human hookworm, these dog worms caused severe inflammation, ulcers, and enteritis.

In Japan, outbreaks of anisakiasis, a worm acquired from undercooked or raw fish, also illustrated this point. The helminth, which was native to seals and dolphins, could cause major symptoms, even life-threatening illness, in people. Clearly, these parasites didn't know their way around the human organism. All of which highlighted how unique *N. americanus*, and other human-adapted parasites, truly were. What could they teach us about how the human immune system operated? Might they divulge a cure, in the form of a secreted protein or enzyme, for the allergy epidemic in developed countries?

These were all titillating questions. At that point in the mid- to late 1990s, however, Pritchard wasn't yet considering deliberate human infection with *N. americanus*. But a series of studies from Africa would change his mind.

IN ETHIOPIA, ASTHMA BEGINS

In the southwestern highlands of Ethiopia, along an old caravan route, sits a town called Jimma. Famous for its markets, the town of 88,000 was growing rapidly during the 1990s. (As of 2007, it had 121,000 residents.) But while increasingly affluent by Ethiopian standards, Jimma remained poor compared with western Europe. Many residents lived in mud-walled houses with corrugated metal roofs. Latrines often stood separate from dwellings. Few houses had electricity. Almost none had running water. Drinking water came from communal wells.

In short, Jimma was passing through an epidemiological transition roughly comparable with parts of the U.S. around the turn of the nineteenth century. And already, the prevalence of allergic disease had shifted. A decade earlier, asthma was almost unheard of in Jimma. But now it was responsible for one of every twenty admissions to the hospital, a high rate for Africa. This clear increase had brought the researcher John Britton and his colleagues from the University of Nottingham.

Theories abounded as to the causes of asthma: pollution, smoking, dust mites, and diet, to name a few. But the sudden and recent appearance of the disease in Jimma suggested that whatever its causes, they began working their influence early in the urbanization process. For a researcher, the temporal proximity was a boon. In Europe, the roots of the asthma epidemic lay decades in the past, maybe even a century. But here, where the condition had just begun to appear, he expected that the causes remained closer to the surface. Britton hoped to identify those factors before the trail went cold.

The first step: to gauge the asthma prevalence in Jimma proper compared with the surrounding countryside. Since Richard Godfrey's work in the Gambia some two decades earlier, others had noted that around Africa, asthma was more common in urban environments, and among the most affluent classes. Here, too, that pattern held. Asthma, Britton and his colleagues found, was three times more common in Jimma, where 3.6 percent of children wheezed, compared with surrounding rural areas.

And while conditions in town might resemble those of late-nineteenth-century Britain, conditions in the country resembled the early Neolithic. Rural Ethiopians lived in round, mud-walled huts with thatched roofs and packed-earth floors. Latrines were uncommon. Most relieved themselves in the bush, or in their fields. "You're living as humans were probably living ten thousand years ago," says Britton.

Now came the comparison. The researchers could, somewhat to their surprise, immediately discount air pollution. Jimma had no major industry and few motorized vehicles. Air quality was roughly equal between town and countryside. They could also dismiss dietary differences. Jimmans ate food made from animals and crops raised in the surrounding countryside. They hadn't yet transitioned to factory farms or preprocessed, packaged foodstuffs. Everyone ate the same traditional Ethiopian cuisine.

Two decades earlier, Australian scientists working in the Fore region of Papua New Guinea had explained the recent increase in asthma there— it had appeared right after sustained contact with Westerners—on the advent of Western-style bedding. They thought that by providing ideal nesting conditions, synthetic blankets enabled dust mites to propagate wildly. Increased exposure to the insect then caused asthma.

Indeed, Britton found that people with better housing and synthetic bedding had a slightly elevated risk of wheezing. And dust mites played a central role. In town, sensitization to mites increased the risk of asthma by a factor of 10, strongly supporting the prevailing dogma: We were right to blame the asthma epidemic on these tiny spiderlike arthropods.

Or maybe not. Observations in rural areas seemed to absolve dust mites. The bugs were actually more numerous in country dwellings, while asthma was less prevalent. Rural folk were more often sensitized to dust mites—determined by pricking the skin, exposing the small wound to dust mite protein, and watching for a wheal. Yet this sensitization failed to predict allergic disease, as it would have in London. In a rural setting, the two phenomena were disconnected. To Britton, this suggested that either something in the countryside was good for you, or something exclusive to town was bad for you. "It could work either way," he said.

Again, Britton and colleagues compiled a list of variables, and began ruling them out. They eliminated measles exposure, which was similar in both groups. Smoking, more prevalent in rural areas, couldn't explain the difference. Neither could hepatitis A infection, sometimes observed to decrease the risk of asthma. Insecticide exposure, often thought to increase asthma risk, had no effect. Only one variable had a consistent inverse relationship with asthma: hookworm infection. *Necator americanus* halved the odds of wheeze in both urban and rural settings. Parasite infection also explained why sensitization to dust mites failed to cause wheezing in a rural environment. By subverting the host's immune response, worms prompted tolerance to third-party proteins. Hookworms incidentally taught people to tolerate dust mites.

What did it all mean? First, as worms disappeared in Ethiopia, you could expect more dust mite allergies, and probably more asthmatics. Second, maybe you could reintroduce worms, and stamp out allergy. That's what Pritchard, who was following this work closely, and who was coauthor on some of the studies, thought. "We got fed up with these stories and decided to test it," Pritchard told me.

In 2000, he gathered a supply of hookworm on Karkar Island and squirreled it away in his luggage. A colleague named Alan Brown infected himself with a backup supply. Who knew how customs might respond? Then the scientists boarded a plane for Britain. In the coming months, Brown, who later estimated that he'd amassed a three-hundred-strong hookworm colony, was delighted to see his lifelong hay fever nearly disappear.

REMOVE WORMS, AND ALLERGIES APPEAR

West of Ethiopia on Africa's Atlantic coast, a Dutch researcher named Maria Yazdanbakhsh had arrived at her own crossroads: Either she would attempt to disprove the reigning immunological dogma, or she would do nothing, and allow ideas she thought inaccurate to continue propagating.

Yazdanbakhsh studied children's immune response to blood flukes in Gabon, a small francophone country in west-central Africa, with the goal of developing a vaccine. Richard Godfrey's work thirty years prior notwithstanding, the prevailing belief among immunologists remained that worms caused allergic disease. Any infection that elevated the allergic antibody immunoglobulin-E would, the thinking went, increase the risk of allergic sensitization.

But this paradigm failed to explain what Yazdanbakhsh observed daily in Gabon. First, children chronically infected with worms didn't have more allergies; they seemed to have fewer. Second, while immunologists assumed that the two modes of immune response, Th1 and Th2 (responses countering microbes or large parasites, respectively), were mutually exclusive, Yazdanbakhsh noted that both were suppressed in children infected with filarial worms and blood flukes. These two immune responses were envisioned as sitting on opposite ends of a seesaw. But everything Yazdanbakhsh observed suggested that a triangle, with the pivot point somewhere in the middle, more accurately approximated reality. In other words, immunologists routinely ignored some mysterious but all-important third player in the immune response.

A simple comparative study would support or rebut the idea that worms worsened allergy. And so working with her colleague Anita van den Biggelaar, Yazdanbakhsh examined 520 Gabonese schoolchildren for blood flukes and filarial worms, measured allergic reactivity to dust mites, and quantified circulating IgE. If worms made allergies worse, children harboring helminths should be more allergic to dust mites than those without. But when she tabulated the results, Yazdanbakhsh observed precisely the opposite. Both groups produced equal amounts of mite-specific IgE; both were sensitized to the allergen; but children who carried blood flukes had about one-third the reactivity to dust mites compared with children who didn't. Where it counted, allergy as measured by skin prick tests, worm-carrying children scored much lower.

How did this happen? What allowed a child to be sensitized but never develop allergic disease? Yazdanbakhsh compared sera from both study groups. A single immune-signaling molecule was elevated in the group with less allergy. Children with blood flukes had much more circulating interleukin-10, an anti-inflammatory cytokine. Here was evidence of that mysterious third arm of the immune system that tempered inflammatory responses. When she exposed immune cells from worm-carrying children to dust-mite proteins, they spewed IL-10. They actively tolerated an otherwise potent allergen.

Now Yazdanbakhsh and van den Biggelaar pursued a second question: Would dewormed children become more allergic after losing their parasites? The researchers arranged a randomized, double-blind, placebo-controlled study. No one knew who received deworming medicine, and who fake medicine. They dewormed 317 children every three months for thirty months, periodically testing the children's allergic reactivity to dust mites. The posttreatment allergic response was, they found, dramatically enhanced. After losing their parasites, children were two and a half times more likely to respond to dust mites.

In just over two years, Yazdanbakhsh had produced an increase in allergic reactivity that roughly equaled the difference John Britton had observed between urban Jimma and its rural surroundings. For that matter, by extrapolation, it was similar to the increase in allergy prevalence observed in developed countries in the late twentieth century. And all it took was removing worms.

THE CELL THAT PROTECTS AGAINST ALLERGIES

Far from the tropics, in Edinburgh, Scotland, a parasite immunologist named Rick Maizels watched Yazdanbakhsh's work with growing interest. He'd spent decades parsing the crosstalk between rodent and parasite in a laboratory setting. More than most, he understood that the relationship between parasite and host was quite nuanced—so nuanced that, at times, it looked like cooperation.

By the early 2000s, a new type of white blood cell had moved to the fore in immunology: the regulatory T cells I mentioned in chapter 1. As far back as the 1970s, scientists had hypothesized that "suppressor" cells existed. But research on the cell—and faith in its existence—had largely collapsed during the 1980s after scientists failed to isolate it. Now, however, scientists had new molecular markers by which to identify the cell subset. And recent studies, including those that looked at mutations in FOXP3, had shown the cells to be real, and critical to maintaining balance in the mammalian immune system.

In a field built on martial metaphors, regulatory T cells were notable for what they guaranteed *didn't* happen. They ensured tolerance to one's own tissues; they helped maintain peace with commensal microbes in the intestinal tract; and they offered a new way of conceptualizing immune-mediated diseases, such as asthma or inflammatory bowel disease. These disorders didn't result so much from misbehaving rank-and-file attack cells—cells that mistook friend for foe, or that relentlessly pursued harmless ragweed pollen. Rather, they stemmed from a deficiency or absence of peacekeeping cells. Not too much yang, but too little yin.

Maizels suspected that these T-regs were critical to understanding the worm-allergy puzzle, and Yazdanbakhsh's observations. To demonstrate the link experimentally, he sensitized mice to dust mites, and then infected them with a millimeters-long, corkscrew-shaped parasite called *Heligmosomoides polygyrus*, the mouse hookworm. Now when he exposed the mice to mite protein, the worms squelched the preexisting allergy. He'd more or less re-created what Britton and Yazdanbakhsh had observed in Africa. Worms stopped the allergic response to third-party proteins irrespective of sensitization.

To prove that T-regs were behind the protection, Maizels transferred regulatory T cells from the worm-infected mice to allergic mice without worms. The recipient mice also lost their allergies. When he removed the T-regs, the protection disappeared.

Here was the lesson: Although T-regs made up a mere 10 to 15 percent of the body's circulating T cells, they were crucial in maintaining a balanced immune response. Provided you had a well-developed anti-inflammatory network—which in this case meant T-regs induced by worm infection—you wouldn't respond in a sneezing, coughing, eye-gunking way to proteins that, judging by your antibodies, you were allergic to.

Why weren't you born with T-regs of sufficient strength and quantity to automatically prevent allergies? Clearly, many people were. But just as clearly, many weren't. Which brings us to an odd conclusion: To tolerate harmless proteins, and maybe even our own tissue, some of us needed a booster shot from organisms that were, by all other measures, health-depleting, life-sucking opportunists. David Pritchard's "hangover" metaphor was perhaps more apt than he realized. Some of us were physiologically addicted, it seemed, to parasites.

THE RED QUEEN:
RUNNING FAST BUT GETTING NOWHERE

Early in *On the Origin of Species,* Charles Darwin explains life's "struggle for existence"—the constant battle against competitors, against members of one's own species, against parasites, and against extreme weather, among other challenges. Seemingly to reassure his Victorian-era readers, Darwin finished the chapter like so: "When we reflect on this struggle, we may console ourselves with the full belief, that the war of nature is not incessant, that no fear is felt, that death is generally prompt, and that the vigorous, the healthy, and the happy survive and multiply."

In other words, no one worried, losers didn't suffer much, and winners won big. But what did triumph really mean, aside from the privilege of struggling another day? More than a century later, the evolutionary biologist Leigh Van Valen asked that question. He looked at the seemingly endless arms race between organisms—gazelles that, ever fleeter, were pursued by cheetahs that ran ever more quickly—and noted that because of continuous adaptation among competitors, nothing really ever changed. With a hint of it's-all-so-pointless ennui, he dubbed this frozen state of affairs the Red Queen's Hypothesis. "No species can ever win, and new adversaries grinningly replace the losers," he wrote in a 1973 paper that, initially, no one would publish.

He'd borrowed the Red Queen from Lewis Carroll's *Through the Looking-Glass*. At one point in the fantastical tale, the protagonist, Alice, says, "Well, in *our* country . . . you'd generally get to somewhere else—if you ran very fast for a long time as we've been doing." To which the Red Queen replies: "A slow sort of country! [. . .] *Now, here,* you see, it takes all the running *you* can do to keep in the same place."

In the decades since, evolutionary biologists have confirmed that our world does indeed resemble the Red Queen's more than Alice's. Van Valen's idea has itself evolved into an influential domain of evolutionary biology. And no relationship better encapsulates the phenomenon of running to get nowhere than that between parasite and host. During the 1990s, as Joel Weinstock, David Pritchard, and others mustered the courage to propose that parasites had indelibly shaped the human immune system, evolutionary biologists were finding that very little about how animals looked, behaved, and reproduced *didn't* bear the mark of parasites.

One simple reason for this outsize impact related to the very different reproductive paces of host and parasite. Parasites typically went through many generations in just one procreative cycle of the host. They could therefore evolve more quickly. As a result, the host could never solve the parasite problem the same way thicker fur could stem the dissipation of heat in a cold climate, for example. Parasites simply adapted to the host's adaptations. And because any given species was essentially condemned to host some parasites, the question became not if, but how many?

Animals developed parasite-limiting behaviors. Perhaps they migrated long distances like the ungulate herds of the African savanna, or the many bird species that fly from the tropics to the Arctic each year. Or maybe they groomed themselves regularly, even obsessively, like many birds and mammals. They might develop a mutualistic parasite-cleaning relationship with another organism, like oxpeckers perched on wildebeest's backs, or cleaner fish in coral reefs. Or maybe they evolved intraspecific grooming behavior—I'll scratch your back if you scratch mine—like many primates, including ourselves.

Some of these adaptations may seem trivial. But, in fact, the necessity of limiting parasites is so dire that many species' most salient characteristics evolved in response. Take sex. Sex—two individuals combining half their genome to create a third—has long puzzled evolutionary biologists. Given other possible modes of reproduction, such as cloning, sex seems remarkably inefficient. Only half the individuals of any given sexual species, the females, actively procreate. Nonsexual species, on the other hand, can reproduce at twice the speed. So why choose this slow strategy?

The answer: to escape parasites. Consider a tiny, extruded spiral-shaped snail that inhabits the freshwater lakes and rivers of New Zealand. The snail can reproduce both sexually and clonally. What determines the strategy it chooses? In lake shallows, where trematode relatives of the schistosomes that infect humans abound, the snail reproduces sexually. In that environment, parasites quickly overrun non–sexually reproducing snails. Their lack of genetic diversity makes them easy prey; the same trick works on all of them. But that's not the case with sexually reproducing snails. Each is distinct, and this difference helps them resist infection. In the relatively parasite-free depths of the same lake, however, clonally propagating snails dominate. Absent the selective pressure of trematodes, duplicating oneself is the most efficient way to procreate. Only parasites, in other words, make sex advantageous.

And with the invention of sex—of male and female—came sexual dimorphism, the different appearance of males and females of the same species. Why bother? Why does one gender, usually the male, so often display excesses of one sort or another—prancing, singing, and bright feathers among birds; horns, tusks, and manes among mammals?

Again, scientists impugned parasites. The male peacock's shimmery feathers signaled, "Look what I can manage, even as lice and worms devour me. Look what my genes can do *despite* the bloodsuckers."

There was a wrinkle, however. Secondary sexual characteristics emerged only in the presence of sex hormones such as testosterone. Testosterone slightly suppressed the immune system. So a buck elk, with its bulging muscles and oversized rack of horns, broadcast not only its fighting prowess, but also its ability to survive in a parasite-ridden environment with a slightly depressed immune system—and to do it all while expending considerable energy locking horns with other males.

This rule, biologists found, extended to primates: Dominant males in chimpanzee troops were, it turned out, also the most parasite-ridden. That was due in part to elevated testosterone. Social dominance, in other words, signaled more than just the ability to pound and intimidate other males; it demonstrated a male chimp's ability to pound rivals while hosting more parasites than they did.

The colorful extravagance of some birds, the deliberately excessive horns, manes, and tusks of some mammals, the complex courting behavior of many animals, the very existence of sex itself—and by extension the love songs on the radio, and all the love-addled poems ever written—all likely evolved because of parasites, because all organisms must run as fast as they can just to stay in place.

In the context of the Red Queen Hypothesis, Pritchard and Weinstock's conjecture that parasites had shaped our immune system seemed a foregone conclusion. The only question was, where to begin looking for direct evidence.

IS ASTHMA A WORM ADAPTATION GONE AWRY?

For Julian Hopkin at the University of Wales in Swansea, the obvious starting point was a gene called STAT6 (short for signal transducer and activator of transcription 6). The gene encoded a molecule that relayed inflammatory signals in the lungs. Two observations suggested that it contributed to both asthma and parasite defense. If scientists inactivated this gene in mice, the rodents completely lost their resistance to worm infections. Parasites overran their bodies. But mice with the deactivated gene were also, scientists noted, invulnerable to experimentally induced asthma. No matter how hard they tried, scientists couldn't make these mice wheeze.

Humans had more than 150 variants of the STAT6 gene, running the gamut from elevated to muted sensitivity. Would any given version produce the same outcome in different environments?

Beginning in 2002, a multinational team of scientists began a comparative study. They traveled to a rural area of Shanghai Province called Xing-Chang, where asthma was rare but parasites ubiquitous. Farmers used their excrement as fertilizer, an ancient and quite sensible practice—why waste valuable nutrients?—that was common until the early twentieth century in the West. (In the U.S., the chamber pot's contents were euphemistically called "night soil.") As a result, these rural farmers inadvertently exposed themselves to endless reinfection with *Ascaris lumbricoides,* the giant roundworm. The community displayed what parasitologists consider a classic "20/80" infection pattern. Of six hundred children tested, one-fifth carried a heavy worm load. The remaining four-fifths had relatively few.

Now Hopkin looked at STAT6 variants and how they correlated with parasite load. In Xing-Chang, the STAT6 gene that enhanced the immune response correlated with lower worm burdens. But in the U.K., where such parasites were a distant memory, people with the very same STAT6 variant had an increased chance of developing hay fever, asthma, and eczema. The inverse was true as well. Those who carried less inflammatory versions of STAT6 had less asthma in Britain, but more worms in China.

Scientists working in rural Mali upped the ante by looking at STAT6 in combination with another variant that increased production of a signaling molecule called interleukin-13. Both were associated with allergy in the developed world. In rural Mali, however, these genes protected against infection with the blood fluke *Schistosoma haematobium*. The protection was additive: Those who possessed both gene variants had the fewest flukes of all.

The geneticists Matteo Fumagalli and Manuela Sironi at the Eugenio Medea Scientific Institute in Bosisio Parini, Italy, took a bird's-eye view of the human-worm relationship. They hypothesized that peoples with the greatest exposure to parasites and pathogens in the recent evolutionary past should have the largest parasitic footprint on their genomes. Rather than comparing specific gene variants, they looked at human populations around the world.

Their first study, published in 2009, supported Joel Weinstock's idea that worms were deeply entwined with inflammatory bowel disease. Six gene variants associated with celiac disease and inflammatory bowel disease were more frequent in populations exposed to lots of viral and bacterial infections in the past, as well as myriad worms. The two selective forces had operated in concert. By chronically depressing the immune system, worms made their hosts more susceptible to bacterial and viral opportunists. But whereas a worm infection wouldn't necessarily kill you on the spot, a viral or bacterial infection could and quite often did. The human immune system countered by bolstering defenses against opportunistic microbial invaders over evolutionary time. Those genes that held the fort against microbial pathogens had evolved to do so in the context of chronic worm infection. When you removed the worms, the genes misfired, increasing the risk of IBD.

A second worldwide study looked at genes underlying asthma. Twenty gene variants strongly correlated with the number and diversity of helminths in the recent evolutionary past, they found. Twelve of these twenty genes were directly implicated in asthma and allergy. Some affected regulatory T-cell development. Others boosted eosinophils—cells that help expel worms from the intestinal tract. Still others increased the sensitivity of cells primed with allergic antibodies, such as mast cells and basophils—the cells that drive the misery of hay fever season.

The scientists concluded that pathogens, and especially worms, constituted the greatest selective force on the human genome of any examined so far—more than diet or climate. We'd evolved light skin to deal with low sunlight and vitamin D deficiency at high latitudes. We'd evolved a lactase enzyme gene that allowed us to drink milk from animals. But all these adaptations paled in comparison with what we'd evolved to deal with viruses, bacteria, protozoa, and especially worm infections. These hangers-on had orchestrated more human genetic diversity than any other factor. And their legacy included gene variants associated not just with celiac and inflammatory bowel disease, but also type-1 diabetes and multiple sclerosis.

Fumagalli and Sironi's studies had acknowledged weaknesses, chief among them the assumption that where pathogens were now plentiful, they'd always been abundant. But together with the comparative studies on single gene variants, they painted a portrait of a genome dramatically sculpted by worms, with the most parasitized populations also the most predisposed, genetically speaking, to developing allergy, asthma, IBD, and perhaps autoimmune disease.

And that raised an awkward issue.

The asthma epidemic began among the people who were, genetically speaking, probably the least prone to developing allergy—people of mostly European descent from higher latitudes where parasites, while not absent, were historically scarcer than in warmer climes. The enrichment of problematic genes in populations more exposed to parasites and pathogens explained, in an uncomfortable it's-in-your-genes way, why the children of immigrants with a recent evolutionary past in the tropics often had higher rates of allergy in countries such as Australia, the U.S., and the U.K. Not only were they exposed to more cockroaches and mites by virtue of socioeconomic realities; their genetically enhanced antiworm defenses made them more vulnerable to overreacting to these proteins.

WHAT MAKES AN ALLERGEN?

What of allergens themselves? Why did humans tend to become allergic to certain proteins, but not others? What about peanuts, eggs, and dust mites, for example, was so provocative compared with, say, chicken or potatoes? In 2007, the British food researcher John Jenkins conceived a rule of thumb: If a protein was more than 63 percent identical to a human protein, one our own bodies manufactured, it wouldn't incite an allergic response. Take, for example, one called tropomyosin. Bird tropomyosin, which was 90 percent homologous to the human version, didn't trigger allergies. But tropomyosins from cockroaches, dust mites, and shellfish were among the most potent allergens known. Why the distinction?

That same year, the Swedish biochemist Michael Spangfort observed that proteins allergenic to humans were absent in bacteria. The vast majority of life on this planet, in other words, lacked the potential to provoke allergies. Only eukaryotes—plants, fungi, and, of course, other animals—produced allergenic substances.

A short time later the British chemist Colin Fitzsimmons pointed out that, of some ten thousand recognized protein families, just ten contained almost half of all known allergens. These ten families overwhelmingly came from invertebrates, especially endo- (internal) and ecto- (external) parasites. The substances that most provoked allergies, in other words, resembled the worms, lice, ticks, and fleas that fed on us during our evolution. Our immune system had an inborn sensitivity to these proteins. "You've got the host response thinking it's recognized a parasite, but it's just a dust mite," says Fitzsimmons.

Why the overreaction? Dust-mite proteins had no powers of persuasion, that crucial component of the host-parasite interaction that, at some level, benefits both. By contrast, real parasite infections actively induced tolerance—those T-regs that Maizels measured, the elevated IL-10 that Yazdanbakhsh observed, and another antibody type called IgG4 that blocked, and partly disabled, the allergy-promoting IgE. Dust mites and cockroaches didn't engage these regulatory circuits. They prompted a reaction meant to expel worms without the temperance of said reaction in its natural context.

These revelations cast the allergy epidemic in a new light, especially in the American inner city, where exposure to dust mites and cockroaches closely tracked allergy prevalence. Evolutionarily speaking, the epidemic might not have emerged from excess exposure to invertebrate proteins, substances that had surrounded us since time immemorial. Rather, it perhaps stemmed from a world full of wormlike proteins where actual worms, which taught us to tolerate these proteins, were suddenly absent.

This was the Red Queen's legacy: You had run hard just to stay in place, and your parasites had run right along with you. They tried their hardest to convince you to ignore them, while you tried your hardest to expel them. From a distance, the struggle appeared to have arrived at a stalemate, but the apparent standstill belied a fraught tension. And when parasites withdrew—or, more accurately, were expelled—the parasite-detecting machinery spun out of control.

EVERYONE LIVES, BUT NO MIRACLE

The first lesson of David Pritchard's dose-ranging trial was that J. A. Turton, the scientist who'd allegedly self-infected with 250 hookworm larvae in the 1970s, either overestimated how many larvae he'd applied to his skin or was simply one tough fellow. The single volunteer who, unawares, received 100 larvae developed a terrible rash. She suffered severe bouts of diarrhea and vomiting. After treatment with antiworm medication, she withdrew from the trial.

Even 50 larvae, the quantity Pritchard found that he'd received, proved difficult to tolerate. Abdominal pain and diarrhea also prompted him to kill off his hookworm colony. The remaining eight participants finished the twelve-week study. Those with the lowest dose—10 larvae—had few symptoms.

The scientists proceeded with a 10-worm dose. They recruited thirty hay fever sufferers for a randomized, blinded, and placebo-controlled safety trial, and infected half of them. Those not receiving worms got histamine to mimic the famous hookworm itch. ("Like nothing on this earth," Pritchard described it.) Participants tolerated the worms fine. Their allergies didn't worsen. And Pritchard observed the very beginnings—a mere hint—of the regulatory immune profile Maizels saw in mice, and Yazdanbakhsh saw in Gabonese schoolchildren. But the changes had no effect on actual symptoms.

Next came a double-blind, placebo-controlled study on thirty-two asthmatics. Hookworm migrates through the lung, and there was some concern that the passage would worsen asthmatic symptoms. The volunteers' asthma didn't intensify, however. Pritchard observed very minor improvements, so minor that random noise couldn't be discounted as the cause. Yet participants raved about their hay fever disappearing. When the trial concluded, many opted to keep their worms.

Overall, while demonstrating safety, the study results were underwhelming. Had they not hosted the worms long enough? Had they used too few?

In Papua New Guinea, the average person carried twenty-three adult worms—and not necessarily the same twenty-three. Individual hookworms were continually cycling through the body. Perhaps Pritchard needed to mimic this constant reinfection to gain benefits.

"That may have been a mistake," John Britton said about the low, one-time dosage. "With hindsight, I regret not using twenty-five."

It may also be that modern-day, worm-naïve Britons simply can't tolerate the number of worms necessary to change their immune functioning. Where hookworm was endemic, people had grown up with constant reinfection. And in animal experiments that clearly showed benefit, rodents received doses that would likely kill a person. As Fitzsimmons, who wasn't involved in the study, put it: "We might not be able to tolerate the medicine." Another possibility: Worms can't fix already established allergic disease. The animal studies aside, the research on worms and allergic disease in humans most strongly suggests that, if anything, parasite infections prevent allergic disease from emerging. They don't necessarily cure it once it is established.

The unconvincing experimental results weren't enough, however, to discourage an underground movement of do-it-yourselfers—people desperate to treat their mostly untreatable autoimmune and allergic diseases, a group I joined when I made that trip to Tijuana. Using higher doses, many have achieved remarkable success. More about them in chapter 13.

But first, let's jump to a parallel line of research. Parasite immunologists aren't alone in thinking that the modern epidemic of allergic diseases stems from a perturbed human ecosystem. We've coevolved with more organisms than just helminths. As Joel Weinstock says, "It would be a miracle if it turned out to be just worms. Then life is so simple."

CHAPTER 6

Missing "Old Friends"

It is now widely appreciated that humans did not evolve as a
single species, but rather that humans and the microbiomes
associated with us have co-evolved as a "super-organism,"
and that our evolution as a species and the evolution of our
associated microbiomes have always been intertwined.
—William Parker, Duke University

By the late 1980s, the German epidemiologist Erika von Mutius was pretty
sure that she was failing as a scientist. She'd spent two years investigating the
relationship between air pollution and croup, a hoarse cough that mostly
afflicts young children. But the project had slipped into disarray. Data collection methods weren't consistent between study areas. And comparisons
of the sort that allowed meaningful conclusions—that air pollution correlated with croup, for example—were proving impossible to make.

"I made so many mistakes," she says. "I was young."

So when her boss urged her to begin another project, von Mutius, filled
with dread at the prospect, applied for what she supposed would be a prohibitive sum—more than 1 million Deutschmarks, or somewhere near
$2.5 million today. To her chagrin, she won the grant.

And then history intervened.

On an early November day in 1989, an East German official made a
surprise announcement during a live press conference: "Today the decision was taken that makes it possible for all citizens to leave the country
through East German crossing points," said Günter Schabowski. Within
hours, thousands of East Germans had gathered at the Brandenburg Gate
in the Berlin Wall, the twelve-foot-high concrete mass that symbolized
both Germany's partition since World War II, and the tense, decades-long
stalemate of the Cold War. A flag-waving, trumpet-blowing throng of
West Berliners awaited on the other side. The Eastern Bloc had begun to
crumble. And Germany, divided for more than forty years, moved toward
reunification.

For von Mutius, these developments presented an opportunity. She was intensely interested in asthma. Unlike croup, which was usually caused by an infection that eventually resolved, asthma was chronic and lifelong. East Germany, where coal was still routinely used for fuel, was far more polluted than West. Industry was also poorly regulated. The Saale River, which ran through the East German industrial city of Halle, had once turned violet with chemical waste. In line with the prevailing pollution-causes-asthma dogma, she'd long suspected that asthma was more prevalent in East Germany. Now she had the opportunity to test the hypothesis.

Working with East German colleagues, von Mutius found that the greater pollution in the East did indeed cause more lung irritation. Bronchitis was about twice as common in East Germany compared with West. But surprisingly, asthma wasn't. Both Germanys had about the same prevalence. And stranger yet, another allergic disease was strikingly less prevalent. In more polluted, poorer East Germany, between one-third and one-quarter as many people had hay fever compared with West Germany. Von Mutius didn't know what to make of the find. Had she made a mistake? Were the results an artifact of differing diagnostic methods? Skin-prick tests confirmed the pattern. East Germans were far less allergic than their West German counterparts, and this despite exposure to more dust mites and molds.

In 1992, she began a yearlong fellowship at the Respiratory Sciences Center at the University of Arizona, Tucson. There she conferred with Fernando Martinez, another scientist keenly interested in asthma. He was puzzling over his own counterintuitive findings. After following a cohort of nearly eight hundred Tucson newborns, he'd noted that the more lower-respiratory-tract infections children had at a young age, the less their chances of developing asthma later. The pattern contradicted the prevailing wisdom. Everyone knew that respiratory infections exacerbated asthma; they didn't prevent it. Now came evidence from an unrelated population in a completely different environment that also countered the prevailing assumptions.

"It dawned on me that we had the wrong paradigm," Martinez says. "There could be exposures that could be paradoxically protective, that we thought were negative for us." Martinez and von Mutius weren't the only ones coming to this conclusion. In November 1989, the same month the Berlin Wall fell, an epidemiologist named David Strachan had published a short paper in the *British Medical Journal* entitled "Hay Fever, Hygiene, and Household Size." The paper had stuck in Martinez's mind like a burr.

Strachan had examined the records of more than seventeen thousand British children born during a single week in 1958. He'd tracked them to adulthood, and then tried to identify early-life factors that correlated with the later development of allergy. A single variable, he found, most correlated with an individual's odds of having hay fever or eczema at age twenty-three: how many older children were in the house at age eleven. The greater the number of older children around in childhood, the lower an individual's risk of allergic disease in young adulthood.

The effect was quite pronounced. Twenty percent of first borns had allergies. But among those with two older siblings, just 12 percent did. And for the group with four or more older siblings, a mere 8 percent were allergic. That was a two-and-a-half-fold difference between first- and fourth borns. Strachan discounted socioeconomic factors—that poorer people had more children, say, and that something about poverty protected against asthma—as driving the phenomenon. Even when controlling for the father's social class, the "sibling effect" held. What could account for these dramatic differences in vulnerability to allergic disease?

Strachan thought that early-life infection most likely explained the pattern. Older siblings increased the odds of contracting colds and other infections. Crowded households were generally more conducive to contagion. His hypothesis also neatly explained the recent increase of allergic disease in the developed world. The relative smallness of the late-twentieth-century family, and the unprecedented cleanliness of modernity in general, had decreased the infectious burden during childhood, he argued. The absence of robust immune challenge early in life was, by still unclear mechanisms, making people allergic.

For von Mutius and Martinez, Strachan's hypothesis provided a framework by which to interpret their own results. Crowding was the common thread.

Martinez continued studying children in Tucson, and von Mutius returned to Munich University. Following Strachan's lead, she noted that not only were living conditions in East Germany much more crowded compared with West, in the former Socialist Democratic Republic, young children more often attended day care while their mothers worked. Seventy percent of East German youngsters had spent time in day care, compared with just 7.5 percent in West Germany. If infections protected as Strachan proposed, then East German youngsters had almost certainly encountered more at a younger age compared with West Germans.

In some sense, von Mutius had discovered a window into the past. Living conditions in East Germany approximated those of Germany and Greater Europe before World War II. And the allergy prevalence in present-day East Germany also resembled that of early-twentieth-century

Germany, before the allergy epidemic. East Germans had remained in that immunological landscape while the West pushed forward into new, more allergy-prone territory. If she could identify the factor that the West had lost or gained in the decades between partition and reunification, she'd have a fix for the allergy epidemic.

In 2000, Fernando Martinez's group at the University of Tucson published the strongest study yet showing that day-care attendance early in life protected against allergic disease. The researchers had followed more than a thousand children from birth to age thirteen. They matched socioeconomic rank, the number of siblings, and other factors. Attending day care in the first six months of life lessened one's chance of wheezing later by nearly three-fifths, they found. Crowded, presumably contagion-enhancing environments early in life definitively warded off asthma. So which infection was it?

SEEKING THE BUG THAT PREVENTS ASTHMA

The popular press dubbed this evolving collection of observations "the hygiene hypothesis": we were too clean for our own good. Scientists used the by-now-familiar seesaw model of the immune system to explain the phenomenon. The immune system had two mutually exclusive responses. One, called T-helper-1, or Th1, attacked microbial invaders, such as bacteria and viruses. The other arm, called Th2, pursued larger interlopers, such as worms and lice. A misdirected Th2 response—an attack mistakenly directed at birch pollen, cat hair, or cockroach proteins—typified allergic disease. The modern hygienic environment, and dearth of childhood infections, had deprived the Th1 arm of stimulation, they argued. Absent that counterbalancing force, the Th2 arm responded overzealously.

You may notice that this explanation is the inverse of Joel Weinstock's thinking in the 1990s. In his model, inflammatory bowel disease stemmed from an overexuberant Th1 response. Add a worm, stimulate Th2, and you'd turn down the Th1. Let me dispel the apparent contradiction: The epidemiology never supported the seesaw model of immune function. If Th1 and Th2 were mutually exclusive, you'd expect allergic and autoimmune disease to occur in opposition to each other, and in different populations. But generally speaking, the same populations—and often, as in my case, the same individuals—suffer from an elevated risk of autoimmune *and* allergic disorders. Scientists now mostly agree that the third arm of the immune system—the suppressor arm—matters most in avoiding both.

But the notion that infection with some virus or bacterium might protect against allergic disease was, in some ways, heartening. It raised the prospect of developing a vaccine against asthma in much the same way scientists had created vaccines for polio or measles. Once the protective bug was identified, weaken it so it was no longer dangerous, put some in a syringe, inject it, and voilà! You'd just cured allergy.

Early on, measles seemed a promising candidate. Nearly everyone contracted the virus before a vaccine became available in 1963. But studies produced contradictory results. Among six thousand British children born in 1970, two years after the measles vaccination became available in the U.K., both natural measles infection and the vaccine seemed to protect against allergy, but only for children with older siblings. The protection disappeared, and even reversed, in firstborn children who'd contracted the virus or received the vaccine.

A look at half a million Finns, meanwhile, suggested that measles infection slightly increased one's chances of developing eczema. And a study of nearly two thousand children in Scotland concluded that while measles protected from asthma somewhat, in general, the more infections one contracted, the greater one's chances of developing allergy. This was not the clear signal you'd expect if measles was truly responsible for the strong inverse relationship between allergies and crowding early in life. So what could explain the pattern?

THE PREEMINENCE OF THE OROFECAL

From Italy came a break in the case. An epidemiologist named Paolo Matricardi looked at allergic disease in more than 1,600 Italian air force cadets. Those infected with hepatitis A, he found, had half the risk of allergy of those who'd never encountered the virus. Rather than focus on the virus itself, Matricardi concentrated on the kind of pathogen it was. Hepatitis A traveled the orofecal route: Widespread infection meant that stool regularly ended up in food and water. You could be sure that people in these environments routinely imbibed other microbes as well. Hepatitis A infection might simply be a marker for other orofecal infections—or even, possibly, innocuous fecal microbes—in general.

A larger follow-up study again emphasized the primacy of orofecal exposures. Infection with chicken pox, herpes virus, mumps, rubella, and measles, which are airborne, didn't protect against allergies. But exposure to *Toxoplasma gondii,* a single-celled parasite native to cats, the hepatitis A virus, and *Helicobacter pylori,* a corkscrew-shaped bacterium that inhabits the stomach—all of which are orofecal—consistently protected. The effect was additive. Infection with one bug lowered the chances of allergy by nearly one-third. Infection with two or more lessened the odds by half again. In all, there was a nearly threefold difference in allergy risk between cadets exposed to more than one of these parasites compared with those exposed to none.

Would this relationship hold in other populations? Matricardi parsed the records of nearly 34,000 Americans gathered during the Third National Health and Nutritional Examination Survey, or NHANES, a periodic study conducted by the CDC. In short, the answer was yes: *T. gondii* and hepatitis A protected against allergy in the U.S. as well. And the greater time depth of the NHANES data permitted a new insight.

The Italian cadets were all roughly the same age. The Americans, however, belonged to various age groups. Comparison of people born decades apart revealed that while hepatitis A infection protected in current environments, for those born before 1920, exposure to the virus conferred no relative advantage. Everyone born in the early twentieth century, both hepatitis-positive and -negative, had the same odds of allergy—which is to say, less than half the risk of the generation born during the 1960s. Generally, about 2.7 percent of those born in the first two decades of the twentieth century developed hay fever, compared with 8.5 percent of those born in the 1960s. But for those exposed to hepatitis A, the prevalence remained steady at 2 percent in all decades.

Like East Germans, Americans exposed to hepatitis A appeared to inhabit a different immunological landscape, one that we all resided in around the turn of the nineteenth century regardless of our exposure to the virus. And then, while those infected with hepatitis A remained in that low-allergy landscape, other Americans moved on into new, more allergic territory.

WHY DON'T FARMING CHILDREN SNEEZE?

While Paolo Matricardi studied cadets, and Erika von Mutius puzzled over the discrepant allergy prevalences in East and West Germany, a chance conversation prompted the Swiss epidemiologist Charlotte Braun-Fahrländer to include a certain question in her asthma survey of rural Swiss children. At the recommendation of a local doctor who'd noted that the children of pig and cow farmers rarely had allergies, she tacked on a question about farming. It turned out to be a critical addition.

Not only were farming children one-third as allergic as their nonfarming rural counterparts, but the more farming they did, the less allergic they were. Children from full-time farming families had half the allergies of those from part-time farming families. And they were one-fourth as likely to have allergies compared with rural children who never farmed. More than a century after the Manchester physician Charles Blackley first noted that farmers, who lived coated in pollen, never got hay fever, Braun-Fahrländer had rediscovered the same phenomenon.

Von Mutius corroborated the relationship in Germany with a survey of more than ten thousand rural Bavarian children. Farmers' children there had half the hay fever of nonfarmers in the same rural area. Protection depended on the frequency and duration of exposure to livestock. Even regularly working with farm animals, but not living on a farm, reduced the odds of allergy by more than half. The effect held when controlling for a family history of allergy. You might have first-degree relatives living in town who sneezed as much as anyone else, but if you milked cows regularly, you'd have less hay fever and asthma. Farmers weren't, in other words, a self-selected group of people genetically invulnerable to allergies.

By 2002, numerous studies from around the world had documented "the farming effect"—in Finland, Denmark, Austria, France, and Canada. The protective farms weren't industrial-sized concentrated feedlots, a work environment associated with its own set of health hazards. They were generally small, family-run operations. What about them protected? Most obviously, cowsheds, pigpens, and stables were brimming with microbes, a wealth of bacteria from manure, animal feed, and mud. So around the turn of the millennium, researchers fanned out across Germany, Austria, and Switzerland to quantify the microbial exposure on farms. They collected dust from houses, kitchens, and barns. They vacuumed beds. They used sucking contraptions to test stable air.

The scientists had settled on a substance called endotoxin as their marker for environmental bacterial load. Endotoxin is a molecule that bacteria with two cellular walls (versus one) use in their outermost membrane. Importantly, it elicits a strong immune response from the mammalian immune system. And sure enough, stables were chock-full of endotoxin, much of it wafting about in the air and inhalable. Farming homes had nearly four times the endotoxin of nonfarming homes. Farmers' bedding had more than five times as much. Even children who had regular contact with animals, but didn't live on farms, brought microbes home on their clothes, their shoes, maybe even in their hair. And the more endotoxin they encountered, the less their chances of allergic disease.

The earlier the exposure began, the better. Children who accompanied their parents to the stables during the first year of life had less allergy compared with children who began working on a farm at school age. Indeed, you could predict a child's odds of developing allergy by measuring the endotoxin in her mother's mattress. Mothers whose beds were relatively saturated with the stuff had children with far less allergy.

The immune system has two major wings. The adaptive immune system—the wing engaged by vaccines—learns over a lifetime. The innate immune system, on the other hand, needs no instructions. From the moment you're born—and even before—innate immune cells can recognize patterns in bacteria, parasites, and viruses. This wing of the immune system has learned, over evolutionary time, that some aspects of the microbial world never change. And it has permanently stored this information in our genes.

Scientists observed a major difference in the innate immune system of these farming children. A protein called CD14, which helps recognize endotoxin, appeared at double concentrations in farmers' blood. And a microbial sensor called toll-like receptor 2 was three times more abundant.

The same way a musician's brain can distinguish between notes and tempos inaudible to a nonmusician, the farming immune system appeared to have a heightened ability to sense the microbial world. Paradoxically, this enhanced microbe-sensing machinery didn't intensify the inflammatory response. Quite the opposite, in fact. When Braun-Fahrländer mixed endotoxin directly with immune cells extracted from farmers' children, and compared the response with that of nonfarming children from the same area, she found that farmers' immune cells responded less vigorously. They tolerated what provoked their nonfarming neighbors. The constant contact with microbes had, it seemed, taught them a certain immune serenity. And that translated to less allergic disease.

THE HYGIENE HYPOTHESIS CONVERGES ON THE COWSHED

The "farm effect" provided a unifying theory for the hygiene hypothesis. Exposure to innocuous microbes could explain the protection of day care, the former East Germany, pet ownership, orofecal pathogens, and older siblings better than infections. (Von Mutius confirmed that farmers didn't have more hepatitis A or *Toxoplasma gondii* than nonfarmers.) The microbial richness that co-occurred in infectious environments—and in day care, homes with dogs, and large families—modified the risk of allergies, not, it seemed, the childhood infections that David Strachan had first proposed.

The issue was one of evolutionary norms. Humans evolved in conditions that more resembled a moldy, hay-filled, manure-dusted barn than a clean, modern apartment in Munich or Zurich. Had the immune system come to expect this microbial wealth? "Our immune system doesn't develop spontaneously," Fernando Martinez told me. "There's no system that develops spontaneously."

From a distance, the U.S. inner city, with its inexplicably elevated prevalence of asthma and allergy, seemed to directly contradict this research. In New York City, neighborhoods with 7 percent childhood asthma prevalence (the Upper East Side) sat within blocks of neighborhoods with 19 percent prevalence (East Harlem). Here were people living in relatively unhygienic environments, exposed to rodents and cockroaches. Yet, despite this "dirty" environment, the mostly African American and Latino populations had an asthma risk higher than just about any other in the U.S.

In the early 2000s, the Columbia University researcher Matthew Perzanowski set out to disprove the hygiene hypothesis in New York City. But even here, if you compared like with like—if you matched ethnicity and socioeconomic class—children raised in homes that had more endotoxin had a slightly reduced risk of eczema. And older siblings protected younger.

"I don't know if I'm failing as a scientist here," he says. "But we keep coming up with data somewhat supportive of the hygiene hypothesis."

There was another thing. For decades, allergen avoidance—clearing out ragweed, fumigating cockroaches, mite-proofing mattresses, avoiding nuts—had served as a cornerstone of allergy management and prevention. And surely, allergic people should avoid whatever makes them sick. But farming children in Germany, Austria, and Switzerland inhaled about five times more dust-mite dander than other rural dwellers, and yet they were far less allergic to it. The same was true of pollen. Farmers inhaled orders of magnitude more grass pollen, but had far less allergic sensitivity to it.

Citing these pollen- and mite-coated farmers without allergies, Braun-Fahrländer and others proposed that allergic disease didn't result from excess exposure to allergens, but from limited exposure to microbes. The allergy-promoting environment was one bereft of bacteria, not one overwhelmed with dust mites and pollens. Indeed, some intervention studies showed that avoiding allergens failed to prevent asthma or allergy, and in some cases actually increased the risk of allergic sensitization.

A study in Colorado encapsulated the problem. The researcher Andrew Liu found that air-conditioning more than halved the endotoxin in Denver-area apartments. Not owning a pet, on the other hand, failed to keep some dog and cat dander from wafting into the house. The end result: Pet-free homes contained animal dander at concentrations high enough to sensitize children, but insufficient microbes to prevent allergy. That about summed up the twenty-first-century predicament: Allergens remained abundant, but the organisms that might help us tolerate them had disappeared.

As you'll recall, this idea parallels one explored in chapter 5—that allergy emerges when we're faced with wormlike proteins, but no worms. The organisms invoked are different (worms in earlier chapters, microbes here), and some immunological details differ, but the greater lesson is consistent: External stimuli help us learn tolerance. Without it, the immune system spirals into dysfunction.

If we really wanted to head off allergic disease, the emphasis on allergens seemed somewhat misguided. Yes, if you're allergic to dust mites, you should avoid them. But if you wished to prevent allergic sensitization altogether, you needed to intercede further upstream. Forget the proteins. The coup de grâce of allergy prevention consists of teaching the immune system tolerance from an early age.

REPRODUCING NATURE IN THE LAB

And so came a battery of animal experiments, each one adding nuance to the central observation that microbes prevented allergic disease. A rat exposed first to endotoxin, and then to egg proteins, wouldn't develop allergy, scientists found. But if the rat first encountered the protein and then the endotoxin, the allergic inflammation worsened. Translation: Early-life encounters with bacteria were important. Conversely, if you already had hay fever, a job as cowhand probably wouldn't fix it. In fact, the enriched microbial environment might make things worse.

Timing was similarly important. Endotoxin could reverse an established allergy in rats if given within four days of exposure to an allergen. If given after, however, it exacerbated the now-established allergy. This meant that if you pranced through a field of ragweed, you'd have a limited window of opportunity to quash the imminent allergic sensitization. Don't miss it.

Work by Anthony Horner, an immunologist at the University of California, San Diego, underscored the importance of chronicity. An allergen accompanied by house-dust extract—a potpourri of microbes and other detritus he'd collected from homes around San Diego—could, if administered to mice in a single large dose, make them quite allergic. But if divided into seven smaller doses given daily over a week, not only did these mice fail to develop allergy, they also became inured to allergic sensitization thereafter. When he again deluged these mice with the potion, they resisted sensitization. Translation: The background noise—the level of basal immune stimulation—was critical in preventing allergic disease from taking root.

A scientist named Meri Tulic tested endotoxin on tiny plugs of sinus tissue removed from allergic children. (They'd undergone surgery for other reasons.) Ragweed pollen alone elicited the typical allergic flurry. But pollen accompanied by endotoxin provoked an allergy-protective response, including about four times as much anti-inflammatory IL-10 compared with the allergen alone. The children's immune cells changed after exposure to bacterial products, sprouting microbial sensors like those seen in farmers. Adult immune cells, however, lacked this plasticity. They didn't transform in the presence of endotoxin. The lesson: Early-life exposure to microbes is extremely important. The adult immune system is disappointingly set in its ways.

A single question drove this research: How to produce a therapy from bacteria? How to put the European cowshed in a bottle? Dale Umetsu, a researcher at Stanford University, erased peanut allergies in dogs with injections of heat-killed bacteria and peanut proteins. Ostensibly, the bacteria accompanying the proteins trained the canine immune system to respond differently.

Others experimented with a substance called CpG-oligo for short (in long form, cytosine guanine phosphodiester oligodeoxynucleotides). The immune system recognized CpG-oligo as bacterial DNA, and responded in kind. Asthmatic rhesus macaque monkeys that inhaled CpG-oligo periodically over thirty-three weeks saw mucus production in their lungs halve, and allergy-promoting white blood cells decline. Their bronchial membranes grew thinner and less inflamed.

Scientists at the University of Manitoba, Winnipeg, moved to blinded, placebo-controlled human studies. Participants received injections of ragweed pollen protein attached to bacterial DNA motifs. Their immune response to the pollen shifted slightly, but during the first ragweed season, they continued to suffer from allergies. Only during the second hay fever season, more than a year after beginning treatment, did they show real, if minor, improvement. A second double-blind, placebo-controlled trial at Johns Hopkins also showed a small benefit, less hay fever in twenty-five allergic adults who'd received six injections weekly of the CpG-ragweed cocktail over six weeks.

But a third study, on forty asthmatics, didn't show much gain. Was the dosage too small? The Canadian study used one-fortieth the dose given to rodents, which was understandable. No one wanted to accidentally trigger inflammatory disease of a different kind. Another possible confounder was age: The immune system was most plastic in childhood, but the study participants were adults. Or maybe complexity itself mattered. What if the immune system required not just one type of stimulation, but several?

Questions like these broached a deeper, epistemological issue. The scientific method that had proven so useful in defeating infectious disease was, by definition, reductionist in its approach. Germ theory was predicated on certain microbes causing certain diseases. Scientists invariably tried to isolate one product, reproduce one result consistently in experiments, and then, based on this research, create one drug. But we'd evolved surrounded by almost incomprehensible microbial diversity, not just one, or even ten species. And the immune system had an array of inputs for communication with microbes. What if we required multiple stimuli acting on these sensors simultaneously? How would any of the purified substances mentioned above mimic that experience? "The reductionist approach is going to fail in this arena," says Anthony Horner, who'd used a mélange of microbes in his experiment. "There are just too many things we're exposed to."

BACK TO THE LAND: MUD, SOIL, AND WATER

Graham Rook peers at me, arched eyebrows pushing into a lined forehead, over rectangular, silver-rimmed glasses that seem permanently fastened low on his nose. When I visit Rook at University College London one day late in May, it's a few months before he'll retire to southern France. Pictures of his retirement house, a manor of hewn stone, full of sun and greenery, fade across the screen of a new laptop on his desk.

Rook has become a kind of godfather to the hygiene hypothesis. Early on, he championed ideas that have since become cornerstones of the field. In the late 1990s, he insisted that the then-dominant model of immune function—two immune response types, the pursuit of microbes and the repulsion of parasites, cross-regulating each other—was incorrect. A third peacekeeping arm, which prevented both autoimmune and allergic disorders from arising, was key—a view that has since achieved orthodoxy. And, bemoaning the compartmentalization of medicine, Rook urged cross-fertilization. Allergy researchers needed to talk with autoimmunity researchers, he argued, and everyone needed to confer with scientists studying human evolution.

Finding the emphasis on "hygiene" misleading, Rook rechristened the hygiene hypothesis the "old friends hypothesis." Major infections don't help the immune system, he argues. If anything, acute inflammation makes things worse. A very specific group of organisms meets the "old friends" criteria—organisms that have accompanied us since the Paleolithic. That includes worms, cowshed-type microbes, lactobacilli, and our own fecal bacteria. It doesn't include, however, measles and your everyday cold virus. Evolutionarily speaking, these are latecomers. They arrived after the domestication of animals, and after humans had aggregated in crowds sufficient to sustain them. That's probably why the research on measles and allergy ultimately arrived at a dead end. They have no evolved relationship with the human immune system.

Rook has spent decades studying one group of these "old friends"—mycobacteria. The mycobacterial family has two famous members: *M. tuberculosis,* the cause of consumption; and *M. leprae,* which causes flesh-eating leprosy. Rook thinks that *M. tuberculosis* protects against allergic disease—not active tuberculosis, but the latent infection that occurs in nine of ten people who encounter the bacterial parasite. And studies in Japan, Estonia, and South Africa, among other places, support the idea. Children who test positive for latent TB also have a lower risk of allergies and asthma.

But really, Rook is more interested in nonparasitic mycobacteria and their role in educating the human immune system. They're called saprophytes, and before the era of paved roads and treated water, we imbibed them with every sip of water, ate them with every bite of fruit, and inhaled them with every breath of dirt-tinged air. These bacteria, which survive by breaking down organic material, coated us inside and out during our evolution. Rook calls them "pseudocommensals"; they don't establish permanent residency in the human body, but by virtue of their constant flow through the digestive tract, and their continuous presence on mucosal surfaces, the immune system treats them as residents. What does that mean? At some level, we tolerate saprophytes. If we didn't, we'd have melted in a spasm of inflammation long ago. And so Rook thinks that environmental saprophytes played a huge role in teaching the human immune system tolerance.

The ubiquity of these bacteria in the muddy, feces-filled world of yore, and their potential importance in our immune functioning, came to Rook's attention in a roundabout way. Ever since the development of the tuberculosis BCG vaccine, scientists have wondered why it protected some, but not others. In parts of Africa, such as Malawi, BCG vaccination hardly helped. But in the U.K., it reduced the odds of contracting TB by 80 percent.

Seeking to explain these discrepancies, scientists' attention eventually fell on environmental bacteria that, to the immune system, resembled the bacterium in the vaccine—Rook's saprophytes. In places where people drank untreated water and lived in houses with packed-earth floors, they were constantly exposed to nonparasitic mycobacteria. Contact with these microbes either served as a natural vaccine, scientists realized, sometimes boosting immunity to TB; or less desirably, it acted as tolerizing immunotherapy. Sustained exposure convinced the immune system to tolerate BCG, and that tolerance reversed the vaccine's protective effect.

In the early 1970s, Rook's mentor and collaborator, a microbiologist named John Stanford, and his wife, Cynthia, traveled to Uganda, where BCG vaccination worked particularly well. If he could identify the bacterium that improved the vaccine's effectiveness in the country, Stanford thought, he could develop a booster vaccine. And from the shores of Lake Kyoga, in what Rook calls "hippo mud," Stanford isolated a nonparasitic mycobacterium called *M. vaccae.*

Back in Britain, the Stanfords tested the bacterium on themselves to prove safety. And the oddest thing happened. Cynthia had suffered from an autoimmune disorder called Raynaud's syndrome, a sometimes painful condition that chokes blood flow to fingers and toes. The winter after her first injections, however, the symptoms disappeared. It seemed as if the bacterium had corrected the immune dysfunction underlying the disorder. And if the bacterium corrected autoimmunity, what other immune malfunctions might it address? *M. vaccae* injections helped their young daughter's asthma, it turned out. Doctors testing *M. vaccae* on tuberculosis in India observed that some patients' psoriasis, a scaly autoimmune disorder of the skin, cleared up.

Rook and Stanford formed a company to formally develop "the dirt vaccine." Immunotherapy with *M. vaccae* snapped the lazy immune system to attention, the thinking went, and corrected ongoing malfunction. But when the first human trial on cancer showed no benefit, the enterprise fell apart. (More about this approach to cancer in chapter 12.) Trials on allergic disease were also ultimately inconclusive. After promising early results, a larger, double-blind trial on eczematic children showed no difference between placebo and treated groups. Both improved by 50 percent. A second randomized placebo-controlled trial on asthmatics showed no improvement—at least initially. Subsequent reanalysis suggested that, after correcting for variation among patients, those who received two large doses of *M. vaccae* had actually improved significantly. And that about sums up Rook's forays into human testing—contradictory results that, upon review, looked promising.

"We were completely incompetent when it came to designing clinical trials," Rook tells me. Now, as a matter of course, Rook, ever self-deprecating, avoids human studies. But he's forged ahead studying *M. vaccae* in animals. And in these much easier-to-control circumstances, he's found that, as you'd expect from organisms that must be tolerated, the bacterium engages the regulatory arm of the immune system. Mice treated with *M. vaccae* have more T-regs, and more circulating anti-inflammatory

IL-10 and TGF-beta, immune-signaling molecules that protect against allergic disease. When he transfers regulatory T cells from treated mice to nontreated, the recipients become resistant to allergy. Importantly, oral treatment protects as well as injections. Immunotherapy directly on the gut has a systemwide effect. Meanwhile, studies around the world—in Vietnam and Ethiopia, among other places—have confirmed that people who drink untreated water from surface sources, water presumably teeming with saprophytes, have less allergic disease.

Rook summarizes his thinking in several tidy dictums: "Coevolution leads to codependence"; or alternately, "evolution turns the inevitable into a necessity." He means that if you can't escape some aspect of the environment, you'll adapt to it. Over time, you'll incorporate the inescapable into your day-to-day functioning. Thereafter, you'll require it. "It's so blindingly obvious that you write into your genome things that are present in the environment," he says. "And you become dependent on them."

He illustrates the principle with a case study of primates and vitamin C. Vitamin C is so important for cellular processes—it's a potent and necessary antioxidant, among other functions—that the great majority of animals manufacture their own. Insufficient vitamin C causes scurvy—swollen, bloody gums and wounds that never heal. Notably, primates and guinea pigs can't make their own vitamin C. What happened?

At some point in our evolutionary past, Rook argues, as we gorged on a vitamin-C-rich diet, our own vitamin-C-manufacturing genes became nonfunctional. But in that environment, the genes were redundant anyhow. So losing the ability to make the vitamin bore no cost. At that moment, the primate lineage outsourced vitamin C production to plants. A new dependency emerged.

Now transpose that model to immune functioning. Contact with another organism—saprophytes, say—develops your immune regulatory circuits. Over evolutionary time, the ability to regulate immune functioning yourself dulls or disappears. Losing this capacity incurs no immediate cost, however. Saprophytes are ubiquitous, and contact with them is unavoidable. Nonetheless, you've just outsourced your immunoregulation to microbes. Now you're dependent on them.

When it comes to the immune system, Rook emphasizes the inevitability of redundancy. We wouldn't outsource the entire job of teaching tolerance to just one organism. "In an evolutionary sense that would be extremely stupid, wouldn't it?" he says, his legs dangling over the armrest of his chair. "You wouldn't want to be dependent upon one bug which one day might get killed off by some peculiar virus or something, and suddenly leave humans with no immunoregulation."

In a manner of speaking, however, that's what has occurred in the past century—not the removal of one crucial organism, but the eradication of an entire critical repertoire.

THE KARELIAN QUESTION

For centuries, the Finns have had to navigate the competing interests of two geopolitical powers: the kingdom of Sweden to the west, which for hundreds of years ruled much of the Baltic rim; and imperial and then Soviet Russia to the east, which ultimately controlled a continent-spanning swath of territory stretching from the Pacific Ocean to the Baltic Sea.

Over the centuries, what's now Finland—historically called the Grand Duchy of Finland—has alternately fallen under Swedish or Russian control. And since the end of World War II, a piece of what was historically Finnish territory has remained on the Soviet, and after its collapse, Russian side. That area, bound by the Gulf of Finland to the west and girded by large freshwater lakes and rivers, is called Karelia.

In Finnish political circles, this state of affairs is known as the Karelian Question. Karelians are linguistically, culturally, and genetically related to Finns. The question is when, if ever, will Russian Karelia, now home to many ethnic Russians, reunite with its Finnish half?

For scientists interested in the hygiene hypothesis, however, the region has furnished what might be called the Karelian Answer. The constant rejigging of the border—a single people divided and redivided—has incidentally produced the sort of experiment that epidemiologists salivate over: two genetically similar populations living in very different conditions, often within one hundred miles of each other. They inhabit the same latitude; they wile away the same interminable winter nights, and enjoy the same endless summer days. And yet the two Karelias present dramatically different prevalences of allergic and autoimmune disease.

Finland mysteriously has one of the highest rates of autoimmune disease in the world, ranking first for type-1 diabetes. (Sardinia is second.) And the Finns suffer from as much allergy and asthma as any other industrialized nation. On both counts, however, Russian Karelians score dramatically lower—so much lower, in fact, that Finnish scientists must go back in time to poor, mostly agrarian 1940s Finland to find statistics that are even remotely comparable.

Despite equal carriage of gene variants that increase the risk of autoimmunity, Russian Karelians have one-sixth as much type-1 diabetes as their Finnish counterparts. Celiac disease is also comparatively rare on the Russian side. Although they consume just as much wheat—often blamed for *causing* celiac disease—Russian Karelians have about one-fifth as much celiac disease as their Finnish counterparts, 1 in 535 compared with 1 in 107 in Finland.

One oft-blamed culprit for immune-mediated diseases, from asthma to cancer, is vitamin D deficiency. But both populations inhabit the same latitude, between 62 and 66 degrees north. (As a frame of reference, Anchorage, Alaska, lies at 61 degrees.) And Finnish scientists have confirmed that both Finns and Russians have the same amount of circulating vitamin D. Pollution is also similar in both regions; there isn't much. And urbanization, which so often predicts allergies and other immune-mediated disorders, is actually greater on the less allergic Russian side.

Finnish scientists stumbled on this "living laboratory" in the 1990s. For much of the decade, they had conducted surveys on chronic diseases in both Karelias. Beginning in the late 1990s, they included questions about allergy. That's when they noted that hay fever was four-and-a-half times more common in Finland than in Russia, and that asthma was two and a half times more common.

They rushed to quantify the differences between Finnish and Russian Karelia. The most obvious was affluence. Stepping across the border, the scientists passed from one of the richest parts of Europe to one of its poorest. GDP fell by a factor of seven. Finland boasted a per capita GDP on par with Sweden, Japan, and Germany. It headquartered globally recognized corporations such as the cell-phone company Nokia, and had a reputation for efficiency, little corruption, and good quality of life.

Russian Karelia, on the other hand, resembled Finland before World War II. People lived in crowded conditions. Families often kept a cow for milk, or chickens for eggs. And the large wealth differential matched a notable disparity in infectious burden. Russian Karelians were rife with infections that, with each passing generation, had grown scarcer in Finland. One in five Russian Karelian children had *Toxoplasma gondii*, the parasite native to cats; nine in ten had *Helicobacter pylori*; and eight in ten had contracted hepatitis A at some point.

By comparison, one in fourteen Finnish children had *T. gondii*; one in four had *H. pylori*; and one in ten had antibodies to hepatitis A. On the Russian side, the child whose immune system hadn't handled one or more orofecal infections was a rarity. On the Finnish side, that child was the norm.

When the Helsinki University scientist Leena von Hertzen correlated allergy prevalence with these infections, she found that, as a group, they explained almost half the difference in allergy and asthma between the two Karelias. *H. pylori* alone, usually invoked in connection with ulcers and stomach cancer, explained the single largest chunk, one-third of that relative protection. (More on *H. pylori* in chapter 8.) What explained the rest? Noninfectious microbes.

Primary schools on the Russian side drew water with minimal treatment from Lake Ladoga, the largest lake in Europe, or one of the area's many rivers. On the Finnish side, however, drinking water came from municipal sources, where it was treated with alkalization, which kills microbes by altering acidity, and ultraviolet radiation.

As a result, Russian water contained nine times more living microbes than Finnish drinking water. And the number of microbes a child imbibed correlated inversely with her vulnerability to allergic sensitization. Four of every ten Finnish children were sensitized to tree pollen. But fewer than one—0.8 to be precise—of every ten Russian children who imbibed this water were, a fivefold difference. Overall, half of all Finns were sensitized to something. One-sixth of Russian Karelians were.

Even within Russian Karelia, allergic sensitization correlated with the number of microbes in drinking water. The more microbes a given child gulped at school, the lower her chances of allergy. Anything above 1 million individual cells per milliliter, roughly one-fifth of a teaspoon, lessened the chances of allergy by two-thirds compared with those who quaffed fewer.

What were these microbes? They included some coliform bacteria, evidence of raw sewage seeping into the water supply. But these bugs didn't constitute the major difference between Finnish and Russian drinking water. Rather, saprophytes like those Graham Rook thought protective, bacteria that lived in water and soil, abounded in the Russian water. In Russian Karelia, water drained from boreal forests, bogs, and farmland, and it contained about forty times more bacterial products than Finnish water. With every swig, Russian children received an immune stimulant at forty times the dose of what Finnish children received. And administration of this naturally occurring probiotic began early in life, with the first drink of water.

The quality and quantity of microbes in homes on both sides of the border also differed substantially. Russian Karelian homes contained far more microbes overall. But it was the type of microbes present that distinguished the Russian environment. In Russian Karelian homes, microbes associated with animals and soil dominated. In Finland, plant- and human-associated microbes held sway. Most striking, the microbial diversity in Finnish homes was relatively impoverished. Many of the microbes that thrived in Russia simply didn't exist in Finland. In sum, Russian Karelians daily encountered more microbes of a greater diversity than Finns, and a greater proportion came from animals and soil.

With differences as stark as these, scientists sought to confirm that a diverse microbiota protected against allergy elsewhere. They conducted a comparative study in Finland proper. In Finnish homes, no other factor, including tobacco smoke, parents having allergy, the presence of a dog or cat, explained protection against allergy save one: Children exposed to the most microbially diverse house dust had one-quarter the risk of allergy compared with children exposed to the least diverse dust. Diversity itself had immunological value.

There was one more important development. Within Finland, barn dust contained a hundred times more bacterial cells per nanogram than urban dust. Urban dust was relatively dead, a few living cells surrounded by debris from human-adapted microbes. To quantify their different effects on the immune system, scientists sprayed mice with both dusts. Those that received barn dust had an allergy-protective response, but rodents treated with urban Finnish dust had a response that veered toward allergy. Not only did Russian-like microbes prevent allergy, in other words, the depleted microbial environment of the average modern Finnish home actively *promoted* allergic disease.

ALL SIGNS POINT TO MICROBIAL DIVERSITY

By the late 2000s, scientists could say with growing certainty what kinds of farms protected against allergy, and by how much: Frequently hanging out in an animal shed lowered chances by 29 percent; haying halved the odds of asthma, as did silage, a method of fermenting animal feed; pig farmers were 57 percent as likely to have asthma as nonfarmers; and drinking unpasteurized milk was consistently protective, although whether this was due to high microbial content, a wealth of health-promoting fatty acids from grass-fed dairy cows, or some other feature that, absent homogenization and pasteurization, remained intact, no one was sure.

Microbiologists, meanwhile, were dissecting the makeup of barn dust, which so consistently protected against allergy. A few bacteria predominated. One was called *Acinetobacter lwoffi*. Pulsing human immune cells with this bacterium made them incapable of initiating an allergic-type response. By extrapolation, if you spent your early life inhaling them while haying or milking cows, sensitization to dust mites and other allergens would simply not occur.

Of course, it hadn't escaped anyone's notice that cowsheds contained many substances other than microbes—namely, hay itself. About 13 percent of barn dust consisted of a plant fiber called arabinogalactan. Many plant fibers interacted directly with the immune system, scientists were finding, often engaging the same receptors as bacteria. (Another was inulin, a fiber derived from onions and garlic, and now a mainstay of the probiotics industry.) And when the microbiologists tested these arabino-galactans, they found that the stuff not only protected mice from asthma, it also induced an antibody type called IgG4. If you had IgG4 specific to ragweed pollen, say, it meant that you *didn't* have an allergy to ragweed. Exposure to this grass carbohydrate actively staved off allergies.

Meanwhile, the methods and technology used to probe and measure microbial communities grew cheaper and more powerful, bringing what was protective about farms and microbe-rich environs in general into clearer focus. Diversity took center stage, not just bacterial, but fungal as well. The intensity of exposure to diverse microbial communities protected both on and off the farm.

The immune profile of people inured to allergic disease also became clearer. Again, diversity was the trump card. The least allergic farming children had the most robustly developed innate immune system. And we're not talking just one or two receptors. If you compared innate immune cells from farming children with your average burgher's, you'd note that the former were comparatively bristling with the full range of possible microbial sensors. Compared to their nonfarming rural counterparts, farmers' immune sensors were in full blossom.

Despite this progress, and the sense that scientists were inexorably closing in on a recipe for an antiallergy potion (derived from pigsties, cowsheds, and animal feed), they could still only guess at how microbial diversity actually fended off allergic disease. In a 2011 article in the *New England Journal of Medicine*, Erika von Mutius and colleagues hypothesized that the continuing engagement of the innate immune system by many different types of microbes—an immune version of growing up in the cultural mélange of New York City, say—changed the tenor of immune functioning, strengthening one's ability to tamp down on inappropriate inflammation, and preventing allergic disease. Another intriguing possibility was that constant engagement by a diversity of microbes enhanced one's defenses. When viruses that might cause asthma arrived, farming children handily fought them off. But everyone else was prone to infection.

As von Mutius pointed out, however, explaining how microbial diversity worked presented a challenge. The innate immune system has a limited number of potential inputs. Humans have just ten toll-like receptors, the sensors of the front line. A small number and variety of microbes could theoretically hit them all. We carry enough microbes in our gut alone—again, in theory—to constantly activate those receptors many times over. Something more nuanced had to be at work.

ESTONIA AND THE MICROBES WITHIN

Late in 1991, the Swedish microbiologist Bengt Björkstén found himself overcome with a distinct feeling of déjà vu. He'd just arrived in Tartu, an old university town in Estonia, a Baltic state that had formally declared independence from the crumbling Soviet Union in August of that year. Until then, Estonia had been something of a mystery to Björkstén. But now that he'd set foot in the country, what struck Björkstén was not Estonia's otherworldliness, but its odd familiarity.

Tartu conjured up memories of Helsinki, Finland, Björkstén's birth-place. Both cities had spent centuries under Swedish rule, which showed in the similar architecture. But really, the city's rundown state brought to mind the Finnish capital a half-century earlier. Occupied by the Nazis and then bombed by the Soviets during the war, Helsinki was in a sorry, rubble-strewn state by the war's end. Tartu hadn't been recently bombed, of course, but nearly forty years as a Soviet satellite had left the country poor and in disrepair.

Precisely these differences had drawn Björkstén to Tartu from his post at the University Hospital in Linköping, Sweden. When the Iron Curtain had suddenly drawn back, he'd leaped at the opportunity to train scientists in Estonia. Estonian is to Finnish about what Portuguese is to Spanish, so language wasn't a major concern. He intended to spend the coming years instructing his colleagues at the University of Tartu about parsing and analyzing datasets, and the protocols of writing for scientific journals. Little did he know that Estonia would forever change the direction of his research, and furnish a crucial piece in the mosaic of observations referred to collectively as the hygiene hypothesis.

Like Erika von Mutius, who was working concurrently in Germany, Björkstén assumed that the relatively heavy industrial pollution in Estonia would exacerbate allergies. Estonians, he thought, would be far more allergic than Swedes. But his first study, a comparative survey of the allergic landscape, found just the opposite. Nearly one in three Swedish children had allergies. By comparison, just one in ten Estonian children did. That was a threefold difference from one side of the Baltic to the other.

Björkstén thought he'd made a mistake. But when he caught wind of von Mutius's studies—less allergy in more polluted and crowded East Germany—and David Strachan's observation that later-born siblings had less hay fever in Britain, he knew he'd stumbled upon a variant of the same phenomenon. Estonians lived in more crowded conditions compared with Swedes. Estonian households averaged 1.5 people per room. In Sweden, on the other hand, the mean was 0.9 person per room.

But whereas Strachan and von Mutius, both of whom were epidemiologists, initially looked to infections to explain the association, Björkstén, a microbiologist, had a slightly different idea. He thought to examine the bacteria at our very core, the microbes that inhabit the human intestinal tract.

He began with a broad comparison of infants in the two countries. Differences were immediately apparent. Estonian children harbored more lactobacilli than their Swedish counterparts. Swedish infants, on the other hand, more often harbored the sometimes diarrhea-causing pathogen *Clostridium difficile*.

The finding might be neither here nor there, but one historical fact provided context. From the late 1950s through the 1970s, West German scientists had methodically categorized infant microflora. Rereading these papers, Björkstén saw that the present-day Estonian microflora resembled that of German infants from the 1960s. The key observation was that the microflora had shifted thereafter. During nearly two decades of profiling infant microbial communities, German scientists had noted a change: Fewer bifidobacteria colonized infants over time. In its gradual depletion of bifidobacteria, the German microflora became more like the present-day Swedish microflora. And somewhere in that transition, the asthma and allergy epidemic had gathered momentum.

Bifidobacteria and lactobacilli were similar: They both produced lactic acid; they were both thought to promote health; and they were both indigenous not just to the human gut but also to the mammalian gut in general. Could the disparity between the Swedish and Estonian microflora explain the different national propensities to develop allergic disease?

Björkstén compared allergic children with nonallergic children within each country. If the makeup of the microflora mattered, he should see differences between groups that resembled what he'd observed between countries. Sure enough, allergic two-year-olds in both countries had fewer lactobacilli.

The working hypothesis went like this: Microflora with fewer lactobacilli and other critical microbes predisposed one to allergy. For reasons that were unclear, Westernization depleted these important microbes. But in Estonia, the nearly half century behind the Iron Curtain had kept the microflora relatively stable and healthy.

Hints of ancestral microflora—and of premodern colonization patterns—came from outside Europe. Like Estonians, Ethiopian newborns hosted far more lactobacilli than Swedes. And in Pakistan, another country with a relatively low allergy prevalence in the 1990s, infants were on average colonized by 8.5 distinct types of enterobacteria within the first six months of life. By comparison, Swedes hosted just one or two strains for their entire life. (Not coincidentally, Pakistani newborns also had more diarrhea.)

Of course, Björkstén couldn't be sure that allergy didn't alter the microflora, and that he'd simply observed the results of allergy, not its causes. To discount this possibility, he launched a prospective study. He followed children in both Sweden and Estonia beginning at birth. He periodically gathered stool samples to determine what microbes colonized them and when. And after two years, he found that children in both countries who were less often colonized by enterococci during the first month of life, and by bifidobacteria during the first year, had a greater risk of developing allergy later.

By now, other groups were reporting similar finds. Finnish newborns who went on to develop allergy hosted more clostridia and fewer bifidobacteria before their allergies appeared. And whatever bacteria arrived first seemed to set the tone for the developing immune system. Early colonization by a bacterium called *Bacteroides fragilis,* for example, reinforced production of the antibody immunoglobulin-A, which protected against allergy.

Why the human microflora should be so different in former Eastern Bloc countries wasn't clear. Estonian house dust had about twice as much endotoxin compared with Swedish, evidence of more resident bacteria. But while increased exposure to endotoxin protected from allergy in Sweden, the correlation fell apart in Estonia. No matter how much endotoxin Estonians encountered in their homes, they had fewer allergies. What protected them?

Björkstén suspected that different food-production methods played a role. Swedes consumed more processed food. Their fresh fruits and vegetables had undergone the semisterilization necessary to enhance shelf life. By contrast, Estonians still ate locally grown produce; there was no other kind. In Sweden, he could buy an apple year-round—they came from as far away as Tasmania. He could put the fruit in his cupboard, forget about it, and rediscover it fresh and shiny weeks later. In Estonia, on the other hand, he could buy the fruit only when in season, and he'd have to eat it within a week or two. If not, it would rot. Fruits and vegetables retained their natural microflora in Estonia. Add to that the custom of fermenting your own vegetables, also still practiced in Estonia, and you had several avenues of exposure that had closed in Sweden.

This constant exposure to a wealth of microbes, what he called "microbial pressure," guided Estonian immune development, Björkstén thought, and steered it clear of allergic disease. In its natural state, the human microflora existed in a kind of "stable chaos." New strains moved in; old ones departed. But the Western microflora had lost much of this dynamism.

It was, in the words of another scientist, "abnormally stable." Björkstén summed up what had gone wrong as the "microbial deprivation hypothesis": Westernization had limited exposure to a variety of microbes that either directly colonized the gut or simply passed through it. And this microbial impoverishment predisposed Westernized populations to allergic disease.

A DIFFERENT IMMUNE FUNCTION IN ESTONIA

Björkstén never confirmed that Estonian food really did supply additional microbial pressure, but in a way, he didn't need to. Wherever it came from, he saw ample evidence of greater microbial activation in Estonian immune functioning. From a very early age, the Estonian immune system followed a different developmental trajectory compared with the Swedish. While Estonian children churned out allergic antibodies (IgE) to proteins in the environment just like their Swedish counterparts, and the quantity of IgE circulating in their blood increased steadily until age five, the welt that formed from skin-prick tests decreased over the same period. Their allergic antibodies failed to translate into clinical allergy. In other words, the Estonian resistance to allergy stemmed less from a lack of sensitization, and more from better-developed regulatory mechanisms. In the first two years of life, Estonians progressively learned to ignore allergens. Swedish children, on the other hand, grew more and more responsive to them. What could explain this divergence? Björkstén's attention turned to the infant's first source of immunological cues: Mom.

For decades, scientists had known that breast milk carries antibodies that confer passive immunity to nursing children. Scientists were now learning that in addition to transferring protection against pathogens, breast milk also relayed a message written in immune-signaling molecules. These signals instructed the infant on how ready to be, and for what, in the current environment. And when Björkstén compared how this information differed in Estonia and Sweden, he found Estonian breast milk conveyed quite a different picture than Swedish.

In Estonia, the very first milk to emerge—the colostrum—carried more anti-inflammatory IL-10 and allergy-preventive interferon-gamma than in Sweden. The Swedish colostrum, on the other hand, carried relatively more of the allergy-promoting molecule IL-13. After the colostrum, Estonian breast milk was also relatively saturated with IgA, direct evidence of elevated microbial pressure.

Could you reproduce the immune profile of Estonian breast milk with an intervention? Björkstén gave expectant Swedish mothers a lactobacillus strain called *L. reuteri*. (Bacterio-prospectors originally isolated this strain from the breast milk of a rural woman in the Peruvian Andes.) The immune profile of the treated Swedish mothers shifted—a little more anti-inflammatory IL-10 and a little less TGF beta—and two years later, children who imbibed milk from treated mothers had less allergy than controls. Notably, however, Björkstén's probiotic intervention didn't elevate IgA. He couldn't completely reproduce the Estonian profile of breast milk, and it's worth exploring why.

A decade of trials with probiotics, bacterial strains touted as promoting health and well-being, had preceded Björkstén's experiment. Early on, several studies found probiotics helpful in preventing infant eczema. Lactobacilli given to expectant Finnish mothers with allergic tendencies, and then to their newborns, halved the risk of developing eczema by age two. A follow-up at seven years of age observed continued protection against the skin disorder. But by then a troublesome downside had emerged: The treated group now had more asthma and nasal allergies than the nontreated group.

German scientists also reported somewhat alarming results. They treated both pregnant mothers and newborns for six months, only to note an increased risk of wheezing in the treated group. And Australian scientists observed more allergy to cow's milk among probiotics-treated children.

Why the divergent results? The studies used different bacterial species. It's entirely possible that each has a slightly different effect. The treatment protocols also varied. Some scientists treated just the mother, others both mother and child, and others just the infant. There's also an inherent unpredictability in using a live organism, one reason that the pill approach—isolating just the key molecule that communicates with the immune system, and purifying it—remains attractive.

But the conflicting outcomes may also indicate a larger problem with the probiotic approach. Probiotic formulas usually contain one or a few bacterial strains. And while it would certainly be fantastic if scientists could isolate *the* bacterium necessary to rebalance the human immune system, the science has repeatedly pointed to the importance of variety.

"To believe that one bacterium is the panacea is probably slightly naïve," Björkstén tells me. "Sometimes you find bifidobacteria are associated with less allergy, sometimes lactobacilli. But what's consistent in all studies is that the diversity is less in allergic children."

The new gene-sequencing technology that became available in the mid-2000s permitted Björkstén to see this diversity with a clarity impossible before. He caught up with the original cohort of Swedish children, now five years old, whose microbial colonization he'd contrasted with that of Estonian children. Analysis showed that the least allergic children in the group consistently hosted the greatest diversity of bacteria early in life. They also excreted the most IgA in their saliva.

Exposure to a wealth of microbes could even, apparently, reverse the momentum of a gathering allergic storm. Children sensitized to allergens early in life generally had a poor prognosis when it came to allergic disease. But Björkstén found that if they escalated production of salivary IgA by one year of age—evidence of having acquired a more diverse microflora— their chances of wheezing at age four fell dramatically. In Estonia, of course, children regularly began churning out lots of salivary IgA earlier than in Sweden, evidence of the greater microbial pressure in that country.

A unifying theory was emerging. In Sweden, the children with the most diverse microbial communities were also generally exposed to more endotoxin at home. They tended to have had more infections, which Björkstén and others now thought was incidental to the true protective factor—microbial enrichment. And they often came from large families. Perhaps the "sibling effect," the "farm effect," and the Karelian Answer (as I called it) not only reflected exposure to a greater array of microbes, but also acquisition of a richer microbial ecosystem within.

GENOTYPING AND PERSONALIZED MEDICINE

Before you run off and join a farming community in Bavaria, there are a few things you should know. As we saw in chapter 3, the genes of the immune system are among the most diverse in the human genome. We've attempted to evade pathogens and parasites by diversifying, by being unpredictable. As a result, everyone won't respond to microbial pressure in the same fashion. Depending on your genotype, farming may be bad for you.

Think of our microbial receptors as gas pedals. Sometimes you get in a car, barely touch the pedal, and find yourself lurching forward. In other cars, however, you can push the pedal to the metal and the car moves forward only lethargically. In much the same way, carriers of different variants will respond differently to the same stimuli. For those with relatively sensitive gene variants, a little microbial pressure may produce a huge response. For those with less sensitive variants, the same pressure barely registers. Eventually (we hope), scientists will produce an allergy-preventive therapy from this research. So understanding these genotypes takes on new urgency.

In theory, the greater one's ability to recognize microbes, the less microbial exposure one requires to properly "train" the immune system. On the other hand, if you have less-sensitive microbe-sensing machinery—a microbial myopia, if you like—you'll require more exposure to avoid allergic disease. These variations underscore why pediatricians can't just bathe children in manure as a precautionary measure. Too much stimulation can push some genotypes into the red zone. For them, excess exposure to microbes could cause a new set of problems.

Consider the case of children, farms, and the microbial sensor CD14. Depending on which version of the CD14 gene Danish farming children had, their risk of allergy was either decreased by one-third—what we'd expect given everything we've just seen—or increased two-and-a-half-fold. Why? Rather than stimulate in a way that protected against allergies, the microbially enriched environs pushed this second group into overdrive and made things worse. They registered too much activation.

The same principle applied to day care. For children with one version of a receptor called TLR2, attending day care protected against allergies. But having another version increased the risk of wheezing in the day care setting.

And then there's Barbados, a Caribbean island whose population descends mostly from West Africa. Asthma has recently increased on the island, but not for all genotypes. Those with two sensitive versions of the CD14 gene mentioned above had one-fourth as much asthma as those without. They fared fine in relatively microbe-free modern conditions, but there was a catch: In surroundings with a greater intensity of microbes, pockets of which persisted on the island, this genotype had a nearly twelvefold increased risk of asthma. They were pushed into the red zone.

This gene variant was comparatively uncommon among Barbadians. Just 9 percent had it, compared with 28 percent in Manchester, and its relative scarcity offered another lesson. The gene happened to increase the risk of septic shock, an organism-wide failure that can occur in response to infections. For those with the sensitive variant, the fury of the immune response was a major liability. In environments where microbial assaults were more or less constant, they were more likely suffer an immune system meltdown.

Tropical West Africa, the Barbadians' ancestral homeland, was just such a place. Too much microbial sensitivity was dangerous there, which probably explained the gene's infrequency among the islanders. But now the very same rarity was making this population more prone to developing asthma. On the whole, they had a harder time registering enough stimuli to ward off the lung disease. Evolutionary pressures had delivered them a relatively insensitive gas pedal.

These intricacies prompted Donata Vercelli, a scientist at the University of Tucson, to imagine a microbially triggered "switch." Light exposure to microbes, less than we'd evolved to expect, might lead to allergies. Just the right amount prevented allergy. And too much could cause other problems, such as chronic obstructive pulmonary disease. The force required to flip the switch hinged on genetic variation. And the gene variants that predominated in a given population—sensitive or insensitive—depended in part on past evolutionary pressures.

If nothing else, these nuances complicated the prospect of a one-size-fits-all microbial fix to the allergy epidemic. Not everyone would respond to bacterial stimulation the same way. Some might worsen. "We have to be very careful," says Martinez. "We're still with this paradigm that there's one solution for everybody. But it could be bad for some people."

On the other hand, these complexities also suggested that personalized medicine—therapies tailored to your particular genetic makeup—are closer than generally thought. You can imagine a pediatrician genotyping your unborn child, and then making recommendations. "Ah, your son has variant X," the doctor might say. "We know that gene greatly benefits from exposure to livestock. So I recommend a chicken coop and a pigsty. Raise this child as a swineherd!"

RETHINKING THE IMMUNE SYSTEM

We began in a reunited Germany, pondered orofecal infections, toured the microbe-rich farms of middle Europe, explored a still-divided Karelia, and ended looking at the different microflora on either side of the Baltic Sea. What needs to happen seems clear: Somehow, we must reproduce what occurs naturally in those environments. How that's going to happen remains, unfortunately, completely nebulous. Nonetheless, these case studies have already changed (for those paying attention, at any rate) how scientists conceive of the immune system. The roughly two decades of comparative research that began with the collapse of the Eastern Bloc has forced a reimagining.

Scientists already knew that the immune system is adaptive, part sensory organ and part cognitive apparatus. It detects the microbial world, decides how to react, and remembers what it has encountered. What they understood less clearly are the requirements of this particular adaptive system. To develop properly, the immune system needs stimulation from the very thing scientists traditionally thought it had evolved to vanquish: microbes and parasites.

In other fields, that adaptive systems need input during development is accepted doctrine. Consider the brain. Kittens raised in a room whose walls are painted with vertical lines, for example, lack the ability to perceive horizontal lines in adulthood. Without the particular pattern of nerve stimuli representing "horizontalness," their visual cortex never develops the neuronal connections to perceive it. Children born with cataracts who don't have them removed while young will, if they remove the blockages later, fail to perceive anything more than vague shapes. Their neurons have failed to make necessary connections during a critical window of plasticity. Their eyes work fine, but their brains can't "see."

Romanian children raised in state-run orphanages during the 1980s, where they were largely deprived of affection and physical contact, have provided scientists with a dispiriting natural experiment of the same sort, an answer to the question What happens to infants when they don't receive basic human stimuli? These children have lower IQ, underdeveloped motor and language skills, and attachment disorders—sad proof of what we already intuit. We require love and affection to develop properly.

The field of physical health has perhaps most deeply absorbed these lessons. Regular exercise, along with a healthy diet, has emerged as the single "medicine" that best staves off a number of degenerative conditions, including cardiovascular disease, some cancers, dementia, and even depression. How? The list of benefits from exercise, including a stronger, more efficient heart, neuronal generation, and serotonin release, is long. But the simpler reason is evolutionary. Adaptive systems need stimuli by the very forces and factors to which they're meant to adapt. Muscles need strain in order to know where and by how much to grow. Bones need stress to determine where to increase density. Hearts must beat quickly for extended periods in order to pump efficiently. Even our eyes, scientists find, require sunlight—not fluorescent lighting, but the particular wavelengths emitted by our star—to avoid myopia.

Fittingly, our first forays off our home planet brought this rule into stark relief. When NASA began sending astronauts into space in the 1960s, scientists quickly learned what happens when a body that evolved under the constant pull of gravity suddenly finds itself weightless. Hearts grew weak. Bones lost density. Muscles wasted away. Some astronauts wouldn't recover lost bone density for years.

The atrophy that occurred during space travel highlighted another rule: Use it or lose it. Muscle and bone need the constant strain of gravity simply to maintain homeostasis—sameness. Not only are adaptive systems constantly growing in response to stimuli, in other words, they're constantly pruning and consolidating in the absence of stimuli. A muscle that's never used shrinks. The dendrites on neurons that never fire are pruned. The immune system that finds itself without microbial pressure grows jumpy (allergies), and turns against the self (autoimmunity).

In this case, what looks like increased aggression really stems from a kind of atrophy as well, a withering of the ability to regulate immune responses. This is no metaphor. Rodents raised in germ-free conditions are more susceptible to asthma and inflammatory bowel disease than their germy counterparts. Absent the commensal flora, inflammatory white blood cells crowd into lungs and guts—and find no one telling them to stand down. Crucially, recolonizing these animals with microbes corrects the problem only if they're young. For adult mice, the window of malleability has shut.

As we begin to contemplate potential therapies, how long early-life immune plasticity lasts and when it begins have become pressing questions. As it happens, our immune system begins developing long before we actually set foot in the world. Our propensity to develop allergic disease, scientists are finding, begins in the womb.

CHAPTER 7

Mom Matters Most

Genes are not Stalinist dictators. . . . They live in a democracy,
and what they do is conditioned by what else is going on
around them.

—David J. P. Barker

Around the turn of the millennium, Erika von Mutius and her colleagues
included a new question in their surveys of farming and nonfarming children in rural areas. They asked about Mom's activity while pregnant: Had
she worked with animals? And if so, how regularly?

The scientists thought the question important for two reasons. First,
while exposure to the cowshed in the first year of life protected most
against allergic disease, children who spent time in stables before their
first birthday—babies in bassinets, essentially, who accompanied their
mothers during chores—were also children whose mothers tended to
have worked the farm while pregnant. So the question was this: Which
exposure was more important in preventing allergies, your mother's while
she carried you in her belly, or yours as an infant?

The second factor related to the earliest appearance of allergic disease.
Often, allergy showed up so early in life, by age one or two, that it seemed
as if some children were born allergic. But immunologically speaking, that
was impossible. The conventional understanding of allergy held that allergic disease resulted from a learned mistake. Somewhere along the way, the
adaptive immune system accidentally confused tree pollen, dust mites,
peanuts, and other proteins for deadly invaders. The consequences were
lifelong sneezing, wheezing, and perhaps vomiting, among other symptoms. But you weren't born with this error hardwired into your immune
system. How could you be? At birth, your adaptive immune system was,
by definition, a blank slate.

And yet scientists everywhere observed that allergies could appear
remarkably early, especially eczema. Indeed, the rash often served as the
first sign of what allergists had dubbed "the allergic march": What began

as an itchy skin condition at age one might advance to severe peanut allergies and debilitating asthma by age fifteen. Sometimes, eczema appeared even before doctors could detect allergen-specific IgE—before a child was measurably allergic. All of which suggested that allergy wasn't just a learned malfunction. A tendency toward inflammation seemed to precede allergic disease.

When the surveys that asked about Mom's activity while pregnant came back, they threatened to reframe the previous decade of work on the farm effect. Had they been inadvertently measuring what happened to the mother, not the child, all along? Expectant mothers who regularly spent time with five or more animals—in the stable, the chicken coop, the hog pen, and so on—had the least-allergic children of all. Remarkably, this relationship held true regardless of the mother's history of allergies. An allergic mother could, provided she encountered lots of animals while pregnant, reduce the odds of "transmitting" the disorder to her child. How did it happen? Having worked with animals while pregnant seemed to develop the fetus's innate immune system, eliciting robust expression of those allergy-protective microbial sensors. With each additional animal Mom fed, milked, and otherwise tended, expression of these genes in her child increased by 10 to 16 percent.

"[M]aternal farm exposure might reflect a natural mode of immunotherapy . . . shaping a child's immune system at an early stage," commented Bianca Schaub, a scientist at University Children's Hospital in Munich. The broader indication was that when it came to allergic disease, a critical period of immune plasticity—that phase of receptivity to environmental cues—occurred before the child ever directly encountered an allergen or infection. It began during one's nine-month development in the womb.

EPIGENETICS—ECHOES OF WHAT GRANDMA DID

In *On the Origin of Species,* Charles Darwin argued that the immense variety of life on Earth, from tortoises to finches and dogs to humans, resulted from the process of natural selection. Key to that process was random variation. Minor differences arose in each new generation of any creature. Aided by these differences, some individuals in each cohort met the challenges of survival at that particular time and place better than others. They would have more offspring. More of the next generation would carry the peculiarities of their successful parents. Over the long term, this selection produced new attributes and eventually new species.

Some sixty years before the publication of Darwin's opus in 1859, the French naturalist Jean-Baptiste Lamarck presented a different idea of evolution. He posited that organisms could change in a single lifetime, and then impart these hard-won modifications to their offspring. Giraffes had long necks because, after straining to reach food, a giraffe predecessor had lengthened its neck. That ancestor then transmitted the gain to its brood. Its longer-necked offspring kept stretching, acquired a few more inches, and passed them on in turn.

In sum, Darwin and Lamarck both agreed that evolution occurred, but differed in their explanations of how adaptations were transmitted over time. Darwin thought traits were predetermined and immutable in any given individual. Lamarck thought them mutable, and transmissible to offspring. With advances in the understanding of heritability and then genetics in the late nineteenth and early twentieth centuries—one gene from each parent for any given "trait"—Lamarckian thinking fell by the wayside, and Darwinian theory dominated.

But in the late twentieth century, a British epidemiologist named David J. P. Barker began chipping away at the dogma of genetic immutability. A mother's privation during pregnancy, and a newborn's low birth weight, strongly predicted an adult's risk of heart disease during middle and old age, he found. What and how much your mother ate while pregnant, in other words, mattered just as much as, if not more than, what you ate as a youngster and adult. Moreover, many disorders of adulthood, including diabetes, high blood pressure, obesity, schizophrenia, and some cancers, also correlated strongly with conditions in the womb.

How did this work? Not by genetics, but epigenetics. Signals from the environment could turn genes on and off. Lamarck wasn't entirely wrong, in other words. (And neither was Darwin.) Life experiences could, by muting or amplifying gene expression, transmit across generations.

Many initially scoffed at the Barker hypothesis. But a number of studies have since moved the fetal origins hypothesis, as it's alternately known, closer to an article of faith. Consider the Dutch Hunger Winter. In the winter of 1944–'45, Nazi forces blockaded the western Netherlands, causing mass starvation. Women who were pregnant during the famine tended to have smaller babies, about 300 grams, or 0.66 pound, underweight. When these children reached middle age, scientists found, they had an elevated risk of obesity and cardiovascular disease. Why? Genes that coded for important metabolic regulators, such as a hormone called insulin-like growth factor 2, were, scientists could see, cranked to "high." The Hunger Winter generation had prepared for a world of scarcity, and instead encountered one of excess. The discrepancy between what their epigenome had prepared for, and conditions as they actually were, predisposed the cohort to disease.

Immune assaults also had transgenerational consequences. Children born to mothers who, while pregnant, had survived influenza infection during the Spanish flu pandemic of 1918 were more likely to develop high blood pressure, cancer, and heart disease in their sixties and seventies compared with children of mothers who hadn't contracted the virus.

Not only did what happened to Mom impact one's risk of disease, so did what happened to Grandma. Epigenetic changes could transmit across three generations. For Swedes born in the late nineteenth and early twentieth centuries, for example, the risk of death from diabetes increased in direct proportion to how much their grandparents ate before puberty. The more your grandfather gorged, the more likely you were to get diabetes. The best-fed Swedes had grandchildren with quadruple the risk of diabetes compared with Swedes who ate less.

Animal models confirmed that these studies measured what they purported to measure. Feeding pregnant rats a low-protein diet, for example, changed the expression of genes involved in metabolic regulation in their offspring. These rats, in turn, passed these epigenetic changes to their progeny. Depriving a pregnant rat of protein could predispose its grandchildren to metabolic disease.

For allergy researchers, these studies were eye-opening. They explained how allergic symptoms could appear in children at such early ages. The critical variable might not be what happened, or didn't, to allergic children, but what happened—or failed to happen—to their mothers.

EASE MICROBIAL PRESSURE ON MOM, AND BABY WHEEZES

Seeking to illuminate how these disorders were transmitted epigenetically, the scientist Bianca Schaub compared the immune response of babies born to allergic mothers with that of infants born to nonallergic mothers. She gauged the divergence by looking at cells extracted from the umbilical cord just after birth. Even at that early stage, the differences were profound. When stimulated, white blood cells from "allergic" cord blood responded with comparatively less anti-inflammatory IL-10, and less antiviral IFN gamma. "Allergic" cord blood also contained fewer regulatory T cells. This translated to a combination of, on the one hand, impaired viral defenses (the low IFN-gamma), and on the other, an inability to squelch inflammation (the paucity of IL-10 and T-regs).

Cord blood from children born to farming mothers, on the other hand, consistently veered toward an allergy-protective immune profile. It contained plenty of regulatory T cells, and these T-regs were especially effective at suppressing responses to harmless proteins. Their ability to regulate allergylike reactions correlated with the number of animals Mom had encountered while pregnant. The more animals she was exposed to, the more effective her child's T-regs. The FOXP3 gene so important for T-regs was also measurably different in farming children—not the gene itself, but epigenetic modification of the gene. Like a book opened to just the right page, in farming newborns the FOXP3 gene was *extremely* available for transcription.

Together, these findings offered a tantalizing glimpse into the epigenetics of the farm effect: The high microbial pressure on farming mothers altered the expression of immunity genes in the developing fetus. Via contact with Mom's immune system, cowshed microbes were preprogramming the unborn child to tolerate allergens—at least, that was the interpretation.

Just to be sure, the researcher Harald Renz sprayed the bacterium *Acinetobacter lwoffii*—the same one found in high abundance in cowsheds, and already shown to protect rodents against asthma in experiments—into the noses of pregnant, asthma-prone mice. Low-grade immune activation followed in the mice's lungs. But in the placenta, the organ that contains the developing fetus, immune activation *decreased*. Bacteria that incited mild activation in the lungs, in other words, triggered anti-inflammatory mechanisms around the fetus—"soothing signals," in the words of the scientist Patrick Holt. And when the mice were born, although they belonged to a lineage that was genetically prone to develop asthma, they resisted developing the disease.

These results are worth pondering: Bacteria of the sort that were likely more abundant in past environments, and that these days persisted in cowsheds, could change how our genome, a set of unchangeable instructions, translated into blood and flesh.

"[W]e seem to be at the dawn of a new incarnation of the hygiene hypothesis," wrote Holt. "[T]he pregnant woman's inflammatory response is crucial to determining the child's likelihood of developing allergic disease."

And microbes weren't the only agent acting on Mom's inflammatory tone.

MOM'S PARASITES PREVENT BABY'S RASHES

The town of Entebbe, Uganda, sits on a peninsula on the northwestern shore of Lake Victoria, the source of the Nile River. The climate is tropical, rainfall frequent, and the vegetation lush. Here, as Schaub plumbed the epigenetics of allergy among European farmers, a scientist named Alison Elliott noted that as she dewormed expectant mothers in the hopes of improving anemia, their children had more eczema—about four times more than children born to mothers who still harbored worms.

The findings were only the most recent of several she'd made that challenged the consensus on the horrors of parasite infection. Some had expressed concern that, by suppressing the immune system, worm infections might accelerate the progression of HIV to AIDS. But Elliott found that deworming HIV-positive Ugandans didn't affect disease progression. Others worried that worm-infected mothers had children who didn't respond to vaccines, but Elliott in fact observed a heightened immune response in these children to the tuberculosis vaccine. And then there was the question of anemia. Anemia, which results from a shortage of oxygen-carrying red blood cells, harms both mother and child, and in the young especially it can retard development.

Elliott found that two-fifths of some 2,500 pregnant women studied suffered from mild to moderate anemia. But she observed no association between anemia and worm infections. (There was a strong association with malaria, however.) What's more, deworming during pregnancy didn't improve anemia, birth weight, mortality, or congenital deformities, all arguments—and sensible ones, generally—for deworming campaigns directed at expectant mothers in the developing world. Essentially, Elliott and her colleagues were finding that in the well-nourished Ugandan population, worms caused little measurable harm.

And now it appeared that removing them might prompt a new disease.

The first study was small, just one hundred women. So Elliott launched a much larger follow-up investigation. In the meantime, surveys indicated that parasite infections were common around Entebbe, but generally not heavy. Two-thirds of women carried one or more parasite. Like elsewhere in Africa, those with helminths had far fewer allergies. Women who hosted filarial worms, for example—a threadlike helminth transmitted by biting midges or mosquitoes—had one-seventh the risk of allergic sensitization compared with those without. Asthmatics were about one-fourth as likely to host hookworm as nonasthmatics.

And then came the results of the double-blind, placebo-controlled deworming trial on 2,500 women. Removing an expectant mother's worms during the second or third trimester increased the odds of her child developing allergic eczema by 82 percent. And the more intimate the worm—the blood fluke *S. mansoni* versus the gut-dwelling hookworm, for example—the greater the apparent cost of evicting it. Mothers treated for blood flukes had children with nearly three times the risk of eczema compared with untreated mothers. And they were nearly 60 percent more likely to wheeze.

"The risks and benefits of routine anthelminthic treatment in antenatal clinics may need to be reconsidered," wrote Elliott in 2011. And she echoed concerns expressed by others over the unintended consequences of cleaning up. "It is possible that, as low-income countries develop, anthelminthic treatment programmes will contribute to an epidemic increase in allergic disease similar to that experienced in affluent countries during the 20th century," she noted.

Mothers tolerating parasites, in other words, gave birth to children primed for tolerance. Mothers not tolerating parasites, however, had children who didn't tolerate quite as well. And what of mothers who were themselves inflamed?

INFLAMMATION BEGETS INFLAMMATION

At least since the late 1990s, scientists have known that when it comes to a child's risk of developing asthma, Mom's asthma matters more than Dad's. The observation suggests that conditions in the womb have an outsize influence on a child's subsequent risk of wheezing. If expectant mothers had vaginitis, for example, an inflammatory condition caused by, among other things, an imbalance in the vaginal microflora, the odds of her unborn child developing asthma increased by 40 percent. A fever episode while pregnant increased the risk of an asthmatic child by 65 percent in one study, and doubled it in another. Contracting the flu during pregnancy nearly doubled the child's chances of developing asthma. The lesson wasn't necessarily that viral infections caused asthma, but that a mother's inflammatory response could imprint on her child and increase the risk of this inflammatory disease.

Other observations also highlighted the importance of maternal inflammation. An extremely preterm birth—between the twenty-third and twenty-seventh week in what's normally a forty-week pregnancy—increased the risk of young-adult asthma two-and-a-half-fold, even compared with preterm children born a month later. A reasonable explanation might be that their lungs were immature, thus the asthma. Another possibility, however, was that a single phenomenon drove both the premature expulsion from the womb and the inborn hypersensitivity of the child's lungs: maternal inflammation.

Scientists garnered direct evidence supporting the latter idea by studying a condition called chorioamnionitis. Sometimes commensal bacteria sneak past the cervix, mildly inflaming the placenta. Rajesh Kumar, a scientist at Children's Memorial Hospital in Chicago, found that children whose mothers developed this condition were almost five times as likely to have asthma later in life. For African Americans, the effect was even more pronounced.

Work by the Maastricht University scientist Boris Kramer again underscored that the revelation in this research wasn't necessarily that low-grade infections cause asthma—although they may—but that the immune milieu in the womb imprinted on the fetal immune system. Where Schaub had examined cord blood, and Kumar the placenta, Kramer looked at the fluid surrounding the fetus. Seeking to reproduce the effect of chorioamnionitis without a live infection, he introduced endotoxin into the amniotic fluid of pregnant sheep. He saw a burst of inflammation, the mirror opposite of those "soothing signals" observed in pregnant mice that inhaled barn bacteria. Amniotic fluid bathes all surfaces of the developing fetus, including lungs and intestines. And Kramer noted the beginning of airway remodeling, a thickening of bronchial tubes that's a hallmark of chronic asthma, in the treated newborn lambs' lungs.

That wasn't all. The inflamed amniotic fluid also interfered with intestinal development, depleting regulatory T cells that normally line the gut at birth, and leaving it more porous than it should be. Altered permeability is characteristic of inflammatory bowel disease, celiac disease, and other autoimmune disorders with no evident connection to the intestine, such as type-1 diabetes. It may also play a role in food allergies. More generally, the prenatally inflamed lambs had half as many circulating T-regs compared with their untreated brethren. These sheep were essentially born with an inflammatory bent, an inability to restrain attack cells. From the outset, they were wired to overrespond to the slightest provocation.

It didn't take long for scientists to replicate some of these finds in humans. Danish researchers following a cohort of more than four hundred newborns observed that, among children who'd developed asthma by age seven, deficits in lung function were already evident at birth. About 40 percent of the observed airway thickening had occurred prenatally.

Meanwhile, Australian scientists launched an ambitious study with the goal of determining, once and for all, how the immune response of children who developed allergies differed before symptoms appeared. They followed 739 children from birth to age five. They collected cord blood at birth, and white blood cells periodically thereafter. And they gauged the immune response along the way.

After five years, they compared the records of thirty-five children who'd developed allergies with thirty-five who hadn't. Children who developed food allergies, in this case to egg protein, had a less-effective T-reg response at birth, they found. (They also produced less of that antiviral cytokine IFN-gamma, suggesting impaired defenses at some level.) And their white blood cells sprouted comparatively few of those microbial sensors that appeared in such abundance in nonallergic farming children.

What did these differences add up to? A radically altered trajectory of immune-system development. The normal immune response slowly ramped up over the course of five years, as if unhurriedly waking and stretching after a long nap. By contrast, these children who went on to develop allergies came out swinging. And then, as their nonallergic counterparts grew more responsive, they grew less reactive. Eventually, the responsiveness of allergic children fell far below that of nonallergic children. Allergic children essentially came out in attack mode and then fell into a kind of catatonia. But by then, their immune systems had learned the bad habits of allergy.

The scientists had preserved the placentas of the mother-child pairs. Now they examined the tissue for differences. Placental tissues from children who developed allergies—those kids who came out swinging—had about one-third less expression of regulatory T-cell genes, they found, precisely the opposite of what scientists had observed in farming children. "We may actually be born allergic," says Meri Tulic, lead author on the prospective study. "If that's the case, we have to change how we think about allergic disease."

Taken together with Bianca Schaub's work, the findings reinforced the notion that allergic disease began in the womb. If genes are instructions, allergic disease started with a failure to read and apply instructions on immune regulation. In chapter 1 we saw the consequences of garbling those instructions altogether: a total immune-system meltdown, autoimmunity, and death. The children in the Australian study at least possessed a clearly written manual. Their allergic proclivity instead stemmed from a failure to adequately implement these directives—to translate them into living cells. And conditions in the womb partly provoked that failure.

Before becoming too forlorn, however, it's important to remember that by definition, epigenetic changes can be reversed. The newborn immune systems, even of these allergic children, weren't beyond repair. Ostensibly, they remained quite plastic. One possible fix, says Tulic, is to give at-risk newborns the fight they apparently expect: expose them to lots of microbes early, and prompt them to develop those otherwise defective regulatory circuits. That seems to happen naturally with children who attend day care.

Communication between mother and fetus wasn't unidirectional either. Just as the tenor of Mom's immune functioning affected the fetus, pregnancy changed Mom's immune system. In fact, the fetus engaged many of the same tolerance-promoting immune circuits as a parasite. And why not? The fetus is essentially a foreign organism lodged inside the mother. One study found that with each pregnancy, allergic moms became less allergic. A mother might have hay fever at age eighteen, say, and, after bearing several children, no longer sneeze by age forty.

This raised intriguing questions about the sibling effect—the repeated observation that later-born children had less allergy than earlier-born. Was it really the microbe-enriched environment younger siblings encountered that prevented their allergy, or the modified womb environment they encountered as later-born children? Had their older siblings made their mothers less inflamed, less allergic, and more tolerant—an immune profile then imprinted on them in utero? For that matter, if microbial enrichment was, in fact, responsible for the sibling effect, were the microbes acting directly on these later-born children, or via Mom's immune system while she was pregnant—or both?

And what of the epidemiology of allergic disease since the Industrial Revolution? How much could be ascribed to women having fewer children? Fertility declined dramatically from the nineteenth century to the twentieth. Women once had double and triple the late-twentieth-century average of 2.06 children. The first women to buck the high fertility trend in the nineteenth century were, of course, the well-off. As we've noted, it's among the upper classes that immune-mediated diseases first appeared. How much of that increased vulnerability simply stemmed from declining fertility—from a womb environment less often modified by older siblings? In the developed world today, for example, firstborns probably comprise a larger proportion of the population than ever before. Could this demographic shift alone explain the late-twentieth-century increase in allergy?

British scientists looked into the question and, in short, they concluded that the shrinking modern family didn't contribute much to the allergy epidemic. By their calculations, the decline in fertility between 1960 and 2000 should have prompted just a 3 percent increase in allergic disease. The increase in allergy prevalence—between a doubling and tripling—was far larger than that. One wishes, however, that they'd included a longer period of time—that they'd begun a hundred years ago, when, in the U.S. at least, the average household had nearly double the occupants compared with today's. Epigenetic changes take time to show up.

These intriguing issues aside, contemplating a fetal origin for allergic disease raised another question: If maternal inflammation in particular predisposed the developing fetal immune system to allergic disease, then why had allergy and asthma increased so dramatically in a period—the previous sixty years—during which infections, a major cause of inflammation, had declined so precipitously? On the face of it, the epidemiology didn't match up.

The probable answer was that we'd lost "soothing signals," not gained inflammatory ones. Moms these days didn't necessarily encounter too many pro-inflammatory stimuli, but a paucity of anti-inflammatory ones. Studies comparing animals raised in "wild" conditions with those raised in "clean" environments bore this out. Parasite-infested sewer rats have a much more adept and muscular anti-inflammatory response compared with their clean, lab-raised counterparts. Pigs raised outdoors that regularly wallow in mud—as they would naturally—show more expression of genes involved in immune regulation compared with pigs raised indoors and routinely fed antibiotics. And of course, mothers in environments that more resemble the world in which we evolved—farming Bavarian mothers, wormy Ugandan mothers—have comparatively enhanced regulatory circuits.

There were, of course, nonmicrobial influences on our inflammatory tone. Diet was one. And the Western diet, dominated by processed foods, plentiful saturated fats, and "empty" calories, was increasingly recognized to elicit a low-grade inflammation that contributed to other diseases of civilization, like type-2 diabetes, obesity, cardiovascular disease, and some cancers. Food could also modify the risk of asthma. In one study, expectant mothers who stuck to a Mediterranean diet while pregnant—lots of anti-inflammatory omega-3-fatty acids from fish, and abundant fiber from fruits, veggies, and legumes—had children with a reduced risk of asthma.

Likewise, in middle Europe, a mother's consumption of unpasteurized milk, butter, and yogurt during pregnancy consistently protected children against allergic disease. Von Mutius and her colleagues suspected that high omega-3 content partly explained the phenomenon. These fatty acids originate in leafy greens, so grass-fed cows produce milk with a relatively high omega-3 content. Leaving the milk unprocessed might preserve these healthful fatty acids. Indeed, there's some evidence that taking omega-3-rich fish oil while pregnant can reduce an unborn child's future risk of allergy. Raw milk may also act as a prebiotic, selectively cultivating beneficial bacteria that then leave a favorable imprint on the developing fetal immune system.

An epigenetic explanation for asthma and allergy nicely explained one nagging bit of epidemiology: In the developed world, the asthma and allergy epidemic seemingly began decades after the major sanitary improvements were already in place. Clean water, sewer treatment, and garbage disposal were established practices in New York City by 1900. (Although, as we've seen, worms persisted in pockets for much longer.) Here was a plausible explanation for why another fifty years passed before these disorders really flew off the hook. It wasn't what happened, or didn't happen, to the children themselves, but what happened, or didn't, to their mothers.

The fetal origin hypothesis also allowed for a prediction: Allergic disease would likely amplify from generation to generation. In some ways, the epidemiology didn't support this idea. After increasing for four decades in the developed world, respiratory allergies mostly plateaued in the 1990s. But as Susan Prescott, lead scientist on the study looking at T-regs in the placenta, points out, even as the incidence of respiratory allergies has remained steady for a decade, formerly rare allergic diseases have continued to increase dramatically, and new ones have appeared.

A NEW WAVE OF ALLERGIC PROBLEMS

Between 1997 and 2008, the prevalence of peanut and tree nut allergies nearly quadrupled among children and teens in the U.S. One of every seventy Americans now has a nut allergy. For children between the ages of six and ten, the prevalence nearly doubles to one in forty. For nearly half of these youngsters, the allergy is potentially life-threatening. These statistics come from surveys that are prone to skewing by several factors. But objective measures paint a similar picture. Nine percent of U.S. children tested have "allergic" IgE antibodies to peanuts. Circumstantial evidence also supports an increasing prevalence. Ambulance calls for food-related anaphylactic shock tripled between 1993 and 2006, as did hospital visits for food allergy. In Australia, one of the most food-allergic countries on earth, referrals for anaphylactic shock more than quintupled in the same period. And whereas in the past children tended to outgrow their food allergies, recent generations—me included—haven't.

Peanut-free zones in schools have become commonplace. Some airlines no longer carry nuts at all. Mention these developments to a middle-aged person without kids, and you'll likely to get an exasperated eye-roll. Some version of "Peanut allergies have become so fashionable!" may follow. But mention the statistics to a parent, someone with firsthand experience, and you're likely to hear anxious disbelief. "My kid, or my friend's kid, has to carry an EpiPen at all times," this parent will say. "Little Bobby nearly died from his throat closing. No one had these problems when I was young."

Prescott calls food allergy "the second wave of the allergy epidemic." Genes clearly haven't changed in one or two generations, but how they're translated has. Parents with more hay fever and asthma had children with more food allergy. And new allergic diseases have appeared. In the past two decades, a condition called eosinophilic esophagitis has emerged to baffle pediatricians. Eosinophils are white blood cells that normally help expel worms and other large parasites, and that also crowd into allergy-plagued lungs, sinuses, and guts.

In this case, the eosinophils move into the esophagus, or the food pipe, causing it to swell and constrict. Over time, the esophageal lining thickens and becomes corrugated like a washboard. Unable to swallow food, the most severely afflicted require feeding tubes, or they'll starve.

Scientists still debate whether the condition is really new, or if its seemingly sudden appearance really reflects new diagnostic criteria—criteria clarified only in the 1990s. An Ohio State University study that looked retrospectively at more than a decade of preserved biopsies found the increase real enough. Since 1992, the condition had increased tenfold, from 0.3 percent to 3.8 percent of patients undergoing biopsies. Other studies that looked farther back—to biopsies collected during the 1980s—also concluded that the increase was real.

Once microbial pressure is relaxed, how quickly do these epigenetic changes occur? Scientists have assumed that early life exposures are most critical to preventing allergic disease later. A childhood spent in rural areas of the developing world fairly reliably prevents allergic disease in adulthood, even if one moves to London or New York. But detailed studies on immigrants suggest that this invulnerability isn't permanent, even in adulthood.

Asian immigrants to Melbourne, Australia, for example, develop allergic disease in direct proportion to how long they've lived in the city. After a decade, some 60 percent of Southeast Asians—people who never suffered from allergies before—develop hay fever. Fifteen percent begin to wheeze. For Asian teens, having lived in Australia for between five and nine years doubles the risk of asthma. And having lived in Australia between ten and fifteen years increases the risk by three and a half times.

Studies on immigrants to Sweden paint a similar picture. These migrants come from Africa, Asia, and the Middle East. They arrive with few allergies, but they have elevated immunoglobulin-E levels compared with the average Swede's, evidence, perhaps, of exposure to parasites back home. Two things happen during the ensuing years. That elevated IgE gradually subsides until it resembles the typical Swede's. And the immigrants gradually become more allergic. After 2.5 years, about 16 percent are sensitized to birch pollen. After 10.5 years, more than half are.

One interpretation of these patterns: Absent the allergy-protective immune stimulation of their homeland, the immigrant immune system becomes prone to the same mistakes that characterize "locally grown" immune systems. A habit of dysregulation sets in. And while adult immigrants to the developed world generally remain less allergic than natives, their children are quite often more allergic, especially if they originally hail from the tropics.

HAVE WE NO SAY IN OUR OWN
ALLERGIC DESTINY?

All this talk of epigenetics may prompt a sinking feeling. If the exposures my mom *didn't* have are causing me to sneeze, then what hope is there of ever fixing the problem? The answer is, a lot—maybe not for you, but for your children. Now that scientists better understand when, exactly, the critical period of immune plasticity occurs, preventing allergic disease altogether is much closer to becoming a reality. The window during which we've got to turn all the right knobs and crank all the correct levers begins before we set foot in the world.

"In contrast to the challenge of changing the determinism of our genes, the intrauterine environment is much more potentially modifiable by a woman to promote the good future health of her child," writes Duane Alexander, director of the National Institute of Child Health and Human Development. And consider how much cheaper preventing a disease like asthma is compared with a lifetime of management. "[I]nvestments targeting fetal health may have higher rates of return than more traditional investments, such as schooling," notes the Columbia University economist Douglas Almond.

Take the Danish study we saw earlier, wherein children who'd developed asthma by age seven had lung deficits evident as newborns. The scientists attempted to intervene. They treated at-risk children with inhaled immune-suppressants, but the steroids didn't help. Even at that young age, the disease had already gained too much momentum. Effective preventive treatment for asthma, the scientists concluded, would probably have to begin before birth.

Unfortunately, probiotics given to pregnant women have so far yielded lackluster results. And yet the robust preventive effect of working with farm animals suggests that immunotherapy during pregnancy, either with living microbes or something that mimics them, will one day constitute an important avenue of prevention for allergic disease, and maybe other disorders. Rebalance the mother's immune system, and the child will emerge protected from allergies.

That said, women especially may view the fetal origins hypothesis with

142

mixed feelings. If Mom matters most, then she's also the guiltiest, right? Do women really need one more thing to worry about while pregnant? Have women spent a century struggling for social equality only to be waylaid by what looks like biological chauvinism?

Well, as Annie Murphy Paul writes in her book *Origins: How the Nine Months Before Birth Shape the Rest of Our Lives*, "Our growing awareness of the importance of maternal well-being to the fetus should lead us to offer help, not to force compliance or mete out punishment." Furthermore, the research we've explored suggests a direct role for third parties in fetal health: If the environment leaves its mark on the unborn child via Mom's immune system, then everyone in the environment—spouses, siblings, friends, pets, livestock—influences the fetus as well. As you'll recall, the relative invulnerability to allergic disease in former Eastern Bloc countries partly related to human crowding. Mom isn't solely responsible, in other words. And everyone else isn't completely absolved. (Think secondhand smoke.) Insofar as it shapes the microbial ecology of the environment, the entire community around an expectant mother leaves an imprint, however faintly, on the developing fetus.

Now let's revisit an orofecal bacterium that, in study after study, seems to protect against allergic disease—a bug with a nasty reputation called *Helicobacter pylori*.

CHAPTER 8

The Disappearing Microbiota

I will be an enemy to your enemies and will oppose those
who oppose you.

—*Exodus* 23:22

The extinction spasm we are now inflicting can be
moderated if we so choose. Otherwise, the next century will
see the closing of the Cenozoic Era [the age of mammals]
and a new one characterized not by new life forms but by
biological impoverishment. It might appropriately be called
the "Eremozoic Era," the Age of Loneliness.

—E. O. Wilson

In April 1982, two Australian scientists got the break that had, until then,
eluded them. Robin Warren and his young trainee Barry Marshall had
spent a year trying to culture a bacterium they suspected caused disease.
Despite their best efforts, however, the bug refused to grow. Now, after a
long Easter weekend, they returned to their laboratory to find that the
petri dishes meant to incubate for two days had instead sat for five. And
the bacterium that, until then, had resisted their attempts at cultivation
had now proliferated.

Warren had observed the odd, corkscrew-shaped bacterium a few years
earlier in the stomachs of patients with ulcers and gastritis. The stomach,
a churning pool of hydrochloric acid, was generally considered a hellish
environment for life, even microbial life. But this bacterium seemed not
only right at home, but well adapted to its environs. It used four string-
like flagella to propel itself. And it embedded itself in the mucus layer lining the stomach wall. Most suggestively, Warren observed the bug only in
patients with ulcers and irritated stomachs. The thought occurred to him
that the bacterium might cause disease.

The idea that a microbe induced ulcers contradicted the prevailing wisdom. The medical community knew that diet and stress caused ulcers. Even the ancient Greeks had noted that people who lived through wars, exposed as they were to extreme strain, suffered disproportionately from the condition. And by the 1980s, doctors had successfully treated ulcers for years—not with antibiotics, but with antacids and drugs that interfered with acid production. Of course, it must have bothered some that, after patients discontinued use of these drugs, the ulcers often recurred. But they likely chalked it up to a resumption of bad habits—gorging on salty, greasy food and stressing out.

So in 1983 when Warren and Marshall proposed that an infection might cause ulcers, the idea encountered some resistance. Maybe the bacterium inhabited people's stomachs, sure, but Warren and Marshall hadn't proved that it caused disease.

Somewhat crestfallen, the scientists returned to the lab. They had to prove causation. They tried to infect animals, but the bacterium wouldn't take. Frustrated, Marshall decided that, with a microbe this temperamental, he'd have to resort to human experimentation. He'd infect himself. And in 1984, he downed a broth of bacteria originally harvested from an ulcer patient. He'd expected symptoms to emerge after some time. But just five days later, an uncomfortable bloat set in. His appetite waned. Friends remarked on his foul-smelling breath. He started vomiting clear, watery fluid in the early mornings. An endoscopy showed clear evidence of inflammation.

This time, the scientific establishment took note. The bacterium was christened *Helicobacter pylori*—in plain speak, helix-shaped bacterium of the stomach (pyloris). The find overturned accepted wisdom. A bacterial infection could, it seemed, cause ulcers.

From the beginning, however, scientists understood that *H. pylori* didn't act like other infections. It didn't ravage and destroy like the smallpox virus; it didn't multiply exponentially like *Yersinia pestis,* the bacterium that causes the bubonic plague. You weren't immune to it after the first infection, either. *H. pylori* established lifelong residency, continually evading the immune system. And it maintained a steady population. Indeed, Marshall's experience aside, you might not even notice when the bacterium colonized your stomach. Many decades could pass before symptoms emerged—before inflamed lesions became ulcerous and began bleeding.

Nonetheless, scientists quickly determined that the bacterium caused significant illness and suffering around the world. It accounted for nearly all cases of peptic ulcer, save those prompted by medication. It drove the formation of duodenal ulcers—erosions in the intestine immediately downstream from the stomach. And although it didn't prompt ulcerations in the majority of those who harbored it, it almost always incited gastritis, a low-grade inflammation of the stomach.

Scientists also realized that the poor everywhere disproportionately bore the burden of *H. pylori*. About half of humanity harbored the bacterium, but most of this half lived in the developing world. In India, four of five people had the bug; in Denmark, just one of four did. Countries in transition fell somewhere in between, with the upper classes, who usually benefitted from improvements in hygiene earlier than the lower classes, losing *H. pylori* first.

Notably, however, in the developed world, the pathogen had begun disappearing before anyone even knew it existed. In Western Europe and the U.S., more than half of sixty-year-olds had *H. pylori*. But just one-tenth of children did. (By comparison, roughly two-thirds of children in the developing world tested positive by age ten.) That sanitary improvements alone had started eradicating the bacterium prompted a collective sigh of relief. The maladies associated with *H. pylori* continued to pile up. Children who acquired it, scientists found, grew more slowly than their *H. pylori*–free peers. And the lifetime of inflammation elicited by the bacterium also predisposed to stomach cancer, a particularly deadly malignancy. Often, by the time physicians diagnosed gastric cancer, the tumors had grown large and spread. Physicians couldn't treat the malignancy at that advanced stage.

In 1994, the World Health Organization's International Agency for Research on Cancer classified *H. pylori* as a group 1 carcinogen, a category that also includes asbestos, formaldehyde, and radioactive by-products of nuclear fission, such as strontium 90. The following year, the National Institutes of Health convened a meeting to decide on a treatment protocol. The gathered experts formally recognized antibiotics as a therapy for ulcers. Robin Warren and Barry Marshall, "who with tenacity and a prepared mind challenged prevailing dogmas," were awarded the 2005 Nobel Prize in Medicine.

On one level, the saga of *H. pylori* was exhilarating. It suggested that the era of the Microbe Hunters—the nineteenth- and early-twentieth-century bacteriologists who demonstrated that infectious microbes caused disease—wasn't quite over. Bugs remained to be identified, their spread halted, and the diseases they caused prevented. Old-fashioned detective work—the scientific-minded pursuit of microbes—could still improve human health.

"I believe *H. pylori* are very likely the first in a class of slow-acting bacteria that may well account for a number of perplexing diseases that we are facing today," wrote the microbiologist Martin Blaser in a 1996 *Scientific American* article that encapsulated this sentiment.

In retrospect, however, Blaser's words are notable for a different reason. Beginning roughly a decade later, he would make an antipodal argument—not that *H. pylori* caused diseases, although he didn't deny that it could and did, but that its eviction from the human stomach had prompted new disorders to arise, chief among them asthma.

We'll get to the specifics of his idea in a moment, but for now, remember the cardinal rule of coevolution: any commensal tolerated by its host likely engages the host's regulatory immune circuits. And as we've seen repeatedly, strengthening those circuits incidentally prevents inflammatory disorders such as asthma.

CAN THE "GASTRIC DEMON" BE GOOD?

These days, *H. pylori* is blamed for about 63 percent of all stomach cancers. That's about 5.5 percent of all malignancies afflicting humans. Scientists suspect the bug in a host of other maladies, from pancreatic cancer, one of the most deadly types, to Parkinson's and cardiovascular disease. Many consider the bacterium a public health problem, especially in the developing world, where prevalence remains high. One scientist simply calls *H. pylori* the "gastric demon."

So it's with some eagerness that, on a fresh, early September day in 2010, I attend a meeting at Martin Blaser's lab at New York University's School of Medicine. The session is dedicated not to offing *H. pylori*, but to its *deliberate reintroduction*.

The room at the Veterans Affairs hospital in the East Twenties where the meeting occurs is institutionally generic. A raven-haired Austrian postdoc named Sabine Kienesberger projects graphs on a screen at the front of the room. She's working to genetically engineer a strain of *H. pylori* that produces proteins from a much more dangerous diarrhea-causing bacterium called *Campylobacter jejeuni*. The rationale is twofold: a single shot doesn't work to vaccinate against pathogens such as *C. jejeuni*. The immune system either forgets about them or can't see them clearly enough to begin with. As a result, one can repeatedly fall ill from the same bug.

Kienesberger thinks, however, that constant immune stimulation can overcome this cloak of invisibility, an ever-present reminder of what the pathogen looks like. Short of daily injections, how to provide this constant stimulus? Enter *H. pylori*. Kienesberger hopes to install a chimera *H. pylori* strain, a bacterium that looks enough like *C. jejeuni* to constantly remind the immune system about the real thing.

Today, as she presents the results of preliminary testing in mice, the abiding question is, have the *H. pylori*–infected mice developed immunity to *C. jejeuni*?

Blaser, smartly dressed in a starched white shirt and blue tie, and his longtime collaborator, Guillermo Perez-Perez, pepper her with questions. Perhaps she should run a control test with dead organisms, says Perez-Perez. Otherwise, how can she know for certain that she's seeing results from a living colonization? Blaser adds that she'll need a replicating organism for this to work. Can she verify that the bacteria are reproducing? No one directly addresses the paradigm-shifting idea here: inserting a carcinogen into the stomach in the hopes of improving health. (In Australia, Barry Marshall is on a similar tack—engineering chimera *H. pylori* strains that display proteins from the flu virus.)

I ask why *H. pylori* of all bacteria, why not a microbe with a less malignant reputation?

"We truly believe that *Helicobacter pylori* has advantage, that it's not a pathogen," says Kienesberger, referring to the bacterium's talent for surviving where little else can. She's arguing that its dominance of the gastric niche suggests that it belongs there—that we, the hosts, need it there.

How did these heretical-sounding ideas become foundational beliefs in Blaser's lab?

FROM NAGGING INTUITION TO PROOF IN PEOPLE

Blaser has thick black eyebrows that contrast greatly with his neatly trimmed silver hair. He has an uncanny ability to recall specific events and discussions from decades past. And he traces his first inkling that *H. pylori*'s relationship with its human host might be more complicated than, say, that of smallpox, to a conversation he had with his mentor and idol, the microbiologist Stanley Falkow, in the late 1980s.

Falkow strove to think about disease-causing pathogens in a larger evolutionary context. And in the broad view, if a bug had infected humans for many tens of thousands of years, and, as was the case with *H. pylori*, caused disease in only 10 to 15 percent of those it colonized, it wasn't strictly speaking a pathogen—a harmful bug. That relationship bordered on commensalism and, who knows, maybe even mutualism.

To the nervous titters of medical students, Falkow, who lectured at Stanford University, had begun to argue that "disease was a distraction." Our own apprehension about illness kept us from truly understanding biology. Entrenched attitudes shaped by germ theory were, in a sense, blinding us to real understanding.

Over drinks one evening, Falkow and Blaser began discussing *H. pylori*'s unusual biology. The bacterium was remarkably diverse. At that point, no one knew how long humans had carried *H. pylori* in their stomachs, but generally speaking, true pathogens were clonal. They self-replicated wildly like cancer cells, each new generation a copy of the old. *H. pylori*, on the other hand, displayed unusual variety among different human populations. (Eventually scientists would refer to the different strains as "races.") So if the bacterium didn't look like a pathogen, and didn't act like one, what was it really?

After that conversation with Falkow, the question nagged at Blaser. Much remained to understand about *H. pylori* as the carcinogenic, ulcer-inducing blight on humanity, however, and Blaser dove right in. Using his own blood—he had an *H. pylori* infection, it turned out—Blaser and Perez-Perez developed a technique for detecting infection with antibodies, an improvement over the down-the-throat biopsy method that until then prevailed. His work on a cohort of Japanese American men born in Hawaii during the early 1900s helped cement the link between *H. pylori* and gastric cancer. In this study group, those who acquired the bacterium were six times more likely to develop the malignancy.

And in 1993, along with an Italian group, he identified one of *H. pylori*'s virulence factors, a protein the bacterium injected into the lining of the stomach that caused irritation. They dubbed it cytotoxin-associated gene A, or CagA. Strains endowed with CagA incited greater inflammation. Harboring a CagA-positive strain roughly doubled your risk of gastric cancer compared with strains without the gene. If a particularly nasty "race" of *H. pylori* existed, this was it. Coincidentally, Blaser himself had a CagA-positive strain. Although he'd never had any symptoms, he decided to eradicate it.

In short, Blaser had a successful career helping to enumerate the horrors of this gastric fiend. But his doubts about categorically demonizing the bacterium continued to grow. For one thing, *H. pylori*–related diseases were by and large afflictions of old age. Two-thirds of gastric cancer cases occurred in people aged sixty-five or older. But we'd only recently begun living past age sixty en masse. For the great extent of human evolution, in other words, *H. pylori* wouldn't have caused cancer, and natural selection would have therefore been mostly blind to its presence. Which raised the question: In the lifetime of coexistence before disease arose—six or seven decades, in some cases—what was the bacterium doing? Was it benefitting the host somehow?

AN AFFLICTION OF WELL-TO-DO WHITE MEN

The first solid evidence that *H. pylori* was more than just a cold-blooded killer came in the late 1990s. The incidence of gastroesophageal reflux disease, or GERD, known colloquially as heartburn, had increased dramatically during the late twentieth century. First described in the 1930s, the condition resulted from acidic stomach juices that splashed up into the food pipe. The constant irritation altered tissue, leading to a condition called Barrett's esophagus. Barrett's esophagus, in turn, increased the risk of esophageal cancer. And even as gastric cancer had declined in the developed world during the twentieth century, the incidence of these three previously rare conditions had increased. Although esophageal cancer struck relatively few people overall, by the 1990s, adenocarcinoma of the esophagus, a particularly aggressive cancer, was the fastest growing type of malignancy in the U.S.

Epidemiologists noted that the prevalence of Barrett's esophagus was inverse to that of peptic ulcer and gastric cancer. People afflicted with stomach cancer and ulcers tended not to develop GERD, and vice versa. And whereas *H. pylori* had earned a reputation as a poor person's affliction, GERD and esophageal cancer became known as well-to-do white man's diseases. Was it more than just a chance pattern? Did *H. pylori* actively protect against GERD and its associated malignancies?

When Blaser and his colleagues looked, they found that those with *H. pylori* infections had much less GERD and fewer cancers of the esophagus. Intriguingly, the association was strongest with the most virulent, CagA-positive strains. The more the bacterium interacted with your stomach lining, it seemed, the more it warded off GERD and associated malignancies.

How did *H. pylori* prevent these diseases? The bacterium regulated stomach acidity, not out of any sense of beneficence to the host, but in its own self-interest. It wanted enough hydrochloric acid to keep competitor species at bay, but not so much that it, too, couldn't thrive. So when acidity grew too strong, the bacterium sidled up to the stomach lining and ran interference with acid production. Incidental to this manipulation, when gastric juices occasionally splashed backward, the lower acidity wasn't as irritating. (Anecdotally, Blaser says he developed acid reflux within six months of treating his own helicobacter infection.)

In much the same way that Joel Weinstock and David Pritchard suspected that we counted on helminths to develop regulatory immune circuits, Blaser thought that humans had probably incorporated this bacterial regulation of stomach acid into our own day-to-day operations. Now we depended on *H. pylori* to maintain optimal acidity.

This argument presumed, of course, that we'd harbored *H. pylori* since the Paleolithic, an assumption that remained unsupported in the late 1990s. That changed when, in the early 2000s, the scientist Maria Gloria Domínguez-Bello isolated a uniquely Amerindian *H. pylori* strain from a tribe living in the Venezuelan Amazon. Compared with specimens from the mestizo population of Caracas, which looked African or European, this Amerindian *H. pylori* resembled East Asian strains. This Asian provenance strongly suggested that the Amerindians' forebears had carried *H. pylori* over the now-submerged Bering Land Bridge when they colonized North America more than twelve thousand years ago. In other words, the bacterium had inhabited the human stomach before the domestication of animals. We hadn't acquired it from cows, horses, or pigs.

A series of genetic studies then cemented the bacterium's place in the pantheon of hangers-on that accompanied *Homo sapiens* out of Africa. Seven founding strains of *H. pylori* existed, geneticists found, all derived, like humans themselves, from African strains. Scientists could see the great human migrations written in *H. pylori*: the Bantu expansion of farmers originating in West Africa, moving south and east some 4,000 years ago; the Neolithic farmers of Middle Eastern origin moving northwest into Europe beginning some 8,500 years ago; the great Polynesian expansion from present-day Taiwan southeastward across the South Pacific beginning 5,000 years ago; the European colonization of Australia and North America within the past 400 years; and the forcible transport of West Africans to the Americas during that same period.

In some cases, *H. pylori*'s regional variation corroborated history better than human genes. Spain's dominant *H. pylori* strain, for example, more resembled North Africa's than that of non-Iberian Europe. That was presumably due to seven centuries of Moorish rule of the Iberian Peninsula.

Sometimes, *H. pylori* raised awkward questions. In one case, a white Tennessee man harbored an African *H. pylori* strain. Was mixed ancestry deep in his past responsible? Or perhaps the apparent mismatch was a legacy of the long tradition in the south, stretching back to slavery, of African American women caring for white children.

In 2007, scientists put a date on the most recent common ancestor of all human *H. pylori* strains: 58,000 years ago in East Africa. When modern humans dispersed from Africa, we already harbored the bug. Our association with the bacterium was ancient.

Meanwhile, everywhere scientists looked in the animal kingdom— whales, primates, rodents, cows, dogs, pigs, and even a few birds—they found unique helicobacter species. Some further clarified aspects of ancient human evolution. The helicobacter inhabiting cheetahs, for example, derived from the human version. The two strains had diverged some 200,000 years ago. Had African cats dined on *Homo sapiens*? Or was inter-specific coprophagy, one species eating another's feces, to blame?

For Blaser, our longstanding association with the bacterium meant that "gastritis"—the chronic, low-grade, often asymptomatic inflammation caused by *H. pylori*—was likely the evolutionary norm. Now, for the first time in human evolution, most children—more than 90 percent in the developed world—found themselves without this immune activation. The new "postmodern" stomach would, he predicted, have consequences. It had to. So he cast about for other conditions whose prevalence had mysteriously increased during the late twentieth century, as *H. pylori*'s had declined. And his attention immediately fell on the asthma epidemic.

DOES THE GASTRIC DEMON
PREVENT ASTHMA AND ALLERGIES?

It was 1951, the first decade of the Cold War, when the mathematician John Forbes Nash noted that sometimes, against all expectations, competitors end up cooperating. He described a situation in which two players of a game both knew their opponent's strategy, but, because changing their own strategy wasn't advantageous, they stayed the course despite that awareness. This became known as the Nash equilibrium: two or more parties in competition that, at some level, cooperated—because cooperation worked best for everyone involved.

In 1994, Nash was awarded the Nobel Prize in Economics for this and other insights. Blaser thought the relationship between *H. pylori* and its human host was a biological Nash equilibrium. But whereas in game theory the competing players didn't change as a result of understanding their opponents, Blaser knew that any organism that persists in another does so by changing the host. This Nash equilibrium involved some amount of mutual tweaking, and Blaser sought evidence in the prevalence of asthma and *H. pylori* infection.

He conducted three studies, one with the pulmonologist Joan Reibman, and two with the epidemiologist Yu Chen. He looked at urban asthma patients in New York, and he parsed more than fifteen thousand patient records collected during the National Health and Nutrition Examination Survey, or NHANES, a periodic study of Americans conducted by the CDC. (This was the same survey that Paolo Matricardi used to link orofecal pathogens with protection from allergic disease.) In all three studies, those harboring *H. pylori* had less allergic disease. Results from the NHANES studies, which allowed for control of variables such as smoking, body mass index, and socioeconomic standing, were particularly compelling. Youngsters who harbored the bacterium were one-third less likely to wheeze, and nearly half as likely to have sinus allergies. For children under age five, the protective effect of *H. pylori* infection was stronger still—a 40 percent reduction in the risk of allergic disease compared with children who did not have *H. pylori*.

The bacterium might have been a marker for something else, of course—a microbe-filled house, for example, or other orofecal pathogens. But one observation suggested that *H. pylori* itself protected. The more virulent CagA-positive strains lessened the risk of asthma most effectively. The more the bacterium interacted with your immune system, it seemed, the less your chances of allergic disease.

Others corroborated Blaser's findings. In Finland, scientists examined sera collected and stored since the mid-1970s. As the amount of immuno-globulin-E specific to birch pollen and other allergens more than tripled in young Finnish adults over time, those with *H. pylori* remained inured to the greater trend of increasing allergic sensitivity. Cross-sectional "snapshot" studies from the U.K., Germany, and Japan also found *H. pylori* protective against allergic disease.

Some, however, remained unconvinced. David Graham at Baylor College of Medicine in Houston thought that *H. pylori* was just "a surrogate for the hygiene hypothesis," not protective itself. Blaser's softening stance on the bug could cause great harm, he thought, if people began refusing to eradicate a recognized carcinogen. Graham set out to disprove, as he put it, the "erroneous attempts to blame some of the problems facing the modern world on a pathogen that was and is responsible for much suffering, morbidity, and mortality." He headed to Malaysia.

For whatever reason, the people of northern peninsular Malaysia, near the border with Thailand, had a naturally low infection rate. Only one in twenty harbored *H. pylori*. The country also had a relatively low asthma rate: just over one in twenty Malay children had asthma, about one-quarter the prevalence in developed countries such as the U.K. If Blaser was correct, Graham thought, Malays should have a comparatively elevated risk of asthma. That they were in fact less asthma-prone proved Blaser wrong. The "dire consequences hypothesis," as he called it, was baseless.

But really, Graham's study disallowed firm conclusions. He didn't directly contrast those harboring *H. pylori* with those who didn't harbor it—the critical comparison. Moreover, *H. pylori* would, according to Blaser's thinking, protect only populations that had coevolved with the bacterium. Populations that had remained naturally *H. pylori*–free, like these Malays, wouldn't have come to depend on it. More broadly, as we've seen, more than one "old friend" lessens the risk of allergic disease. The population Graham examined was certainly exposed to the other protective factors—parasites, for example, close contact with animals, and sundry microbes in water and soil. Without controlling for these exposures, his would-be rebuttal flew somewhat wide of the mark.

Meanwhile, experimental work began clarifying the mechanisms that *H. pylori* used to establish lifelong infections. The bacterium turned up the dial on the very cells that, in study after study, prevented allergic and autoimmune disease: regulatory T cells.

HELICOBACTER, SCULPTOR OF REGULATORY CIRCUITRY

I meet the scientist Karen Robinson in her spacious office in a gleaming white research building at the University of Nottingham in England. She has curly auburn hair. Black-rimmed rectangular glasses frame her blue eyes. She speaks slowly and deliberately, appearing to choose her words after a moment of silent, internal deliberation. The habit accentuates her aura of extreme patience. But my interest in the potential benefits of *H. pylori*—her area of expertise—has her visibly worried.

From the outset, she's at pains to communicate that even if *H. pylori* prevents allergic disease, and who knows what else, at this point it's all still very theoretical. Stomach cancer is no joking matter. It's very difficult to detect in its early stages. And depending on how far it has advanced, and what part of the stomach it strikes, treatment options, which include removing the stomach, chemotherapy, and radiation, run the gamut from dismal and slightly effective to torturous and totally ineffective. The bottom line: Stomach cancer is a malignancy you want to avoid. A century ago, it killed more American men than any other cancer, and it remains a major killer of men over sixty-five in developing countries. Robinson worries that people reading about the nuances of *H. pylori* infection might refuse to eradicate it.

So let's be clear: If you've got an ulcer and your doctor advises treating your *H. pylori* infection, do it, even if you've read this book, agree with its arguments, and consider the bacterium part of the natural human flora. Listen to your doctor. Even "old friends" can turn against us.

Now, let's get on with it.

When Robinson first began studying *H. pylori* some years earlier, it was apparent that no one really understood the natural response to the bacterium. Scientists had penned report after report on inflammation that, over a lifetime, added up to gastritis, ulcers, and cancer. But plenty of people—

the majority, in fact—harbored the bug without complications. What distinguished the immune response of those who carried it peaceably from those who developed problems?

She looked to asymptomatic patients for an answer. The distinction, she found, was in *how* they responded. Successful carriers greeted *H. pylori* with anti-inflammatory signals, not inflammatory ones. They tolerated the bacterium. Those with peptic ulcer disease, on the other hand, responded with aggression. What determined these different outcomes? Peptic-ulcer patients had 60 percent fewer T-regs than asymptomatic carriers. Ulcers didn't arise, strictly speaking, by *H. pylori*'s hand. Rather, one's own immune response caused the lesion. "Inflammatory responses damage bacteria," says Robinson, "but also, the person." Those who peaceably harbored *H. pylori* did so by unilaterally calling a ceasefire. Without receiving any obvious concessions from the bacterium, they held back.

When she compared asymptomatic carriers with noninfected controls, a key difference was apparent, one critical to understanding how *H. pylori* might prevent asthma. Infected patients had nearly triple the number of circulating T-regs compared with those without the bug. Did these excess suppressor cells protect against allergies? In mice, *H. pylori* infection lowered the risk of allergic sensitization to dust mites by nearly 40 percent. That was the same relative risk reduction Blaser had seen in American children carrying the bacterium.

Outside of the U.K., the protective effect of *H. pylori* was stronger still. Working with scientists in Ethiopia, Robinson launched a prospective study. Her Ethiopian colleagues followed nearly nine hundred children from birth to three years of age. By then, two-fifths had acquired *H. pylori*. Controlling for worms, other bacteria thought to prevent allergic disease, and socioeconomic status, the scientists found that acquiring *H. pylori* early in life reduced the risk of developing eczema and respiratory allergies by more than half.

So what about those ulcers? Why did some responded with inflammation and develop lesions, while others tolerated the bacterium just fine? Genetics likely played a role. When confronting *H. pylori*, different genotypes responded with either more or less inflammation.

A more intriguing possibility than genes, however, related to the timing of the bacterium's arrival. Anne Müller at the University of Zurich found that the earlier *H. pylori* colonized its host, the better the host tolerated it. Mice colonized as newborns suffered far less damage from *H. pylori* than mice colonized later. Presumably, they'd be less prone to stomach cancer in old age as well.

"The outcome of host-pathogen interaction is extremely different depending on the time of colonization," Müller says. Timing was also important in preventing allergic disease. Asthma-prone mice that acquired the bug earliest were the most protected from wheezing later. Essentially, an early arrival vigorously engaged the host's regulatory circuitry, while prompting comparatively little inflammation. With later colonization and more inflammation, the host response became less protective against allergic disease, and more predictive of cancer. The equivalent period in humans, she says, might be the first year of life—the period during which, in many less-developed countries, children still acquire *H. pylori*.

Müller made another important find. When she cleared the *H. pylori* infection with antibiotics, the mice's regulatory T-cell population collapsed, and the invulnerability to asthma disappeared. Mice with no history of asthma began wheezing. The lesson: Asthma-prone mice required constant priming by the bacterium to prevent the disease. By extrapolation, humans with inborn asthmatic tendencies might also need continual stimulation to avoid the lung disorder.

"A single shot wouldn't work," says Müller, contemplating possible therapies. "T-regs definitely rely on regular stimulation." Blaser had been prescient: only colonization with the living bacterium produced the desired benefits.

Müller's work also helped explain another hotly debated phenomenon, the so-called African Enigma.

WHY DON'T AFRICANS
GET STOMACH CANCER?

When it became apparent that half the world carried *H. pylori,* and that most of these carriers resided in the developing world, some marveled that vast swaths of this infected pool of humanity did not develop disease. In 1992, the British scientist C. Holcombe dubbed this the "African Enigma." In sub-Saharan Africa, where most people acquired the bug early in life, the incidence of gastric cancer remained remarkably low— much lower than you'd expect from nearly universal prevalence.

"Above all else, the data from Africa underline the multifactorial nature of the cause of peptic ulcer and gastric cancer," Holcombe observed. "*H. pylori* exerts its influence in concert with other environmental, social, and genetic factors."

One fallback explanation was that Africans didn't live long enough to develop cancer. And certainly, life expectancy in Nigeria at the time was fifty-five, which would support the argument. Gastric cancer generally struck after age sixty-five. But a short life expectancy couldn't entirely explain the mystery. For one thing, although Africans acquired the bug much earlier than Westerners—between two and three decades earlier— they tended not to develop the precursor conditions to cancer.

In South Africa, meanwhile, where before the HIV epidemic life expectancy was relatively high for sub-Saharan Africa—sixty-three years for men—and gastric cancer mysteriously low, scientists observed that Africans responded to *H. pylori* infections differently from Europeans. They didn't mount a full, frontal assault. And that semitolerant response presumably made the bacterium less carcinogenic.

Meanwhile, other helicobacter-related enigmas kept arising. Since Holcombe's paper, scientists have observed an Asian Enigma, an Indian Enigma, and what you might call a "large, free-living feline enigma." Zoologists note that, as in humans, the helicobacter species native to large African cats, called *H. heilmannii,* causes disease in captive cheetahs. Helicobacter-related illness is the leading cause of death in some captive populations. Wild cheetahs, on the other hand, have few problems from hosting *H. heilmannii.* What's the difference?

Anne Müller's work revealed the mechanism by which earlier infection was both less pathogenic and more protective against asthma: a tolerant response to bacterial colonization. Her work also raised the intriguing possibility that, in the past, *H. pylori* didn't cause disease even in the West—that a European Enigma lay buried somewhere in history.

Call it the primordial balance of the human superorganism, that semi-mythical time when not only were all residents of the human organism present, they also arrived on schedule. Coincidentally enough, one scientist glimpsed that period. It seems to have ended with the advent of the Industrial Revolution.

THE EUROPEAN ENIGMA

In a series of detailed studies, the epidemiologist Amnon Sonnenberg found that, in Europe, the incidence of *H. pylori*–associated disease first changed for people born around the turn of the eighteenth century. Working backward from death certificates in several industrialized countries, he observed that the risk of stomach cancer peaked among those born in the mid-nineteenth century and then fell off among those born in succeeding decades.

Did people not live long enough in the eighteenth century to develop stomach cancer? Sonnenberg compared like with like—sixty-five-year-olds with one another—largely eliminating the possibility that he was seeing an artifact of more people reaching old age. And there was another reason to think the pattern was real. The incidence of ulcers followed the same boom-and-bust motif, but with a delay of a few decades. Someone born in 1850s Denmark had nearly six times the odds of getting gastric cancer compared with someone born fifty years later (or earlier). But as the risk of cancer subsided, the odds of gastric ulcer skyrocketed, peaked, and then declined among those born during the early twentieth century.

Sonnenberg traced this same upsurge of gastric cancer followed by increased ulcer risk, plus or minus a decade, in England, Wales, Italy, Japan, Denmark, and Switzerland. In all these countries, industrialization brought some new factor that first increased the risk of *H. pylori*–associated diseases, and then, some decades later, caused it to fall off. One possibility, supported by Müller's work, was that delayed acquisition of *H. pylori* drove the phenomenon. As affluence increased and sanitary reforms went into effect, colonization by *H. pylori* occurred later and later. Delay it a little, and your risk of stomach cancer goes up. Delay it a little more, and your risk of stomach cancer declines, but the odds that you'll develop ulcers increases. Put off first exposure even more—the situation in the twentieth century—and the bacterium stops taking at all.

Enter the allergy epidemic. While it officially began in the mid-twentieth century, a creative reading of Sonnenberg's and Müller's work suggests that the collapse of the human superorganism began at least a century earlier. As you recall, hay fever appears a short time later. Sanitary improvements may have been partly responsible, but the first changes in *H. pylori*–associated disease occurred before major sanitary reforms. Something else altered *H. pylori*'s interaction with its host before sewers and clean water. Changing cultural norms may have played a part. With the body and its secretions increasingly vilified as unclean, once-common practices, such as prechewing an infant's food, likely subsided in Western Europe. The bacterium arrived later, eliciting more inflammation.

Another explanation mentioned by Sonnenberg is that a third-party bacterium disappeared from human stomachs as Europeans urbanized. Absent competition for the gastric niche, *H. pylori* expanded its range and became more pathogenic. This scenario could also apply to *H. pylori* types alone, argues Blaser. If you're exposed to a variety of *H. pylori* strains, you may harbor a more diverse helicobacter ecosystem, one that inhabits your stomach more peaceably.

Yet another possibility, however, is simpler yet: a decline in worm prevalence.

HUMAN, WORM, AND BACTERIUM:
AN ANCIENT TRIAD

In the late 1990s, James Fox, a scientist at the Massachusetts Institute of Technology, wondered if worms might explain the African Enigma: Could helminth infection protect against the stomach malignancies associated with *H. pylori*?

He infected mice with the rodent worm *Heligmosomoides polygyrus,* and then with a helicobacter species called *H. felis.* Mice that received just *H. felis* developed severe gastritis and precursor conditions to stomach cancer. But wormy mice didn't develop this malignant inflammation. The worms' skewing of their immune response protected them from the ravages of infection, and ultimately from cancer. There was a trade-off, of course: wormy mice harbored a larger colony of *H. felis* bacteria than nonwormy mice. But the greater multitude had little notable impact on murine health.

Fox and colleagues followed up with a comparative study of two human populations, one from Colombia's coast, the other from the country's interior. *H. pylori* infected about 95 percent of both populations, but children from coastal Tumaco had twice as many worms as their sierra-dwelling counterparts. And in the wormy coastal population, gastric cancer was far less prevalent.

Genetic differences between the two populations limited the drawing of firm conclusions: the coastal population was of African descent, the interior mostly Amerindian and Spanish. But elsewhere, scientists observed patterns consistent with Fox's interpretation. Nearly all Tanzanians had the bacterium, for example, but inhabitants of the country's highlands, and denizens of the slopes of Mount Kilimanjaro where worms were less prevalent, had much more stomach cancer. Likewise, across the Japanese archipelago, gastric cancer incidence increased with latitude, again the inverse of worm prevalence.

Scientists suspect that *H. pylori* strains of different virulence partly explain this variation, especially in Africa. Diet perhaps accounts for another chunk. Vitamin C protects against stomach cancer, and people living in warmer climates eat more fresh fruit—what David Graham calls the "banana hypothesis." Excess dietary salt, on the other hand, promotes stomach cancer. Sonnenberg notes, for instance, that European governments abolished salt taxes just before the incidence of gastric cancer spiked. Perhaps the timing wasn't coincidental. Nor was it coincidence, some argue, that as refrigerators increased the availability of fresh fruit and veggies during the twentieth century—and decreased the consumption of cured and salted meats—stomach cancer declined on its own in the West.

So we see that many factors may account for the enigmas surrounding *H. pylori*. If the bug were ever put to therapeutic use, these cofactors would have to be clarified. But in some ways, whether it causes cancer isn't relevant to the question of whether it prevents asthma. Both may be true. And given the immune system's requirements for certain stimuli to operate smoothly, the real takeaway here is that one more stimulant has mostly disappeared. If we need all the right buttons pressed, we're fast running out of organisms qualified to press them.

There is, however, one pertinent, unresolved question. Martin Blaser proposes that *H. pylori* benefits the host, but so far these are benefits in the sense that, if you remove the bacterium, the host becomes unbalanced. Biological dysfunction arises. That's really a dependency: normalcy reigns when the bug is present; problems appear when it's absent. But what about direct advantages from the bacterium's presence? If the bug is more mutualist than parasite, how does it actively contribute to the greater whole? What good does *H. pylori* do?

An answer came from the tuberculosis field. Of every ten people exposed to the bacterial parasite *M. tuberculosis,* only one fell sick. Some beat back the bacterium entirely, but most developed latent infections. They kept the bacterium safely walled off in their body. The qualities that distinguished those who developed active TB from those who kept it controlled presented a long-standing mystery of intense interest. And the Stanford University scientists Sharon Perry and Julie Parsonnet stumbled on a plausible answer. They were tabulating the results from a survey of tuberculosis prevalence among immigrants living in south San Francisco when they noted that while people with *H. pylori* were just as likely to carry tuberculosis, they were far less likely to develop active tubercular disease. Did *H. pylori* somehow protect from active tuberculosis?

The scientists confirmed that harboring *H. pylori* protected against active TB in the Gambia and Pakistan. Then they tested the idea experimentally. They exposed macaque monkeys that naturally carried *H. pylori* to *M. tuberculosis.* These monkeys were, they found, one-third as likely to develop disease six months later compared with noninfected counterparts. How did it work? By activating antimicrobial aspects of the immune system—especially interferon gamma, which was also, incidentally, so important in protection against allergy—*H. pylori* helped its host manage the tuberculosis infection. One bacterium helped control a second. (And another infection—worms—may have helped limit damage caused by the first.)

The finding "raises the intriguing possibility that our microbiota can be manipulated to modulate disease risk from *M. tuberculosis,* as well as other common human pathogens," wrote Perry and Parsonnet in 2010. Indeed, German schoolchildren harboring the bacterium were nearly one-third as likely to suffer from diarrhea compared with their helicobacter-free peers. And Israeli soldiers hosting *H. pylori* were less susceptible to diarrheal disease than those without. In the future, before you travel to a region where tuberculosis is endemic, or dysentery common, physicians may colonize you with a specially tailored helicobacter strain.

WHAT ELSE DOES HELICOBACTER PREVENT?

At this point, a picture should have formed of the human superorganism in all its messy glory, a portrait we've cobbled together from the consequences of our having dismantled it piecemeal. So it will probably come as no surprise that, in addition to warding off allergic disorders, *H. pylori* may protect against autoimmune diseases, such as lupus, multiple sclerosis, inflammatory bowel disease, and perhaps even heart disease. In some case studies, autoimmune disease—type-1 diabetes, rheumatoid arthritis, and Crohn's disease—has flared immediately after patients eradicated the bug.

As always, nothing is clear-cut with the bacterium, however. Some evidence suggests that *H. pylori* causes autoimmune disorders and that it contributes to other conditions with an inflammatory component, such as Parkinson's or cardiovascular disease. Given these apparent contradictions, it's tempting to dismiss the idea that the bacterium benefits us at all. Yet a cursory rejection would ignore much of what we just learned: The bacterium can either hurt or help, depending on the greater context of the superorganism. The only rule here is that insofar as the bug elicits damaging inflammation, it will worsen health. But to the degree that it strengthens immune regulation, it will prevent diseases of immune dysregulation. It can do both.

For his part, Blaser thinks that in the future, pediatricians will deliberately introduce *H. pylori* into young stomachs, strains designed to fit the genotype in question. The bacterium will help prevent asthma, allergies, and maybe other inflammatory and autoimmune diseases. Then, come adulthood when the risk of ulcer and gastric cancer increases, doctors will eradicate *H. pylori* with a narrow-spectrum antibiotic. Asthma prevented, cancer averted; benefits reaped, costs avoided.

For allergy, immune-system-modulating drugs based on *H. pylori* proteins are already a possibility. The Italian scientist Mario D'Elios has isolated a single protein from *H. pylori* that, in mice, prevents allergic disease. The protein may one day serve as a drug to prevent allergies.

Meanwhile, Blaser frets most about *how* the bacterium began disappearing, and what its vanishing act really means. The bug started going extinct before anyone knew it existed, and without anyone intentionally trying to eradicate it. A century ago, most Americans had it. These days, just under 6 percent of children in Western Europe and the U.S. do.

Improved public hygiene, larger living quarters, smaller families, the obsolescence of premastication, and everything else that hinders the sharing of saliva and fecal matter likely contributed to the bacterium's decline. But Blaser worries most about antibiotics, bowel-scorching drugs directed not necessarily at *H. pylori,* but other infections. With each course of antibiotics, the odds of losing *H. pylori* fall between 15 and 50 percent. And while *H. pylori* serves as a kind of proof of concept, for Blaser, the pressing question is, what other good microbes are disappearing along with it? *H. pylori* landed on the radar because of its connection with disease. But what of those purely beneficial residents?

Blaser describes this particular anxiety as "the disappearing microbiota hypothesis." We inherit our first microbes from our mothers. Many, like *H. pylori,* live only in humans. Once they're gone, they're irrecoverable. "Each generation could be beginning life with a smaller endowment of ancient microbes than the last," he wrote in a 2011 article in *Nature* entitled "Stop the Killing of Beneficial Bacteria."

Indeed, epidemiologists have for some time noted a correlation between asthma and antibiotic use. The more people use antibiotics early in life, the greater their chances of asthma as adults. Similar patterns have emerged for inflammatory bowel disease. By one count, children exposed to seven or more courses of antibiotics had nearly triple the risk of developing Crohn's disease compared with those who never took such drugs, and double the risk compared with those who'd taken just one or two courses.

These survey-type studies are, of course, prone to errors of reverse causation: Those who develop allergic and inflammatory disease may take more antibiotics early in life because they're infirm to begin with. Indeed, this explanation would seem the most legitimate, except that the combined force of Blaser's, Robinson's, and Müller's work—and that of several others—indicates that *H. pylori* can prevent these diseases. And it does so by strengthening those regulatory immune networks increasingly understood to preempt allergic, autoimmune, and inflammatory disorders.

There are other reasons to take Blaser's worries over inadvertent extinction seriously. What if these heirloom microbes, which you receive from your mother, are uniquely evolved to your genotype? *H. pylori,* for example, recombines and adapts to the unique conditions in each individual stomach. The strain that your mother has, in other words, is likely better suited to your particular makeup than a stranger's.

Why does this matter? If you host your own bacterial "race," perhaps it's less pathogenic. Maybe the symbiosis is tighter, the mutual benefits greater. Maria Dominguez-Bello, for example, has found that Amerindian strains are markedly less damaging than those from the Old World. And among the Venezuelan Warao, where nearly everyone—99 percent—has intestinal parasites, those who also harbored *H. pylori* showed improved nutritional status compared with those who did not. In ways scientists can't yet explain, they benefitted from *H. pylori*'s presence. Is their unique Amerindian strain, coevolved to fit their stomachs, partly responsible?

"No one knows," says Blaser. But in the meantime, for whatever reason, the imported African and European strains are outcompeting the native strains in South American stomachs. Presumably, they're more aggressive. Introduced strains are driving the Amerindian strains extinct.

In parasitology, there's an old rule of thumb: hangers-on that travel from parent to offspring are more likely to confer benefit than parasites that travel horizontally. If parasites need their current host to reproduce so they can find a home for their own offspring, the thinking goes, then their interests and the host's align.

On the other hand, if the parasite skips laterally through a crowd, then its interests don't align with the host's. In fact, horizontal transmission may favor the opposite qualities—virulence, a microbial run on the host's bank. For these parasites, which include most crowd diseases, you're just a meal along the way. They have no stake in your continued existence.

H. pylori can spread both ways, but for most of human evolution, it likely passed from mother to child, and the rules of vertical transmission applied. On some level, its interests aligned with the human host's. The bacterium is just one of many transmitted vertically, however. We inherit a universe of microbes from our parents, our siblings, and our greater environment. Blaser and others worry that we've distorted these little-understood ecosystems, and that this derangement is also contributing to the diseases of modernity.

This microbial community, alternately called the human microbiota or microbiome, is the subject of the next chapter.

CHAPTER 9

Community-Wide Derangement

There is a dimension to human evolution—a microbial evolution—that is likely occurring at a very rapid rate as our societies undergo dramatic shifts in socioeconomic status and cultural norms, redistribution of populations from rural to urban areas, changes in patterns of food consumption, and alterations in our exposures to xenobiotics, ranging from antibiotics that we intentionally take to various potentially toxic compounds that we unintentionally or deliberately ingest.
            —Jeffrey Gordon and Todd Klaenhammer

Envision three groups of people: one lives outdoors in a large messy crowd; a second lives indoors together; and the third lives indoors as well, but its members are isolated in small apartments, each individual regularly taking antibiotics.

Now ask the question: Which group is healthiest? If your assumption is the latter group, the situation that most resembles modern, urbanized life, you'd be wrong. The healthiest individuals come from the first group—from those living in conditions that most resemble life before the Industrial Revolution.

Scientists conducted this experiment not with people, but with pigs. Imke Mulder and Denise Kelly at the University of Aberdeen, in Scotland, reared three groups of swine, one outside in the mud, one inside, and the third, also indoors, housed in isolation cages and routinely fed antibiotics. They noted three major differences between the groups. First, the indoor, antibiotics-treated pigs more often turned on genes involved in inflammation. Second, the indoor group also harbored more potentially pathogenic bacteria, microbes that could cause disease.

And third, the outside animals, whose immune profile leaned most toward tolerance, harbored a very different community of microbes. More than three-quarters of the outdoor pigs' resident bacteria were lactoba-

164

cilli. By contrast, in the intermediate group, just 13 percent of the bacteria came from that family. And for antibiotics-treated individuals reared in isolation, lactobacilli accounted for just 3.6 percent of gut microbes.

We all come into this world sterile. We encounter our first microbes as we pass through the birth canal. We continue acquiring them throughout life. This study showed that the microbes that most promote health weren't, however, evenly distributed among different environments. Pigs that wallowed outside in the mud acquired the healthiest microbes, as reflected by patterns of gene expression. Pigs that were sheltered and continually "cleansed" harbored the least-healthy community.

We've repeatedly explored the idea that genes *aren't* destiny—that genetics alone can't explain allergic and autoimmune disease. Here we see that microbial communities *can* influence fate through gene expression. And environmental exposures alone determined the makeup of those communities. Our surroundings, it seems, can affect our health by seeding our microbial organ.

In October 2010, I attended a meeting in Miami hosted by the American Society of Microbiologists entitled "Beneficial Microbes." Not long ago, most clinicians would have considered the very phrase oxymoronic. Microbes were sneaky and murderous. They'd caused immeasurable human suffering. They'd impeded human progress. The only good microbe, it followed, was a dead one.

The original nineteenth-century microbe hunters didn't necessarily ascribe to this extreme view, however. Ilya Metchnikoff, for example, the Russian scientist who first described cell-mediated immunity—white blood cells that devour invaders—and who was awarded the 1908 Nobel Prize for this work, was obsessed with bacteria he thought improved human health. The fascination stemmed in part from his observation that Bulgarians and some Russians lived unusually long lives, and that they routinely drank fermented milk. He examined the drink, isolated lactic-acid-producing bacteria, and began advocating for their deliberate ingestion. Today, Metchnikoff is often cited as the father of probiotics.

And concurrent with the original germ theorists, botanists were working out the cooperative relationships that underlay so much plant life. Most plants cultivated special fungi on their roots that provided phosphorus, a critical nutrient. They were called mycorrhizae. Legumes—peas, beans, and alfalfa, among other important crops—relied on rhizobia bacteria for nitrogen. It was with relationships like these in mind that, in 1879, the German botanist Heinrich Anton de Bary coined the term *symbiosis*: "the living together of unlike organisms."

So the study of cooperation between life-forms started off strong. But then, perhaps because there was more glory in defeating microbial terrors than identifying allies, the study of microbes that make humans healthy languished, at least in the medical mainstream. That's now changing, partly because the challenges we currently face—antibiotics resistance and degenerative inflammatory diseases—have shifted. The study of human-microbe symbiosis is flowering.

The Beneficial Microbes conference takes place on the ground floor of a polished downtown Miami hotel—a neighborhood of colorful modern towers rising improbably from the swamp-covered, south Florida bedrock. The mood is ebulliently insurrectional. If the people in the large conference room were to protest at the headquarters of germ theory—if such a place existed—they'd hold signs reading WAR NO MORE, HUMAN AND MICROBE: UNITED WE STAND, or HONOR THY SYMBIONTS—all titles of recent papers by respected scientists in major journals.

New, improved, and ever-cheaper technology has enabled the uprising. A little over a decade ago, scientists primarily studied microbes by cultivating them. You could therefore investigate only those microbes you could grow, a small fraction of the total diversity inhabiting the human body. New techniques, however, eschew cultivation, and look directly at what microbes are present, and what they do, by "fingerprinting" their DNA. Freed of having to grow the bugs, we can finally see them with some clarity. And we observe something similar to what Galileo Galilei saw in the seventeenth century when he trained his telescope on the heavens. Just as we revolve around the sun, not vice versa, it's increasingly apparent that microbes do not revolve around us; we revolve around microbes.

A long-overdue cross-fertilization of different disciplines underlies the endeavor. The scientists at the conference represent a mishmash of expertise—entomologists, botanists, microbiologists, and more. One conference organizer, Margaret McFall-Ngai, has spent decades studying the symbiotic relationship between the squid species *Euprymna scolopes* and a light-producing marine bacterium called *Vibrio fischeri*. The squid gathers these luminescent bacteria from the sea floor, sequesters them in a special organ, and uses them as a cloaking device, a way to disguise itself from predators during nocturnal feedings.

Others here focus on insects and their resident microbes. Some study microbial communities in people. What brings them into the same dimly lit room in this downtown Miami hotel? All multicellular creatures—plants, animals, and fungi—retain many of the same sensors for communicating with microbes. McFall-Ngai calls this commonality "the language of symbiosis." No one thinks this retention across kingdoms is accidental. Earth has always been, and always will be, a microbe-dominated world.

To get by, you'd better keep a line of communication open with the ones in charge.

Microbes existed for at least 2.5 billion years before the first multicellular creatures. By the time animals arose during the Cambrian explosion 540 million years ago, they'd already carried on for 3 billion years. In that time, they'd evolved ways, often by forming communities and working together, to colonize every exploitable niche on Earth.

If you're an aspiring multicellular organism lumbering into this world, do you reinvent the wheel or do you defer to the masters with billions of years of experience? The evidence says we used the experts. The "big bang" of multicellular life occurred only because one microbe imbibed another. An amoeba-like cell either consumed or was invaded by a bacterium. The bacterium survived, eventually becoming a critical organelle in the larger cell. They're the energy-producing mitochondria of animals, and the sun-catching chloroplasts in plants. The fundamental building block of our bodies—the eukaryote cell—derives from this ancient mutualism.

And thereafter, new mutualistic relationships continued to appear. Sponges, usually billed as the oldest living proto-animal—our ancestor—often host symbiotic bacteria in their tissues. Termites cannot digest their food, which is rich in cellulose, one of the toughest organic substances on earth, without symbiotic protozoa. Grazing animals have evolved multiple stomachs to house their own fermenting microbial communities. It's not that symbiosis occasionally occurs in nature; it's that everywhere, at every level of complexity, symbiosis makes what we call nature possible. Indeed, Margaret McFall-Ngai's research on the squid and its luminescent bacteria has inspired her to rethink our adaptive immune system entirely. Did it really evolve to fight off pathogens—the dogma—or to form tighter symbioses with a wider range of microbes? Does it really serve the function of a standing army, in other words, or diplomatic corps?

When these scientists look around the room at the two hundred or so faces in attendance, they don't see PhD candidates, postdocs, and professors. They see anaerobic digesting chambers with arms and legs—spaceships for microbes that, wherever they go, leave a trail of living slime. One presenter says as much. Humans "can be regarded as elaborate vessels that have evolved to permit the survival and propagation of microorganisms," says the Stanford University microbiologist Justin Sonnenburg. Every day you put food in one end and excrete microbes from the other.

Let's review the basic facts: In the human body, bacterial cells outnumber human cells by ten to one. One thousand microbial species naturally inhabit the human intestine, including archaea, viruses, and yeasts, reaching about 100 trillion individual cells in total. The information in our collective microbial genome is a hundred times greater than that contained in the human genome.

Importantly, these microbes aren't a random collection. Just four of more than fifty known bacterial phyla, or large families, inhabit the human gut. The narrow spectrum of successful colonizers suggests a very specific coevolution. Meanwhile, just fifty to one hundred bacteria are pathogenic to humans. Compare that with the thousand potential commensal species, and you immediately note that most of our daily interaction with microbes has nothing to do with disease.

The bulk of our resident bacteria live in the large intestine, the final, sweeping loop of bowel before the exit. If you unrolled and hammered flat the human gut, it would cover about 100 square meters, roughly half the area of a singles tennis court. That's quite an interface, and maybe explains why 70 percent of immune activity occurs around our intestines.

This last factoid, it's turning out, is key to understanding the immune-mediated diseases we've discussed. Without bacteria present, the immune system remains half asleep. And depending on what's present or absent, not only does one's immune activity change, so does one's ability to store calories as fat, one's proclivity to form kidney stones, even, scientists have found, one's mental acuity. There's little about how the mammalian body works that *isn't* affected by microbes in the gut.

Which partly explains the palpable undercurrent of anxiety at the conference. We're only now beginning to understand the importance of our microbial organ, but already the overriding question is, have we altered our microbiota without realizing it? And are we suffering the consequences?

Our microbial community is quite plastic. It shifts according to diet, to microbial exposures, to individual makeup, with age over time. This very mutability may be one answer to the question, why have a microbiota at all? An ecosystem of microbes can evolve and shift more quickly than our own comparatively stiff genome. That malleability affords a greater flexibility—to eat a broader variety of foods, for example—than if we relied on "our" genes alone. But as with any evolved codependency, there's a limit to how much either party can change before the relationship collapses. And the dramatic reordering of human experience in the past two centuries may have produced just such a mismatch between the human genome and the human microbiome.

The "microbiota we think of as healthy may be a derived one," says Justin Sonnenburg, one "that predisposes us to Western diseases."

LESSONS FROM RECONSTRUCTING
THE MICROBIOTA

Louis Pasteur, the giant of microbiology who, among other accomplishments, developed the first rabies and anthrax vaccines, once described an experiment he'd like to conduct: He proposed raising an animal with "pure nutritive products which have been artificially and totally deprived of the common microorganisms." He suspected that microbes were absolutely necessary for animal life. The outlined experiment would, he thought, prove that dependency.

Beginning in the mid-twentieth century, following a hundred years of almost miraculous progress in medicine—including the triumph of germ theory, the advent of antibiotics, and the polio vaccine—scientists finally looked into Pasteur's idea. They delivered mice by C-section, fed them sterile food, and raised them in germ-free bubbles. And they found that Pasteur was wrong: animals could survive without microbes.

But they looked really weird. Aside from requiring dietary supplements of vitamins B and K, nutrients normally synthesized by resident bacteria, their physiology was off. One region of their intestines—the cecum—was abnormally enlarged. The total surface area of their gut, however, was reduced by about one-third. They secreted excess mucus, but the contents of their intestine moved at a snail's pace.

Stranger yet, organs distant from the gut appeared malformed. Their hearts, lungs, and liver were shrunken. All the while, these germ-free animals required one-third more calories than conventional mice to survive. This last observation allowed scientists to quantify at least one aspect of the microbiota: its contribution to our energy supply. Although resident microbes clearly took a cut of incoming nourishment, the net effect of their presence was not to drain resources but to enhance the host's ability to extract energy from food. (For humans and pigs, omnivorous animals that subsist on a less coarse diet than mice, the microbiota's contribution is slightly less: our resident microbes furnish 10 percent of our calories.)

As it turned out, Pasteur had proposed a second experiment: adding microbes back one by one until the animal could again thrive. In the early 2000s, a postdoc at Harvard Medical School named Sarkis Mazmanian started this investigation. He thought he'd rebuild the microbiota from the ground up, but he never made it past the first microbe, a bacterium called *Bacteroides fragilis*.

In addition to other abnormalities, germ-free mice displayed major immune deficits. Normally, white blood cells shuttle around the body in the blood stream and gather in lymph nodes, rest stops along the way.

But germ-free mice mostly lacked lymphoid tissue. The nodes were fewer or absent. Their attack cells remained in a state of arrested development. And most important for our purposes, they didn't have as many regulatory T cells.

Introducing *B. fragilis* into this human-created anomaly, Mazmanian found, corrected all these deficiencies immediately. T cells were primed; lymphoid tissue grew; the immune system came to life. *B. fragilis* could also change disease outcomes. Another bacterium, *Helicobacter hepaticus,* was a normal member of the microbiota in wild mice, but could cause disease in lab rodents. What made the difference? If *B. fragilis* arrived first, Mazmanian found, *H. hepaticus* behaved like a contributing member of society. But if *B. fragilis* was absent, then *H. hepaticus* incited chronic inflammation and colitis.

"This raises the possibility that the mammalian immune system, which seems to be designed to control microorganisms, is in fact controlled by microorganisms," Mazmanian, by then running his own lab at Caltech, wrote in 2009 in *Nature Reviews Immunology*. Which is to say that our immune system seemed to have assigned certain functions to certain commensals.

Across the country, New York University scientists happened on another microbe with a different, but equally important role. Dan Littman and Ivaylo Ivanov had purchased mice from three different vendors. All mice were genetically identical, and so they should have possessed identical immune systems. But while mice from two of the vendors were indeed indistinguishable, those from the third were different. They lacked certain pro-inflammatory T cells—labeled Th17—important in defense against microbial opportunists.

The different immune repertoire would bias any experiments the scientists conducted, which was annoying, but the more intriguing question was, what underlay these differences? They had the same genes, so why did they have different immune systems?

The absence of a single microbe, they found, explained the paucity of those Th17 cells. Segmented filamentous bacteria, long whiplike strands anchored at one end to the intestinal wall, were absent in the Th17-deficient mice. And transplanting the microbiota from the other lab mice to these deficient animals immediately boosted their Th17 cells. Cohousing them also did the trick.

Whereas Mazmanian's *B. fragilis* appeared tasked with strengthening the regulatory arm of the immune system, segmented filamentous bacteria prompted attack cells. In the context of autoimmune disease, an abundance of inflammation-inducing cells would seem problematic. But in the real world full of opportunists itching to loot and steal, inflammatory potential was a necessity.

Indeed, when Littman and Ivanov introduced a mouse pathogen called *Citrobacter rodentum,* mice hosting these bacteria—mice with more Th17 cells—better resisted invasion. Ever wonder how some people can dine on street food in any slum around the world without falling ill? Here was an answer: maybe they hosted bacteria that better primed their immune system against incursion.

So what of the immune system that had a tendency to turn against the self? The Harvard scientist Diane Mathis, who studied autoimmune rheumatoid arthritis, a painful degenerative inflammation of the joints, found that these same segmented bacteria could induce arthritis. In germ-free conditions, mice bred to develop the disease remained mostly arthritis-free. But when she introduced just this bacterium, their immune system turned on their joint cartilage. In a second study, Sarkis Mazmanian showed that the bacterium could also prompt multiple sclerosis in MS-prone mice.

These were highly artificial models, of course. And anyway, segmented filamentous bacteria weren't, as a rule, native to the human gut. But the experiments revealed a remarkable facet of the relationship between resident microbes and mammalian host. Each bacterial species might induce a mirror population of immune cells in the host. Those cells affected the potential for autoimmune disease at sites distant from the gut, such as the joints and the central nervous system.

Real-life microbial ecosystems were, of course, umpteen times more complex than these laboratory re-creations. But as a whole, the research suggested that immune-mediated diseases might result from imbalances between, say, the soothing *B. fragilis* and the human version of the rabble-rousing segmented filamentous bacteria. Health might depend on the right ratio of different microbes.

And on the question of balancing pro- and anti-inflammatory tendencies, a breakthrough was in the works.

WHAT KEEPS THE PEACE IN HUMANS?

For decades, scientists had searched for an infectious cause of inflammatory bowel disease. A relative of tuberculosis called *Mycobacterium avium paratuberculosis* continued to draw attention. And "sticky" strains of *E. coli* that adhered to the gut lining also attracted suspicion.

But when Daniel Frank and Norman Pace at the University of Colorado, Boulder, looked at biopsies of Crohn's patients, they were struck not by the presence of these usual suspects but the absence of just two types of bacteria. Bacteroidetes and certain clostridial species, both of which were normal inhabitants, were hundreds of times less abundant in diseased guts.

So striking was the depletion of these bacteria that Alexander Swidsinski at the Charité Hospital in Berlin proposed using their absence as a quick way to diagnose active Crohn's disease. By reading a "punched fecal cylinder" suspended in paraffin wax—rather like the sediment cores climate scientists extract from lake beds, but in this case drawn from feces—one could accurately diagnose IBD, he argued. The absence of these bacteria in the "fecal-mucus" transition zone invariably signified inflammation.

French scientists zeroed in further on a single species with the immense name *Faecalibacterium prausnitzii*. The risk of relapse for Crohn's patients who'd undergone surgery was directly inverse, they found, to how abundant these bacteria were. If you harbored lots of them, your prognosis was better than if you didn't. Swedish scientists, meanwhile, could predict which twin from identical twin pairs had the greater chance of developing Crohn's disease just by measuring who harbored less *F. prausnitzii*.

Still, while these correlations were compelling, they didn't show causation. Proving that these bacteria had an active role in peacekeeping fell to the University of Tokyo scientists Koji Atarashi and Kenya Honda. They took a top-down approach, and slowly whittled away at the mouse microbiota with narrow-spectrum antibiotics, waiting for that moment when the host's population of T-regs crashed. A course of vancomycin, which specifically targets Gram-positive bacteria, was the turning point, the moment the mouse peacekeeping cells collapsed. The scientists focused on those clostridial species that were depleted in Crohn's disease, of which *F. prausnitzii* was one. Now they endeavored to reverse-engineer the microbiota to the point where T-regs rebounded.

They devised a cocktail of forty-six clostridial strains, and introduced them into mice. T-regs increased. Mice routinely eat one another's feces, and when the scientists cohoused nontreated mice with treated ones—mice that didn't harbor these important bacteria with those that did—the nontreated mice also saw an expansion of T-regs. This was immune modulation that was contagious.

To quash any remaining doubt, they colonized young mice with a microbiota that contained those clostridial bacteria. As these mice matured, they had more T-regs. And this muscular ability to control inflammation conferred resistance to both experimentally induced colitis *and* allergic disease later in life. A single group of bacteria prevented both diseases of immune dysregulation.

In the real world, meanwhile—the complex microbiota of free-living people—scientists repeatedly observed that immune-mediated diseases were often preceded by predictable shifts in the microbiome. In 2008, the year Dan Littman published his first study on segmented filamentous bacteria, Finnish scientists noted that before rheumatoid arthritis set in, the microbiota changed. Bacteria with anti-inflammatory potential, including the *B. fragilis* Mazmanian had studied and the bifidobacteria that Bengt Björkstén and others found so important in warding off allergies, declined months before inflammation began to intensify in the joints.

Scientists at the University of Florida, Gainesville, meanwhile, began following a group of children who had gene variants linked to autoimmune diabetes, but no disease—yet. Whereas children who never developed diabetes amassed an increasingly diverse and stable microbiota as they approached toddlerhood, children who later developed diabetes had a relatively nondiverse and unstable microbiota, with some species abnormally blooming before the disease ever appeared. Did this impoverished microbiota predispose to type-1 diabetes?

Yolanda Sanz, a scientist at the Institute of Agrochemistry and Food Technology in Valencia, Spain, observed similar patterns in celiac disease. For mysterious reasons, in some people the protein gliadin prompted debilitating inflammation in the gut. If the irritation continued indefinitely, it could cause osteoporosis, failure to thrive, even occasionally neurological symptoms, and sometimes death. The good news was that unlike other autoimmune diseases, you could control celiac disease by avoiding wheat and other gliadin-containing grains. The bad news was—as we've seen—the disease had increased dramatically during the latter twentieth century. Our ability to tolerate foods we'd eaten for at least ten thousand years, and probably much longer, was suddenly deteriorating. What had changed?

Sanz found that children with celiac disease had a shifted microbial community, more Gram-negative compared with Gram-positive bacteria, fewer bifidobacteria, and a paucity of those clostridial species that boosted regulatory T cells. She observed this difference both in children with active disease and in children in remission. The altered community didn't result from a wheat-free diet.

To measure its effect, she transplanted this "celiac" microbiota from humans into germ-free rats. The rats' intestines became more permeable; the ability of proteins and microbes in the gut to pass through the intestinal wall increased. Clearly, some permeability is always required, or we wouldn't be able to absorb nutrients from our food. But if the intestine is too porous, scientists think, substances that aren't supposed to leak through. In the case of celiac disease, the offending substance might be gliadin, which then prompts more inflammation, which makes the intestine even more sensitive to the protein, and so on, in a vicious cycle.

Some argued that gliadin was naturally toxic to people with celiac disease. Yet Sanz found that simply adding a third microbe, a bifidobacterium she'd isolated from a healthy nursing infant, could blunt the protein's toxicity. With bifidobacteria present, gliadin didn't inflame. What's more, bifidobacteria changed how other residents of the microbiota behaved. In a model of the celiac gut, a native *E. coli* species switched on genes that increased its virulence. But with bifidobacteria present, the *E. coli* carried itself as a cooperative member of society. In game theory parlance, *E. coli* could change from a strategy of cooperation to a winner-take-all approach. And the relative abundance of a third party—those bifidobacteria—determined which strategy it chose.

Ecologists had long noticed these sorts of complex and unpredictable interactions in external ecosystems. Aspen stands had recovered after the reintroduction of wolves to Yellowstone, for example. Willow stands along streams also rebounded. Small birds and other creatures that preferred lush streamside habitats became more abundant. But what did carnivorous wolves have to do with trees? Their mere presence kept grazing elk away from the woods, scientists surmised, allowing the trees to recover from years of overgrazing.

Microbial ecologists were essentially finding that our internal ecosystems followed similar rules, and that they were prone to similarly unpredictable knock-on effects. In wolfless Yellowstone, aspens and willows were food. With the large predators present, however, they remained off-limits. In the human gut, this *E. coli* strain went on a bullying rampage when bifidobacteria disappeared. But when bifidobacteria were present, it behaved itself.

Sanz's research recast celiac disease as a problem of microbial deficiency. The observations on the diabetic microbiota, meanwhile, highlighted the perils of community instability. The research on IBD suggested that certain species were supremely important in maintaining harmony between the host and its microbial residents. These were the types of extinctions—or nonarrivals, as the case may be—that Martin Blaser fretted about, an ongoing loss of microbial diversity occurring at our core. And as we saw in chapter 6, the impoverishment likely began at birth.

INTERRUPTING MOM'S MICROBIAL MASTER PLAN

On the second day of the conference, a scientist named Gregor Reid begins his talk by spastically dancing across the dais to an '80s tune called "Subculture." The performance is meant to attract attention that, he must know, is flagging after a round of presentations. But Reid's performance also ties in with his subject matter. He starts his talk with a question: "What is culture?" he asks in a thick Scottish brogue. "It's transmitted from generation to generation," he answers.

The presentation is entitled "The Vaginal Microbiome's Role in Humanity." I assume from the title that he's being half serious, but as the talk progresses, I see that Reid means it literally: human reproductive success depends entirely on the microbes inhabiting the vaginal canal, he argues. He isn't talking about infections, but rather native microbes in the wrong proportions—what's called vaginosis. This type of imbalance can wreak havoc on the reproductive process, from interfering with conception to prompting premature birth. (In chapter 7, we also saw how low-grade inflammation incited by these microbes could predispose the developing fetus to asthma.)

"To not place a huge focus on the vaginal microbiome is like putting human survival at risk," says Reid at one point. It seems like hyperbole, but this idea—that commensal microbes affect all aspects of health, including reproduction—is rather dramatically supported by our own biology.

Consider the expectant mother. During late pregnancy, vaginal secretions change, becoming enriched in glycogen, a starchlike carbohydrate. The excess glycogen feeds specific lactic-acid-producing bacteria to the detriment of other types. Before birth, in other words, Mom selects a particular community of microbes. The acid they produce impedes pathogens from ascending the birth canal. And when the baby descends, she is also coated with this protective, slightly acidic microbial soup—the first life-form other than her mother that she encounters.

Other changes in Mom's body collude in the endeavor to, apparently, seed the newborn with a specific community of microbes. Colonies of bifidobacteria sprout deep in expectant mothers' nipples, and pass out with breast milk, a starter culture for the infant gut. Scientists have long wondered how these bacteria, which don't much like oxygen, found their way to the outward-oriented milk ducts. But in 2007, a group reported that, at least in mice, white blood cells transported bacteria from the gut, where they were abundant, to the mammary glands. The microbes didn't arrive via an external route; they had internal VIP privileges. Two years later, a Finnish group announced that healthy placentas, long thought sterile, were in fact coated with DNA from lactobacilli and bifidobacteria.

Breast milk also, scientists find, cultivates a distinct community of microbes. It contains a dizzying number of sugars called oligosaccharides—maybe two hundred—that the infant cannot digest. Only bifidobacteria can metabolize these sugars. So Mom's milk both transmits bacteria and provides the food they need to thrive. Why all the effort to ensure this particular microbial colonization? The easiest answer is that friendly commensals prevent potentially nasty bacteria from gaining purchase. They also jump-start the immune system, setting it on a path toward healthy development.

And scientists fairly consistently observe an elevated risk of allergic disease among children born by cesarean section—newborns who've circumvented their mother's inoculum—particularly for children born to allergic parents. In a study of 2,800 Norwegian children, for example, among those born to allergic mothers, the risk of food allergy was seven times higher if born by cesarean. For those born by cesarean to nonallergic mothers, however, there was no increased risk—a critical comparison. Dutch children born by C-section followed to age eight had an 86 percent increased risk of allergy with one allergic parent. Two allergic parents plus a C-section, on the other hand, nearly tripled the risk of asthma.

At least one autoimmune disease followed the same pattern: A metanalysis of twenty studies found that C-section children had about a 20 percent higher rate of childhood-onset type-1 diabetes compared with those born vaginally.

How does circumventing the vaginal canal affect the microbiota? After one month, Finnish children born by C-section still hadn't acquired the number of bifidobacteria and lactobacilli that vaginally born children harbored after a few days. And six months later, they had only half as much *B. fragilis*. Another study found that alterations to the microbiota in C-section children persisted to seven years of age. Your mode of delivery had long-lasting effects on your microbial community. And so did the location of your birth.

In the Netherlands, children born at home had half the risk of allergic disease by age seven compared with children born in the hospital. (Again, this held true only for children with allergic parents.) What was the major difference? Colonization by the pathogen *Clostridium difficile,* which tended to occur more in the hospital.

C-section children weren't, of course, sterile. They were simply colonized by a different set of bacteria. In Venezuela, whereas vaginally born babies acquired microbes from the mother's birth canal, C-section-born children were colonized by skin bacteria from the doctors and nurses who handled them. If evolution had conspired to seed newborns with a specific set of microbes, this wasn't it.

As of 2007, one-third of all infants were born via C-section in the U.S., a 50 percent increase over a decade earlier, and a mass diversion from the evolutionary norm. We don't necessarily have to avoid C-sections to preclude altered microbial colonization. In this instance, you can have your cake and eat it, too. At the conference, rumors circulated (which I couldn't ultimately confirm) that physicians in some parts of the world have become concerned enough by altered colonization patterns that they swab C-section newborns with their mothers' vaginal secretions. That's a nice workaround.

The broader point is not that cesarean sections are responsible for the allergy epidemic, although they may have contributed; it's more that considerable biological effort goes to populating newborns with certain bacteria, and interfering with this conveyance appears to cause problems. The generalizable rule, therefore, is that any perturbation of our microbial communities may cause trouble. And on that front, C-sections are far from the most obvious culprit.

WEAPONS OF MICROBIAL MASS DESTRUCTION

In 2004, Gary Huffnagle and the postdoc Mairi Noverr at the University of Michigan, Ann Arbor, carried out an unusual experiment. Most scientists looked to the lungs—infections, pollution, cigarette smoke, allergens—for the root cause of asthma, but Huffnagle and Noverr suspected that what most considered a lung disease originated in the gut.

As you'll recall from chapter 6, scientists had observed a consistent link between changes to the microflora and allergic disease. At that point, at least four studies—and many more since—had demonstrated a close association between antibiotic use in early life and asthma later. Again, these studies didn't demonstrate that antibiotics *caused* asthma. It might be that people with asthmatic tendencies tended to take more antibiotics. Another possible confounder: doctors might mistake asthmatic wheezing for a bacterial infection, and throw antibiotics at children who really had asthma. If you viewed this sequence in retrospect—a child who imbibed lots of antibiotics, and who then received a diagnosis of asthma—you'd see causality where, in fact, there wasn't any.

Huffnagle and Noverr moved to resolve these lingering questions experimentally. They treated mice with an antimicrobial called cefoperazone. And then they administered what they thought was the coup de grâce: a single dose of the yeast *Candida albicans*.

The yeast was a normal part of the natural human microbiota, but yeast infections, an overgrowth of this commensal species, were increasingly recognized as a problem after antibiotics treatment. Knocking out competitor bacteria with antibiotics gave *C. albicans* free rein. And these yeasts secreted powerful immune-modulating hormones—prostaglandins that in excess could, Huffnagle and Noverr suspected, skew the immune response toward allergy.

Normally, mice resisted colonization by this human-adapted yeast. But after antibiotics treatment, it quickly took hold in the rodent gut. And when these mice subsequently encountered a mold called *Aspergillus fumigatus,* one typical to the average damp home, the inhaled fungal spores triggered asthma. Antibiotics had enabled a yeast overgrowth in the gut, and that overgrowth had altered allergic sensitivity in the lungs. Now a common mold triggered wheezing. By changing the microbial community within, antibiotics could increase the risk of asthma.

Real-world support for what they called the "microflora hypothesis" for allergic disease abounded. Populations that avoided antibiotics had a different microflora and a diminished risk of allergies. In Sweden, the so-called Steiner families adhered to a philosophy called anthroposophy laid out a century ago by the Austrian philosopher Rudolf Steiner. As an article of faith, the families restricted vaccines (only tetanus and polio), used fewer antibiotics, and ate lots of fermented food.

They got measles much more often, but they also suffered from less allergic disease. For every four allergic Swedish children, there were only three allergic Steiner children. (Continuing along that gradient, for every three allergic Steiner children, there were only two allergic Swedish farming children.) The Steiner children had measurably different microflora in infancy: more lactic-acid-producing bacteria than the average Swedish child, and an increased microbial diversity that was, incidentally, directly proportional to whether they were born at home or in the hospital, and whether they used antibiotics. The fewer antibiotics they used, the more diverse their internal ecosystem.

A survey of twenty studies concluded that antibiotics in the first year of life increased one's risk of asthma by 50 percent. Most studies found a dose-dependent relationship; the more you used as a youngster, the greater your chances of wheezing later. And, as we've seen, a Danish study showed the same pattern for inflammatory bowel disease. The greater your intake of antibiotics early in life, the higher the risk of inflammatory bowel disease later.

Huffnagle's experiments aside, early studies looking closely at antibiotics' effect on the microbiota suggested that in free-living humans, it recovered quickly after disturbance. But as technology improved and scientists peered more deeply, they began to discern rather dramatic long-term changes. In one case, scientists caught up with patients two years after they had received a single weeklong course of clindamycin. The diversity of bacteroides species remained impoverished. Others found that repeated insults had a cumulative effect. Les Dethlefsen and David Relman at Stanford University noted that one course of ciprofloxacin, a broad-spectrum antibiotic, only slightly altered the community, but after a second disruption, the community stabilized in a different configuration. Some species disappeared entirely. "It's as if your beneficial bacteria 'remember' the bad things done to them in the past," Relman said. "Clinical signs and symptoms may be the last thing to show up." The outcome was akin to cutting down a pine forest once, watching it regrow, clear-cutting it again, and having oak scrub grow back, with nary a pine in sight.

Another factor emerged as important to recovery: access to the bacteria you'd lost. After administration of an antibiotic cocktail, diversity remained depleted in mice. If scientists housed an untreated mouse in the same cage as the treated mice, however, the treated mice recovered. Recovery after perturbation, it seemed, depended on reseeding from a backup supply of microbes.

The finding raised a new set of anxieties. If, like mice, we depend on our fellow human beings for constant reseeding, and for recovery after disturbances, then the widespread depletion of microbial diversity would have unforeseen consequences. The problem wasn't just that a few people had successively thinned and perhaps permanently altered their microbial communities; the worry was that everyone had. "I think the pervasiveness of antibiotics as a whole in society will likely play a role in the disorders of immune dysregulation," Dan Littman told me. "My guess is that the microbiota in the entire human population in developed countries has been shifted by the use of antibiotics."

MICROBE-KILLERS EVERYWHERE

The first antibiotic, penicillin, became widely available after World War II, a decade or two before the beginning of the allergy epidemic. By 1992, the apogee of the allergy epidemic, children and teens were averaging nearly one course of antibiotics per year, often for minor conditions, such as otitis media, an ear infection. That was half again as many antibiotics as the average child took in 1980. In that same period, the number of broad-spectrum antibiotics, such as cephalosporins and fluoroquinolones—drugs that functioned like sledgehammers rather than scalpels on microbial communities—increased.

By the turn of the millennium, worry over antibiotics resistance had lessened the prescription rate by half. (Doctors also realized that otitis media was a self-limiting condition, and didn't always require antibiotics.)

It's perhaps no accident that the allergy epidemic leveled off by the new millennium as well.

The secret in plain sight, however, was that most antibiotics weren't given to people at all. Roughly 70 percent in the U.S. went to animals, not humans. They were given to check infections in some cases, but mostly, low-dose antibiotics were used to accelerate animal growth. How often they made their way into the food supply remained a largely unexplored question.

Protocols existed to ensure that meat and dairy remained antibiotic-free. Chief among them, livestock producers were to discontinue antibiotic administration an adequate amount of time before slaughter. The problem was, the USDA's Food Safety and Inspection Service—the agency tasked with enforcing these rules—had, in its own words, "serious shortcomings."

A scathing 2010 report by the Office of the Inspector General cited a litany of failures: carcasses known to be tainted that entered the food supply anyway; repeat offenders ignoring multiple warnings; and an inability to discern where, in the complicated web of livestock raisers, feedlot owners, and meat resellers, tainted meat originated. The National Residue Program was "not accomplishing its mission of monitoring the food supply for harmful residues," the report concluded.

Even vegetarianism couldn't save you. Between three-quarters and half of all antibiotics consumed by animals were, after all, excreted in feces and urine. That manure went to fertilize crops. First, it usually sat in pits where, apologists argued, microbes made quick work of any residues. But scientists looking into the question found that antibiotics endured, arriving intact in farmland. And food crops readily absorbed them.

In experiments, lettuces took up florfenicol, levamisole, and tri-methoprim from tainted soil. Carrots absorbed diazinon, enrofloxacin, and florfenicol. Staple crops, such as wheat, corn, barley, and potatoes, soaked up veterinary antibiotics when present. And that was hardly the extent of the problem.

Treated human sewage became fertilizer as well. Just over half of the sludge removed from U.S. wastewater treatment plants ended up in fields. Human-derived slurries contained an even wider array of pharmaceuticals than your average animal manure—antidepressants, antimicrobials, compounds from fragrances and soaps, hormones from birth-control pills—in short, everything people took. Food crops sopped up these molecules as well. Soybeans fertilized with human biosolids readily took up the antimicrobials triclosan and triclocarban. Hydroponically grown cabbage absorbed the antimicrobials sulfamethoxazole and trimethoprim.

These drugs diffused far and wide into the environment. Even waterways distant from manure pits and sewer outfalls contained traces. Of 139 streams tested across thirty states, over one-quarter contained a little trimethoprim. One-fifth, although not the same fifth, contained erythromycin, lincomycin, and sulfamethoxazole. A more recent analysis of drinking water plants found ten different antimicrobial drugs were common enough. Meanwhile, sediments laid down in the 1970s in waterways across the country were full of the antimicrobial triclosan, a common ingredient in hand sanitizers and some toothpastes. The molecule, which, aside from affecting microbial communities directly, degrades into dioxin, a potentially toxic substance that binds to hormone receptors, accumulated in snails, frogs, fish, and earthworms wherever it was present.

To make a long and depressing story short, the world was full of human-made, microbiome-altering substances. The only boon was that concentrations tended to be minute. In the carrot experiment, for example, the skins contained 10 percent of the official daily maximum intake amounts. And in those drinking-water plants, the amounts were even more minuscule, several orders of magnitude below therapeutic levels. But the drugs nonetheless had far-reaching impact. Antibiotic-resistant bacteria showed up in wild seagulls and sharks, among other wildlife.

The concern over antibiotics administered to livestock had traditionally centered on antibiotics resistance, but given that some members of the human microflora were more sensitive to them than others, alteration of our internal microbial communities presented a new reason to worry. A scenario whereby a few crucial allies were knocked out, leaving behind an imbalanced, destabilized ecosystem, wasn't completely farfetched. Scientists demonstrated as much experimentally. Giving broad-spectrum antibiotics to young mice, for example, increased their vulnerability to asthma. Why? Those important clostridial species died off, and the peacekeeping T-regs melted away.

You didn't necessarily need a high dosage to effect change either. Ilseung Cho, a scientist in Martin Blaser's lab at NYU, found that antibiotics given to mice at one-tenth the therapeutic dose measurably altered the microbiota and increased body fat by 20 percent. Antibiotics given at concentrations seen in carrots fertilized by animal slurries, in other words, could change your vulnerability to, among other conditions, metabolic syndrome.

And however the derangement had occurred, comparative studies showed that the postmodern microflora was, indeed, quite different compared to the microbiota of people with premodern lifestyles.

AN ANCESTRAL MICROBIOTA,
VARIATIONS, AND GLOBALIZATION

In 2010, Carlotta De Filippo and Paolo Lionetti at the University of Florence published a telling comparative study. They contrasted the microbiota of rural, village-dwelling children in Burkina Faso with that of Florentine children, and found them to differ significantly. The African microbiota, a stand-in for the human microbiota at the beginning of agriculture, was more complex, and it harbored several species specialized in digesting plant fibers that were completely absent in Italy. It also had a relatively high ratio of bacteria from the bacteroidetes phylum, as well as fewer firmicutes.

Bacteroidetes, of course, were less abundant in guts suffering from IBD. And the scientists surmised that the Burkina Faso microbiota was especially adept at limiting inflammation. The relatively high ratio of firmicutes seen in Italy, meanwhile, was a pattern repeatedly associated with obesity. And there you had it: In Western Europe, the microbiota promoted weight gain without doing much to ward off inflammation, while in rural Africa, it helped digest tough plant material and limited inflammation.

Could genetics explain these differences? Tanya Yatsunenko and Jeffrey Gordon at Washington University in St. Louis addressed this question. They compared the microbiota of Amerindians living in the Venezuelan Amazon, rural Malawians, and North Americans. Although Amerindians and Africans were separated in time and place by 60,000 years, and a migration spanning four continents, including Asia and North America, functionally speaking, their microbiomes more resembled each other than the North American. The African and South American microbiotas specialized in degrading complex plant carbohydrates. The North American microbiota, by contrast, excelled at breaking down simple sugars. Evolutionarily speaking, it seemed, the North American microbiota was the outlier.

Other regional differences came to the fore. By looking at genes and their functions—a portrait of what the microbiome, in aggregate, can do, not just which microbes are present—scientists found that the Japanese microbiota had a unique ability to digest seaweed. *Nature* called it "the sushi factor."

Bacteria can exchange genes by sharing DNA-containing bubbles called plasmids. So it wasn't that seaweed-dwelling microbes had directly colonized Japanese guts; rather, at some point in the past, native bacteria had internalized instructions on how to break down seaweed from bacteria living on seaweed. "Cultural differences in diet may, in part, dictate what food our gut microbiota can digest," observed Justin Sonnenburg. In a roundabout way, the study also highlighted the importance of eating food that contained living microbes. Those bacteria that were, after all, already eating the food you consumed could impart expertise to your resident microbes they might otherwise lack.

More broadly, the find raised the specter of what one scientist called "the globalization of the microbiota": As human experience converged, and diets became more alike everywhere, human microbial communities would suffer the same loss of regional variation that New Urbanists bemoaned in the American landscape, a microbial "mallification" and "chain-store invasion" that would undoubtedly have consequences for human health.

A DIET THAT SIMPLIFIES ECOSYSTEMS

Among the greatest changes in the developed world during the twentieth century was a shift in how we ate. Before the Industrial Revolution, people mostly subsisted on unrefined grains, tubers, occasional vegetables, some dairy from grass-fed animals, and lean meat. In the early stages of the Industrial Revolution—1815 in England—per capita yearly consumption of refined sugars was around 6.8 kilograms (15 pounds). By the year 2000, it had ballooned to nearly 70 kilograms (154 pounds) per person. Whereas before the Industrial Revolution, most people got their calories from one or several nonrefined staple crops—wheat, barley, and potatoes in the West, for example—by 2000, Americans received three-quarters of our energy from refined sugars and grains, vegetable oils, and dairy products. This dietary regimen was, evolutionarily and biologically speaking, unprecedented.

During just the twentieth century, the consumption of vegetable oils, rich in unhealthy fats, quintupled. Concentrated feedlots replaced pasture for grazing, and marbled meat, which rarely existed in grass-fed animals, became standard. Perhaps more important than total fat consumed was the ratio of different fatty-acid types to one another. Omega-6 fatty acids, which incite inflammation, and saturated fats, which do the same, made up an ever-larger proportion of total fat consumed. Anti-inflammatory omega-3s, meanwhile, which come from leafy greens, nuts, and the animals that eat these greens, comprised an ever-declining share of dietary fat.

Scientists had long associated these dietary patterns with an increase in cardiovascular disease, type-2 diabetes, certain cancers, even dementia—the diseases of civilization. They assumed that the effect was direct, and part of it must be. But as they plumbed the microbiota, they found that diet changed our microbial community fairly dramatically, and that these changes themselves contributed to disease. The microbiota, it turned out, sat at the interface between diet and immune function. That is, our resident microbes translated lousy eating habits into obesity and cardiovascular disease.

In one study, scientists fed lean people a junk-food diet, and noted an immediate bloom in firmicutes, and a decline in bacteroidetes—a rough approximation of the Italian microbiota compared to the rural African. The altered community improved the men's ability to harvest calories and store them as fat. Obesity, of course, brought with it an increased risk of metabolic syndrome, a suite of symptoms including insulin resistance and low-grade inflammation that predisposed to further complications— cardiovascular disease and cancer, among them—down the line.

Many assumed that the inflammation was a symptom, not a cause, of the greater syndrome, but Patrice Cani and Nathalie Delzenne at the Catholic University of Louvain, Brussels, found that, actually, inflammation drove the syndrome, and that a shifted microbiota prompted the inflammation.

They fed junk food to rodents, and observed the same shift in the microbiota seen in people. In this case, the altered community increased gut permeability, and the tendency of microbes and their by-products to leak into the circulatory system. As more bacterial products passed through, a low-grade systemic inflammation took hold. The body's cells couldn't respond to hormonal signals while on a slow burn, and as a result, they became resistant to insulin. The mice overate. They grew obese. The microbiota shifted further. Inflammation intensified. The cycle reinforced itself. And, unable to keep up with the skyrocketing demand for insulin, the organ that produced the hormone, the pancreas, eventually failed. The mice now officially had diabetes. (This was the lifestyle-associated type-2 diabetes, not autoimmune type-1.)

This was a new model for metabolic syndrome: A fast-food-like diet altered the microbial community, which prompted systemic inflammation, which precipitated insulin resistance, which then drove overeating, and the cycle continued. As a point of comparison, germ-free mice could eat a high-fat diet without gaining weight, and without developing low-grade inflammation. Only germy mice developed metabolic syndrome when gorging on grease and sugar. Resident microbes were the key to metabolic syndrome.

Amazingly, Cani and Delzenne could disrupt the cascade of events by adding plant fibers to the junk-food diet, oligofructoses that only bifidobacteria could digest. Feeding those critical bacteria, making sure they maintained a strong presence in the gut ecosystem, kept the intestinal barrier intact, prevented systemic inflammation, and precluded the burgeoning diabetic storm.

A similar trick worked in humans. Paresh Dandona at the State University of New York, Buffalo, fed lean men a typical fast-food breakfast—egg muffin, sausage, and two hash browns—and watched as systemic inflammation increased almost immediately. Just adding orange juice to the fast-food meal, however—fodder for protective bacteria—blunted the systemic inflammation.

The lesson was that food could affect immune function by, among other things, altering the microbiota. And the modern, highly refined, calorie-rich diet failed to culture some of the most important residents of the microflora. Eating fast food was akin to dumping sewage onto a coral reef: weedy species bloomed, and specialists declined. An exquisitely evolved, complex ecosystem simplified. Scientists also likened its impact to that of the agricultural runoff washing down the Mississippi River. Every year, the nitrogen- and phosphorus-rich effluent produced a New Jersey–size dead zone in the Gulf of Mexico. Excessive nutrients paradoxically created a wasteland.

Viewing the microbiota through this ecological lens suggested a few easy interventions—not just probiotics, but prebiotics, food for native microbes that made you healthier. "I want to take care of my current microbiota," says Stanford University's Les Dethlefsen. "So I'm going to make sure I give them lots of plant fiber, food rich in plant matter, some of it uncooked, and try to avoid white flour and grains."

There was one final determinant of your microbial community: your genes.

CONFUSED CAUSES: MICROBES CONTROL GENES; GENES SELECT MICROBES

How and why does an otherwise happy-seeming microbiota shift to cause inflammatory bowel disease? That was the question Wendy Garrett at the Harvard School of Public Health sought to answer when she hobbled the immune system of mice by knocking out an important gene called T-bet, a mainline of communication between microbe and host. With that line severed, the modified mice couldn't tolerate their native microbes whatsoever. When colonized with bacteria, they responded with all-out inflammation and quickly developed colitis.

The offspring of these mutant mice also got colitis, which wasn't that surprising. They had the same genetic defect, after all. More remarkable, however, normal mice housed with these colitis-prone mice developed colitis as well. Although they didn't have the genetic defect, exposure to the "inflamed" microbiota was enough to irritate their guts. It was fairly easy to stop the colitis. When Garrett transplanted in T-regs, the inflammation resolved. Introducing bifidobacteria also quashed the inflammation.

Garrett's experiments underscored one of the oddest, and in many ways counterintuitive, lessons to emerge from research on the microbiota. Host genetics "chose" the microbiota, but the microbiota could also change how the host's genes were expressed. Host and microbiota operated on a bidirectional feedback loop.

Other experiments illustrated this dynamic. In one, scientists incapacitated an innate immune sensor called toll-like receptor 5 in mice. TLR5 detected the whiplike flagella bacteria use to propel themselves. Mice with this sensor disabled had a shifted microbiota, and the altered microbial community drove them to overeat. They became obese, developed low-grade systemic inflammation, and metabolic syndrome. And when the scientists transferred the microbiota from the mutant obese mice to regular mice—animals with perfectly functioning TLR5 genes—the recipients also developed metabolic disease.

Again, an immune defect had shaped a microbial community; disease ensued; and when transplanted, the "diseased" microbial community incited the same syndrome in animals without the primary genetic defect. Microbial communities could imprint diseases on their hosts, but genes could also contribute by recruiting an aberrant microbial community.

These studies prompted intriguing questions about the cascade of events leading to autoimmune disorders. How did our own genotype contribute? Did genetic variants increase the chances of disease by changing our immune system directly, or by doing something akin to Garrett's experiment—cultivating a deranged, inflammatory microbiota?

Daniel Frank, the scientist who noted the peculiar absence of those peacekeeping bacteria in people with inflammatory bowel disease, looked into this question. Geneticists had identified variants of a gene called NOD2, which encoded a microbial sensor, as predisposing to IBD. Frank found that people with these gene variants harbored a dramatically shifted community of microbes regardless of whether they had overt disease or not. Not surprisingly, the assembled communities tended to lack those anti-inflammatory microbes.

Yolanda Sanz observed something similar in celiac disease. Infants with celiac-associated gene variants had a shifted microbiota, and these changes were apparent long before the actual disease showed up. If the microbial community was a garden, these genotypes tended to cultivate a disorderly, pest-infested, weed-strewn plot. Variations in our microbial sensors moved us closer to inflammatory disease by accruing an inflammatory microflora.

And yet, while genes might select for certain microbes, how genes were expressed was clearly modifiable by, among other things, microbial exposure. Epigenetic modification could depend, as we saw in chapter 7, on the mother's immune state as the fetus was developing, and on the microbes she encountered while pregnant. That brings us to the most confounding aspect of the host-microbe interaction: If it wasn't a unidirectional chain of causality, how did the relationship ever stabilize?

In ecosystem science, keystone species are defined as players in a community that, because of their outsize impact, greatly determine what the ecosystem looks like. Elephants on the African plains, for example, shape the grassland with their trampling of encroaching bushes, maintaining pasture for all grazing animals. In North America, beaver dams create marshes and regulate stream flow—impacts that have far-ranging consequences, from providing forage to wetland-loving moose to keeping trout pools full during dry season.

Clearly the host is one keystone species in the superorganism, the flask that houses the microbial community. Within the gut, other keystone species may play a similar stabilizing role. Mazmanian's and Littman's work suggested that critical microbes could set the tone for immune functioning. And there were other obvious candidates, such as *Helicobacter pylori,* which uniquely colonizes the acidic stomach far upstream from the colon, or even the comparatively large parasitic worms. Indeed, Vincent Young at the University of Michigan, Ann Arbor, found that infection with the mouse hookworm increased the relative abundance of anti-inflammatory bacteria nearby. Like elephants on the savanna, helminths changed the surrounding ecosystem. Presumably, they either secreted substances that cultivated these microbes (too much inflammation could harm the parasite, after all) or they altered the local immune response directly, and that alteration selected for friendly bacteria. In either case, Young's experiment added a new layer of complexity to Joel Weinstock's original hypothesis (he was coauthor): The helminth's power to sway the host immune response may stem in part from its recruitment and cultivation of certain microbes.

In the bigger picture, that microbes played such an important role in our health was heartening. Imagine the conditions we could treat, from obesity and depression, to cancer and allergic disease, not with drugs directed at our genetic self, but by fiddling with the microbiota—not by modifying our genes, but those of our meta-genome.

ECOSYSTEM RESTORATION ALL AT ONCE:
FECAL TRANSPLANTS

In 2008, Alexander Khoruts, a gastroenterologist at the University of Minnesota, Minneapolis, admitted a starvation-thin sixty-one-year-old woman. She had suffered from chronic, ongoing diarrhea for eight months. The condition, caused by the bacterium *Clostridium difficile*, had appeared after treatment with antibiotics sometime back. She'd taken more antibiotics to eradicate the *C. difficile*, but to no avail. She'd improve during treatment, only to relapse after finishing the latest course.

Now the woman was wheelchair-bound. She suffered frequent and uncontrollable bowel movements that forced her to wear diapers. She'd lost nearly 60 pounds over the course of the ordeal. If the infection wasn't eradicated soon, Khoruts knew, she would likely die.

Khoruts tried several different antibiotics, but the *C. difficile* strain was resistant, a growing problem, and it always bounced back after temporary improvement. He tried probiotics. That didn't work either. At wits' end, he turned to a therapy that, at least anecdotally, had shown great promise: a fecal transplant.

In theory, this meant replacing an obviously defective microbial community with a balanced one. In practice, it required taking a sample of feces from the woman's husband of forty-four years, mixing it with a little saline solution, and administering it deep into her colon.

Two days after the transplant, the woman had a solid bowel movement, the first in months. Khoruts continued monitoring her. He sampled her microflora regularly and noted that bacteroides bacteria, largely absent before, made a triumphant return. About two weeks later, her microbiota resembled her husband's. A month later, it had shifted, becoming more personalized, but it retained her husband's signature. Six months later, her diarrhea remained vanquished, and her health was restored. "I didn't expect it to work. The project blew me away," Janet Jansson, who ran DNA analyses on the woman's microbiota, told the *New York Times*.

The phenomenon of *C. difficile* and diarrhea illustrates one peril of disturbing the microbiota. In the U.S., deaths from *C. difficile* quadrupled between 1999 and 2004. Of the 500,000 cases of *C. difficile* yearly, between 15,000 and 20,000 die. Six percent of people diagnosed succumb within three months. In people over eighty, the mortality rate is more than double that—13.5 percent.

The bacterium is naturally present in more than half of healthy newborns and 2 percent of healthy adults. It generally causes disease only after antibiotic treatment and time spent in hospitals, where patients unwittingly pick it up.

No one is sure why the bacterium has become more virulent in recent decades, but a greater depletion of the microbiota may be at fault. One prospective study found that patients admitted to a Montreal hospital who went on to develop *C. difficile*–associated diarrhea had an altered microbiota to begin with—more firmicutes and fewer bacteroides. If you remember, the comparison of Italian and African microbiotas suggested that the Westernized microbiota had, in fact, shifted in precisely this direction—toward a configuration vulnerable to invasion by *C. difficile*.

Khoruts's approach of restoring the entire ecosystem has emerged as one of the most promising treatments for *C. difficile*–associated diarrhea. A metanalysis, which admittedly included only small, uncontrolled studies, concluded that fecal transplants fixed the problem in nine out of ten cases. The transplants had no adverse side effects.

Given the microbiota's importance in immune and metabolic functioning, scientists have begun asking if "microbial engraftments," as Khoruts calls them, might address these diseases as well. Thomas Borody, a Polish-born Australian scientist who pioneered "fecotherapy" in the 1990s, has successfully treated people with IBD. Enemas containing "human probiotic infusions"—feces from healthy donors—sent six ulcerative colitis patients into remission. And they remained completely disease-free thirteen years later. Borody is proceeding with a trial on Parkinson's disease.

More recently, Dutch scientists tried stool transplants for metabolic syndrome. In a double-blind, placebo-controlled study, they transferred stool from lean individuals into eighteen obese males who'd recently been diagnosed with the disease. (Controls received their own feces.) The recipients didn't lose weight, but they did see improvement: blood levels of unhealthy fats declined, and insulin resistance decreased.

IS FECOTHERAPY
THE WAVE OF THE FUTURE?

One day we may look back on these experiments and deem them crude. By then we'll better understand the intricate interplay between genotype and microbiome. We'll be able to shape the microbiota with narrow-spectrum antibiotics and prebiotics—to add and subtract species, to balance the community. We'll likely have stool-like microbial mixtures just for this purpose, a product already in development. It's not going to be simple, however. One study that looked at human microbial communities in four countries found that we have fecotypes much the same way we have blood types—at least three of them. We can't throw any old microbial community at any old gut, in other words, and expect miracles. The microbes may have to match the individual.

Nonetheless, these transplant experiments, and the many more likely to come, will undoubtedly be lauded as the first attempts, however effective, at treating human disease by tweaking our resident microbes. There's much more to know: What's the best way to dislodge a dysfunctional microbiota and install a healthy one? And as it becomes more apparent that our microbes are species-specific—that, for example, mouse, chicken, and human lactobacilli are not the same—how are we going to preserve the human microbial heritage from extinction?

One can already discern the beginnings of a conservation movement dedicated to saving the human microflora—to "preserving this treasure of microbial diversity from ancient rural communities worldwide," as Lionetti and De Filippo, the scientists who compared children in Burkina Faso and Italy, put it. Given the worries over societywide derangement, one can even envision a gold rush of feco-prospecters—scientists hurrying to places where people continue living the "old" way, such as the Amazon or rural Africa—seeking to gather, catalog, and somehow protect the ancestral microbiota from disappearing.

In the meantime, we've got to wonder about pet ownership and the sibling, day-care, and farm effects—if maybe they're mediated by microbial communities in the gut. Pets, which protect against allergies, also increase the microbial diversity in the home. Large families and crowded conditions generally facilitate microbial transmission. And as we saw in chapter 6, younger siblings, who tend to have fewer allergies, have a more mature and diverse microbiota at an earlier age.

Scientists also know that the diversity of the human microbiota represents just a fraction of that observed in grazing animals. Functionally speaking, gut microbial communities grow simpler the more carnivorous the species, so maybe close contact with animals allows farmers to amass a more complex, stable, health-promoting microbial ecosystem, what ecologists of the external world call an apex community. Contact with animal fodder incidentally ensures a steady flow of prebiotics. Likewise, some suspect that raw milk functions as a prebiotic, cultivating healthy microbes in the gut. The well-developed farming microbiota imprints favorably on the developing fetus. And when farming children are born, they inherit this robust microbiota.

This is all speculation, of course, based on converging lines of scientific investigation. So far, few studies have directly looked at the microbiota of farmers. Intriguingly, one that did found that lactobacilli dominated farmers' internal microbial communities, much as they did that group of pigs raised outdoors. Generally, more microbial diversity is considered better. But in this study, farmers, the least-allergic of all subpopulations in developed countries—and among the least vulnerable to inflammatory bowel disease—had a comparatively simple microbiota. Perhaps, the scientists theorized, having the right "keystone" species—in this case a preponderance of lactobacilli—was more important even than diversity. And by virtue of their close contact with animals, farmers retained access to those keystone microbes.

Clearly, not all microbes in the Bavarian cowshed can colonize the human gut, but experiments with probiotics suggest that microbes don't have to take up residence to change the community. Just passing through can shake things up. As an aside, it's worth noting that some peoples long ago understood that manure had medicinal and nutritive value. The Bedouin, a nomadic people of the Middle East, occasionally ate camel dung to fix stomach trouble. And the Inuit of the Arctic often consumed lemming droppings and half-digested plant matter scooped from caribou insides.

Meanwhile, certain ingrained human behavior appears designed to enhance microbial exposures. As I write this, I marvel at my eight-month-old daughter's desire to jam everything into her mouth. It's difficult to imagine that this instinct, with such obvious dangers, such as acquiring a deadly pathogen, doesn't also provide some benefit. Early exposure to certain bad bugs, like poliovirus or hepatitis A, is clearly preferable to later exposure, which alone might account for the oral fixation. And early acquisition of *Helicobacter pylori* may also be preferable to later. But other potential arrivals, such as dysentery, continue to be major killers of babies around the world. What benefits could possibly make death by dysentery a risk worth running?

What about an early acquisition of a robust microbial ecosystem? The microbes one acquires while young set the tone for the community to come, and for immune functioning. Could the requirement of obtaining a healthy microbial community drive some of my daughter's oral fixation?

We've seen how parasite avoidance influences animal behavior—grooming, migrating, even group size. The evolutionary biologist Michael Lombardo argues that the necessity of acquiring the right symbiotic microbes may account for some animals' social behavior. While parasites drive us apart, purely mutualistic microbes may bring us together. Termites lick their parents' anuses to acquire their microbes. Without them, they can't digest woody matter and they'll starve. Juvenile iguanas eat their parents' feces, presumably for similar reasons. Naked mole rats have a special cry to request parental poop, which they also consume. Birds of all sorts regurgitate food for their young. For some of these species—such as the termite—acquiring the right microbes is a matter of life and death. Are people similarly dependent on certain hand-me-down microbes? Maybe primate kissing even evolved, Lombardo speculates, as a way to transfer the microbiota.

And that causes some worry. In the not-so-distant past, the toys, fruit, armrests, blankets, and other objects that my daughter eagerly devours would have been coated with a very particular biofilm—from the extended family, from the animals we kept, from dirt outside, and so on. They're still full of microbes, but judging by the studies comparing farming homes with nonfarming, and Russian with Finnish Karelia, they contain orders of magnitude fewer, far less diversity, and probably the wrong type entirely. My daughter's accumulating microbiota will only be as good as the one she builds from, and that one has undoubtedly diverged from the evolutionary norm. So as I watch her suck armrests and chew books, my worry is not how filthy my apartment is, and not even whether it's too clean, but whether it's dirty in just the right way.

Follow-up studies on those farm pigs suggest my fretting isn't baseless. This time, the scientists allowed the newborn piglets to remain with their mothers outside for two days and *then* put them in isolation chambers. Although these piglets had acquired all the necessary microbes from their mothers—a full starter culture—without continuous exposure, their microbiota still failed to mature. It never achieved the apex formation seen in outdoor-raised pigs. Moreover, absent those critical microbes that expanded peacekeeping cells, the indoor immune system veered toward inflammation. Upon weaning, these piglets responded to food proteins with an unwarranted ferocity—rather like twenty-first-century children, who are so bafflingly allergic to peanuts and other foods. By extrapolation, we not only require access to the correct microbes—in this case, from others of our kind, soil, and the greater farm environment, we also need that access continually during infancy and childhood.

Given that the postmodern immune system has lost some ability to quash inflammation, the abiding question is how much of this atrophy stems from the postmodern microbiota? One view of the microbial organ's role in the greater organism is as a baseline inflammatory reference for the immune system. Once "good" inflammation has fought off pathogens or repaired tissue, the basal stimulation of resident microbes marks how far to turn back the dial. The Westernized microbiota, sparsely populated to begin with, depleted by antibiotics, and further altered by a sugary, calorie-rich diet, may have moved that reference point up a few notches. As a result, we're more prone to immune malfunction.

There is one more important constituent of the human superorganism, of course, a final player in our immune function: the human virome.

CHAPTER 10

Multiple Sclerosis:
Worms That Never Come, a Tardy Virus,
and a Brain That Degenerates

> If the tree of life metaphor is to be useful in future, we have to remember that viruses have been and still are essential agents within the roots and stem of the tree of life.
> —Luis P. Villarreal, molecular biologist

In the early 2000s, when Joel Weinstock was testing his pig whipworm in Iowa, fate, in the form of economic collapse, handed an Argentine neurologist his own helminth experiment.

We like to think of sanitary progress as unidirectional. You get sewers, running water, and garbage collection, and you've arrived at modernity. There's no going back. But regression can occur. And that's what happened in Buenos Aires for a few tumultuous years around the turn of the millennium.

The Argentine peso went into free fall. Inflation shot up. Unemployment rocketed to 24 percent. Protesters took to the streets. Scores died in riots. Two presidents resigned in quick succession. Prohibited from withdrawing money from banks—and unsure what the collapsing currency was worth anyway—many Argentines resorted to a barter economy. More than half of the population slid below the poverty line. An estimated 30,000 to 40,000 people began collecting cardboard and lugging it to recyclers for a living. They were called *cartoneros*.

For the neurologist Jorge Correale, at the Institute for Neurological Research Dr. Raúl Carrea in Buenos Aires, the turmoil translated into new infections. The clinic provided free care to the city's economically disadvantaged. Some multiple sclerosis patients from poorer parts of town started presenting with a condition called eosinophilia, elevated counts of white blood cells called eosinophils. Eosinophilia usually signifies either allergies or a parasitic infection. None of the patients had a history of allergies, however, so when Correale ordered a stool test, he wasn't surprised to find they'd acquired a range of helminths: a tiny tapeworm native to rodents called *Hymenolepis nana,* as well as human whipworm, giant roundworm, threadworm, and pinworm.

Sanitary conditions in some neighborhoods had clearly deteriorated, a distressing development. Yet in this backslide, Correale, who was familiar with Weinstock's work, saw an opportunity. How might these naturally acquired worm infections affect these patients' autoimmune disease? There was only one way to find out. He sat the patients down and explained the hypothesis: worm infections inhibited allergic and autoimmune diseases in animals; some thought they did the same in people. He mentioned Weinstock's pig whipworm studies. And then he made a proposal: Under close monitoring, the patients might want to keep their worms. Maybe the parasites would help with their multiple sclerosis.

Correale would document their progress with magnetic resonance imaging, or MRI. And whenever they chose, they could easily flush out the parasites with deworming medicine. Twelve patients signed up for the experiment. And so began an investigation that may prove seminal for the treatment of a terrifying and intransigent autoimmune disease—the gradual, inexorable stripping of myelin from neurons, and the creeping, sometimes painful paralysis of multiple sclerosis.

WHEN DID MULTIPLE SCLEROSIS BEGIN?

More than six centuries ago, a fifteen-year-old Dutch girl fell while ice skating, and broke a rib. Her name was Lidwina, and that fall was the first sign of a chronic lifelong illness. She went on to suffer prolonged bouts of dizziness, limb weakness, and blurred vision. By age nineteen, she could no longer use her legs. According to some accounts, she occasionally shed bits of herself—skin, blood, even a length of intestine. Eventually, she became almost completely paralyzed. When she died at age fifty-three, she could move only her left hand.

In the late nineteenth century—five hundred years after Lidwina's fall—the church canonized her Saint Lidwina of Schiedam, the patron saint of skaters. When she was alive, a priest had suggested that the affliction came from God, and that her suffering had a divine purpose. Neurologists remember her for other reasons, however. Judging from the descriptions of her progressive disease, Lidwina's was the first recorded case of multiple sclerosis.

So multiple sclerosis is not, apparently, new. And yet, like hay fever and inflammatory bowel disease, the incidence of multiple sclerosis seems to have increased during the nineteenth century. In 1822, Augustus d'Esté, an illegitimate grandson of King George III, was struck by blindness. He was twenty-eight. His vision problems resolved, but leg weakness, night spasms, and incontinence plagued him for the rest of his life. He died at age fifty-four. From his own meticulously kept diary, scientists proclaim d'Esté the first case study of MS.

The first description of what, biologically speaking, occurred in the disease came in 1838. That's when the Scottish illustrator Robert Carswell sketched the autopsied brain and brain stem of a deceased woman who'd died from paralysis. Brown-colored lesions and scars covered the specimen. A few years later, the French anatomist Jean Cruveilhier noted similar lesions on the brains and spinal cords of two women who'd suffered from limb weakness, blurred vision, and paralysis. Cruveilhier and Carswell, who had seen their fair share of diseased brains, especially syphilitic ones, thought these lesions and scars "mysterious" and "peculiar."

The French neurologist Jean-Martin Charcot usually gets the credit for defining the symptom set of MS. In 1868, he described *"sclérose en plaques"*: inflammation without apparent cause that stripped fatty myelin off neurons. The degeneration led to slurred speech, blurred vision, and coordination problems. Thankfully, it was rare: He observed fewer than forty cases during his entire career.

But in the latter nineteenth century, as snazzy resorts began to cater to the newly hay-fever-smitten upper classes, and the first wave of inflammatory bowel disease baffled British physicians, more and more MS cases popped up—in England, Canada, and the U.S. In 1921, a decade after a similar meeting convened in London to discuss "non-infectious dysentery"—inflammatory bowel disease—the Association for Research in Nervous and Mental Diseases called a meeting in New York City to consider this progressive disease of the central nervous system that had become worrisomely more common.

Whatever its actual prevalence before the Industrial Revolution, during the twentieth century, the incidence of MS definitely increased in the industrialized world. And from early on, the great variation in its prevalence mystified scientists. Broadly speaking, it grew more common the farther one moved from the equator. Even in a single country, the U.S., northerners got it more than southerners, at least in the early twentieth century. Given the latitudinal gradient, perhaps vitamin D played a role. But peoples from the highest latitudes—the Sami of northern Scandinavia, the Inuit of the frozen reaches of North America and Greenland, even the Maori of New Zealand—bucked this rule by almost never succumbing to MS. And when the Iron Curtain drew back, scientists saw that Eastern Europeans developed MS less often than Western, although they resided at the same latitudes.

Maybe ethnicity explained the odd prevalence patterns. A thousand years before, Vikings either came from or settled those countries with the highest incidence, such as Norway and Scotland. And everywhere Scots emigrated, including Northern Ireland, MS rates were relatively high. Perhaps a Viking gene was at fault. That would explain the Inuit and Sami invulnerability to the disease.

As the twentieth century wore on, this, too, seemed less likely. South Asian immigrants to the U.K. had a lower risk of MS than locals, but their decidedly non-Viking children had the same risk as native Britons. What's more, identical-twin studies showed an extremely low rate of concordance—31 percent in one study, and 24 percent in another. This was, it seemed, a disease determined by environment.

Beginning in the mid-twentieth century, scientists began linking multiple sclerosis to hygiene. In Israel, the risk of MS was directly proportional to one's sanitary amenities at age ten. Having had piped-in water, a flush toilet, and fewer than two people per room increased one's vulnerability to MS later in life. Using an outhouse, drinking rainwater, and having more than two people per room in childhood, on the other hand, protected against MS in adulthood. The scientist Uri Leibowitz dubbed this the "sanitation hypothesis" two decades before David Strachan's "hygiene hypothesis" for allergic disease.

Others linked socioeconomic status and infectious burden to the risk of MS. In South Africa, native-born Afrikaners, who were originally of Dutch ancestry, had one-tenth as much as recent immigrants from northern Europe and the U.K. Native-born English-speakers, on the other hand, landed somewhere between. They had about three and a half times the risk compared with Afrikaans-speakers, but a lesser risk compared with recently arrived European immigrants.

The native South African Bantu, on the other hand, seemed invulnerable. "No single case of MS has yet been found among South Africa's 15 million Bantu . . . though there are good Bantu hospitals with well-trained neurologists in the large cities," wrote Geoffrey Dean, a physician who looked into the matter, in 1971. "The disease does occur among the Coloured"—people of mixed ancestry—"and Asian South Africans, but it is very uncommon," he noted.

The pattern mimicked that of paralytic poliomyelitis. As was the case in other places where living conditions varied greatly according to class and ethnicity, native black South Africans, who lived in more crowded conditions and were exposed to more orofecal infections at younger ages, tended not to suffer from paralytic poliomyelitis as often as the white upper classes.

The disparity had little to do with their avoiding the virus. To the contrary, in fact: Blacks were exposed to it more often, and earlier. But in this case, exposure in infancy translated to relatively benign symptoms—a mild fever, and maybe no illness at all. Infants infected with the poliovirus largely avoided the paralysis that could afflict children, teens, and adults.

So it was notable that, in South Africa, all whites weren't equally vulnerable to paralytic polio. It struck native whites, who also got less MS, about half as often as European immigrants. Dean applied this observation to MS. He reasoned that the Afrikaner lifestyle worked to their advantage. They farmed, had large families, and lived with Bantu housekeepers and nannies. Compared with their English-speaking counterparts, Afrikaners presumably had more and earlier exposure to a range of infections, the poliovirus included. They probably had earlier contact with the MS virus, assuming one existed, as well.

PARASITES AND MULTIPLE SCLEROSIS

The virus aside, if these patterns sound familiar, it's because of an uncanny resemblance to Joel Weinstock's studies on IBD and worm prevalence during the twentieth century. One was the inverse of the other. But whereas New York City served as the epicenter of IBD in the U.S., for MS, the nucleus was Olmsted County, Minnesota. Records stretching back a century showed more MS there before anywhere else in the country. In general, as with IBD, in the early twentieth century, northerners got MS more often than southerners, and African Americans everywhere appeared immune. But these disparities began to narrow during the second half of the twentieth century, first between white southerners and northerners, and then between whites and African Americans. Irrespective of geographic location, urban birth and belonging to society's upper echelons consistently increased the risk of MS.

Canadian scientists eventually linked the two inflammatory conditions, one of the gut, the other of the central nervous system. In a district-by-district analysis of Winnipeg, the capital of Manitoba, they found that having MS greatly increased the risk of also having IBD. High socioeconomic standing elevated the risk of developing both diseases. Having orofecal infections, on the other hand, lowered one's chances. Although they lived within blocks of each other, Native American peoples (in Canada, called First Nations) were less prone to both IBD and MS than whites. Perhaps not coincidentally, the great majority of First Nations people also harbored *Helicobacter pylori*.

Did genetics account for the disparity? Were First Nations people immune? It seemed not. Even within the First Nations community, the risk varied greatly according to where one lived. Those who remained on the reservations had one-tenth the risk of developing Crohn's disease compared with their city-living counterparts. Again, the strongest determinants of these diseases seemed environmental, and linked to hygiene.

As an exercise, in the mid-2000s, John Fleming, a neurologist at the University of Wisconsin, Madison, correlated the prevalence of a single helminth species—*Trichuris trichiura,* the human whipworm—with the incidence of MS around the world. Whenever whipworm prevalence fell below about 10 percent, he found, MS shot up. There was no intermediate slightly wormy phase with just a mildly elevated risk of MS. Once you passed some threshold of societal hygiene, the incidence of MS increased abruptly.

In fact, worm prevalence, Fleming observed, was a far better predictor of MS risk than latitude. MS rates in Jewish Jerusalem, for example, were more than double those of the Arab quarter in the same city—the inverse of worm prevalence in the two adjacent communities. And in high-latitude East Asian countries, where MS was still comparatively rare, worm infections were common until relatively late in the twentieth century. Five of every six South Korean schoolchildren, for example, had parasites as recently as the 1970s. On the other hand, Canada, with near-zero worm prevalence for much of the twentieth century, had among the highest MS rates in the world.

Worms that prevented MS were one thing; parasites that treated already established MS were quite another. Fleming's attention turned to Jorge Correale's ongoing study in Argentina. Could worms really halt multiple sclerosis?

THE ARGENTINE MIRACLE

Although the overall trajectory of multiple sclerosis tends inexorably downward, the symptoms of MS can come and go from day to day. Dizziness and slurred speech can appear and then disappear. Numbness can set in and then dissipate. To know for certain that his patients' worm infections were helping, Jorge Correale knew he'd need a long period of observation. And so for nearly five years, he watched. He withdrew blood periodically, measuring what kinds of white blood cells were present, and how they responded to immune challenge. Every six months, he peered into his patients' brains with an MRI machine. And as the years wore on, what he saw happening—or not happening—was remarkable.

On an MRI, the healthy brain has folds, meandering creases, and sharp delineations. The lesions characteristic of MS appear as dull holes eating at this complexity, areas where neurons aren't firing anymore. As the disease progresses, new lesions appear and old ones grow larger.

Compared with uninfected MS controls, Correale's worm-bearing patients saw an almost complete halt to the progression of their disease. Existing lesions stopped expanding. New lesions stopped appearing. All in all, compared with nonparasitized controls, the disease slowed by more than 90 percent in the twelve wormy patients.

In 2007, Correale published his results in *Annals of Neurology*. John Fleming, who wasn't involved, calls the outcome "very important." Granted, the study was small and observational, and the way the patients acquired infections was, from a scientific perspective, lamentably uncontrolled. Nonetheless, says Fleming, "we've never had anything like this in MS, nothing even close to it." The improvement was greater than that seen with any drugs currently on the market, he added. The study prompted Fleming to launch his own trial at the University of Wisconsin using Joel Weinstock's porcine whipworms.

Civilians caught wind of Correale's study as well.

TAKING TREATMENT INTO YOUR OWN HANDS

One morning in October 2004, Dan M. awoke to find the left side of his body numb. "I'm twenty-eight, and I'm having a stroke," he thought. "Ridiculous."

But it wasn't a stroke. After a month of tests, neurologists diagnosed him with multiple sclerosis. Dan, who lived in Santa Barbara at the time, was horrified. "I was picturing myself in a wheelchair in a year," he says.

He brought the flare under control with a course of immune-suppressing steroids. As a long-term strategy, he began injections every other day of a drug called Rebif. No one understood how it helped in MS, but trials showed that it did. And Dan's numbness receded.

But when he got an MRI six months later, new lesions had appeared. He was controlling his symptoms, but the disease was progressing anyway. He started another drug called Copaxone, a molecule that mimicked myelin proteins. How Copaxone slowed disease progression was also a mystery—perhaps by providing a decoy for autoimmune myelin-seeking cells, or by transforming these self-directed cells into regulatory T cells—but in trials, it helped.

And for a few years, everything went smoothly: no new flares; lesions mostly stopped expanding. But at age thirty-one, Dan relapsed. This time, symptoms were more severe. The left side of his body went numb as it had before. But he also lost motor control. He couldn't make precise movements with his left hand. And he began to have difficulty walking.

He fell back on the usual protocol: powerful immune-suppressing steroids to control the flare in the short term. He stayed with his Copaxone. But the left side of his body never fully recovered.

"That scared me way more than the first time," he says.

He began trolling the Internet for alternative long-term therapies. He read about avoiding gluten, various allergens, not eating dairy. And then he came across something called "helminthic therapy": a fellow named Jasper Lawrence was selling hookworm to treat autoimmune and allergic disease. (More on Lawrence in a bit.)

Dan had studied ecology, and the approach appealed to him. He saw it as restoring an ecosystem—as reestablishing an ancient relationship. The Correale paper, published in 2007, swayed him to try hookworm. But when he mentioned the idea to his neurologists, they were dismissive. Instead, they urged him to consider a new MS drug called Tysabri. Consisting of intravenously delivered antibodies, the drug would interfere with the self-directed attack. But Dan demurred. Reports were emerging of fatal brain infections associated with the drug. (The FDA has since included a warning on the label about the infection, called progressive multifocal leukoencephalopathy.)

And so in 2009, with his wife's support, but without his neurologists' knowledge, Dan ordered a hookworm self-infection kit from Jasper Lawrence. A short time later, he pressed a bandage ostensibly containing thirty-five invisible hookworm larvae against his arm. Within a few minutes, he felt the itch.

And that was it.

Then, within weeks, his seasonal allergies disappeared. "I could, like, snort pollen, and nothing would happen," he says. "I haven't had allergies since." He had a few urgent bowel movements in the same period, but mostly no side effects.

And he hasn't had an MS flare in the years since.

In May 2010, he had an MRI, the first in three years. Afterward, his neurologist sent him an excited e-mail. One lesion had shrunk. Another had disappeared entirely. The others remained stable. "Things are actually improved and no new lesions are seen!" the doctor wrote. "This is great news!"

In the meantime, Dan, who now lives in Portland, Oregon, upped his worm population—sixty-five additional hookworms and a dose of one thousand human whipworm eggs. Throughout his ordeal, Dan has stayed on Copaxone. He's fully aware that the drug may be helping him, not the worms—but the drug didn't, he notes, prevent flares before. Or maybe he's just had dumb luck, a longer-than-usual ebb in the inexplicable coming and going of any autoimmune disease. After the hookworm, however—but not Copaxone—he recovered almost full mobility in his left arm and leg.

And there are other compelling reasons to think that Dan's parasite infection helped his MS.

AMONG THE CELLS:
WHAT REMISSION LOOKS LIKE

In Buenos Aires, Jorge Correale kept documenting the parasites' effect on his MS patients. The tenor of their immune system shifted. White blood cells that sought out myelin, the fatty insulation on neurons, were a hallmark of MS, but in his worm-infected patients these self-reactive cells changed. When they encountered myelin, they now spewed anti-inflammatory rather than pro-inflammatory molecules. The cells tolerated rather than assaulted the protein.

More generally, the quantity of circulating regulatory T cells increased—as did regulatory B cells, a newly described member of the immune repertoire. Together, this enhanced regulatory musculature prevented the inflammation that drove MS. Correale was witnessing the squelching of an autoimmune disease in real time. These observations also answered two larger questions. First, during the course of our evolution when parasites were ever-present, was the self-directed attack that drove MS even possible? Apparently not. Second, Weinstock's work had demonstrated that worms could influence their immediate environment, the gut. But what about sites farther away, like the brain? "For us, it's the most exciting thing," Correale told me. "You can see that there's a possibility to have systemic protection even though the parasites are in the gut."

Worms were increasingly seeming like only one protagonist in the MS story, however, albeit maybe the most important one. In the decade that saw Weinstock's first experiments, and Correale's five-year vigil of parasitized patients, scientists had zeroed in on a putative trigger of MS, a virus they suspected provoked the slow-motion meltdown of the brain.

THE MYELIN-STRIPPING VIRAL PARASITE

It was 1941, the early years of World War II, when British troops made landfall on the Faroe Islands, a rain-sopped, windswept archipelago in the North Atlantic between Iceland and Norway. A few years later, new cases of MS began popping up among the native Faroese. Several more waves of MS followed during the British occupation.

Before the British arrival, MS was absent on the Danish-run islands. At least, that was the conclusion of scientists investigating this apparent case of infectious MS four decades later. The British had brought some virus, they argued, to which the natives, living isolated in the sea, had no immunity. Again, paralytic poliomyelitis served as a model: if you got the poliovirus early, you might not even notice; but if you got it late, all hell broke loose. The Brits brought the virus, which was until then absent on the islands, and the adult Faroese immune system, which had never encountered it before, spiraled into degenerative disease.

Other scientists disputed this conclusion. Were the observations of MS accurate? Who was excluded, who included? Had previous MS cases sought treatment in mainland Denmark, removing themselves from the study population? To this day, debate swirls around "infectious MS" on the Faroe Islands. But the idea that a viral infection can instigate multiple sclerosis has gained considerable currency. One virus has emerged to the fore.

Epstein-Barr (EBV for short) is named for its discoverers Michael Anthony Epstein and Yvonne Barr. Nearly all adults everywhere have it. So you see the immediate problem linking it to autoimmune disease: How and why would a virus that's nearly universal cause a degenerative disease in just a few people? And why would those few people reliably hail from the most affluent populations in the most northerly reaches of the world's most developed countries? The answer: It's all about timing.

EBV was spread in saliva, and, like poliovirus, if you contracted it early in life, you might not even notice. But if your first encounter came during adolescence or adulthood, you were at greater risk for what American teens referred to as the "kissing disease," and what physicians called "infectious mononucleosis." Symptoms might include long-term fatigue, slight fever, and swollen lymph nodes. Broadly speaking, the epidemiology of mono matched that of MS. In the developing world, where children generally acquired EBV in infancy—by inadvertently sharing saliva with each other, older siblings, and directly with parents—both mono and MS were rare. In the developed world, by contrast, half of all children still didn't have EBV by age ten.

In the mid-1990s, Danish scientists sought 6,800 people who'd contracted mono as teens in the 1960s and '70s, searching for their names in the national MS database. Those who'd been diagnosed with mono as teens had nearly triple the risk of multiple sclerosis as adults compared with those who hadn't. Others soon corroborated the link. In the U.K., having had mononucleosis quadrupled one's chances of developing MS a decade later. In the U.S., among a cohort of nurses that scientists had followed since the 1970s, contracting mono doubled one's risk of MS. Most notably, compared with those few lucky individuals who somehow escaped EBV infection at all, mono increased the risk of MS twentyfold. Multiple sclerosis, it seemed, didn't occur without prior infection by this virus.

The strong correlation was worrisome for a number of reasons. First, no vaccine for EBV existed. Second, if late acquisition of the virus increased the risk of MS, then, as families grew smaller, the notion of personal space larger, and overall hygiene continued to improve, the pool of people vulnerable to MS would keep expanding. More than one-third of Danish teens, for example, still hadn't acquired the virus by puberty. At the very least, late EBV infection partly explained those patterns of MS incidence along ethnic and socioeconomic lines. In the U.K., for example, the incidence of mono followed class (and so did that of MS): northern Britons got more mono than southern, richer ones more than poorer, and native-born more than immigrant. The apparent conundrum: the liberal idea of a more egalitarian society where everyone enjoyed improved living conditions would also, terribly, create more infectious mononucleosis—and more multiple sclerosis.

HOW DOES A QUIET VIRUS SPARK
A MAJOR MELTDOWN?

One old theory about autoimmune disease in general held that invading viruses and bacteria disguised themselves as human tissue in order to fool immune surveillance. But, never to be outdone, your immune system got a fix on them . . . and then, like a dog chasing its own tail, accidentally pursued some aspect of the self. That was one explanation for the virus's role in MS, but when scientists tried to isolate viral proteins from lesions, they had inconsistent results. Some claimed to have isolated virus particles; others denied they were there.

A more subtle theory held that the virus imbalanced the immune system. After all, self-directed white blood cells—autoimmune cells—were, as we've seen, natural and healthy. They helped catch rogue cancer cells, repair tissue, and defend the organism. Regulatory T cell handlers usually held them in check until needed. Maybe the virus upset this balance. Indeed, EBV was implicated in other autoimmune diseases, such as rheumatoid arthritis, a painful condition of the joints, and the autoimmune meltdown known as systemic lupus erythematosus. If the virus really knocked the immune system off balance, you'd expect other autoimmune diseases like this.

Moreover, EBV's interface with the immune system was provocative—literally. It belonged to the herpes family. In Greek, *herpes* means "creeping." And like its relatives, the chicken-pox-causing varicella zoster, or the cold-sore-inducing herpes simplex viruses, EBV established lifelong infection. But whereas varicella hid in sensory nerves connected to the skin, and herpes simplex viruses silently occupied nerve cells leading to mucosal surfaces around the mouth and genitalia, Epstein-Barr headed straight for the center of the immune system itself.

The virus invaded B cells, the lymphocytes that manufacture antibodies. It established permanent residence in a specific subgroup called memory B cells. These are the cells that normally provide lifelong protection after a vaccine, say. They remember old foes, and beat them back when they reappear. EBV hid out in these long-lived sentinels, changing their architecture so they'd resist self-destruction. And then, as other immune cells prowled about nervously, aware that something fishy was afoot but unable to pinpoint the cause, the virus went quiet—and stayed mostly silent for decades.

You could see how this incursion at the heart of the immune system might upset the greater balance. But, strictly speaking, this model didn't finger EBV as the *cause* of autoimmune disease. The virus didn't *create* those self-reactive cells (although some argued that it did). Rather, it expanded, strengthened, or somehow changed them. That's an important distinction. While worms and other "old friends" enhanced regulatory aspects of the immune system, EBV appeared to strengthen aggressive tendencies.

As you might expect, this observation prompted some to wonder if the relationship was mutualistic. Herpes viruses were quite host-specific, and EBV had likely been with us since before we dispersed from Africa. It continued to infect more than 95 percent of the world's population (by age thirty-five). Might a virus this universal in time and place (except among teens and children in the developed world) benefit us somehow?

JUMP-STARTING THE IMMUNE SYSTEM

By the turn of the millennium, most scientists had abandoned the notion that the classic childhood infections—measles, mumps, and various respiratory viruses, among others—explained the "hygiene hypothesis," the observation that later-born siblings and children who attended day care had fewer allergies. A consensus of sorts had emerged that acute infection, especially with respiratory viruses, in fact increased the risk of allergic disease. By the early 2000s, attention had shifted to nonpathogenic microbes in cowsheds, crowded apartments, and perhaps day care. Then came evidence that parasites could protect against allergies.

Of course, we also had a long history with a suite of viruses. And in the mid-2000s, Caroline Nilsson and colleagues at the Karolinska Institute in Sweden noted that children infected with Epstein-Barr virus early in life had about one-third the risk of allergy compared with their virus-free counterparts. The association was too strong to ignore, and Nilsson began a prospective study. She and her colleagues followed 219 children from birth to age five. The goal was to determine how their immune systems changed if, and when, they acquired EBV.

Nearly two-fifths had acquired the virus by age five, which meant the majority still hadn't. After controlling for breast-feeding, day-care attendance, allergies in the family, and other variables implicated in allergy, the scientists arrived at the same conclusion: Children infected with EBV before age two were, as previously observed, one-third as likely to have allergy at five years of age compared with uninfected controls. The prospective nature of Nilsson's study brought to light another remarkable aspect of infection: If children contracted the virus after age two, their risk of developing allergy nearly quintupled. EBV infection in late toddlerhood and childhood *increased* the risk of allergic disease dramatically.

One could always argue that EBV wasn't the important agent, that it signified another bug that was plentiful in saliva-coated homes. But Nilsson also noted that immune functioning changed after infection. It appeared to be engaged in managing the virus. As a result, regulatory circuits were engaged. Nilsson surmised that, depending on when the immune systems experienced this upsurge, it could either prevent or cement allergic disease. For young children whose immune systems hadn't yet started to remember dust mites and other allergens, the viral stimulation prevented that memory from forming. But for those whose immune systems were beginning to recall proteins in the environment, the virus etched the memory more deeply. To borrow Martin Blaser's term (which he borrowed from the mid-twentieth-century bacteriologist Theodor Rosebury), EBV was *amphibiotic*: It could both hurt or help depending on when it arrived.

It remained to be seen how the altered risk of allergic disease at age five would play out later—if, for example, those who acquired the virus late would still have an elevated risk of allergy as adolescents—but Nilsson's findings were stunning for what they implied about the rise of allergy in the developed world. Over the course of human evolution, infants likely acquired the saliva-borne virus early from their mothers, siblings, and fathers. But as families shrank, mothers stopped prechewing food, and hygiene improved everywhere, EBV likely arrived later and later, eventually not arriving until adolescence or adulthood. How much had this late entrance not only deprived the maturing immune system of an expected developmental cue, but encouraged incipient allergic disease and sparked evermore MS?

WELCOME TO THE VIROME

Scientists estimate that herpes viruses, which are ubiquitous in the animal kingdom, are at least 400 million years old and possibly much older. They infect mammals, birds, reptiles, fish, and a few mollusks. And they're host-specific. Humans have eight native herpes viruses. All of them establish lifelong residence, and although they're occasionally associated with cancers and blisters, mostly herpes infections don't do much of anything. This distinctive M.O. puzzles virologists. They arrive, and then promptly go to sleep, remaining latent for decades. We might recognize this as a parasitic strategy especially tailored to small groups of people—a strategy from the Paleolithic. And, as parasitologists did with large, multicellular parasites, a few intrepid virologists began asking what advantage these "silent commensals" might confer on their hosts.

Nature offered some intriguing examples of utility. Viruses could serve as agents of biological warfare working in their host's favor. The American gray squirrel, for example, which was introduced by human hand to the British Isles, carries a certain poxvirus. The American squirrel tolerates the virus with few problems, but the native red squirrel finds it deadly. Partly due to its viral hanger-on, the American squirrel has steadily and inexorably advanced across Britain, driving the red squirrel into refuges such as the Isle of Wight.

This pattern resembles what followed the European arrival in the Americas. By one estimate, 90 percent of Native Americans succumbed to diseases, brought by Europeans, to which they had no immunity. In Darwinian terms, once Old World diseases became part of the New World landscape, Europeans had a competitive advantage over the relatively pristine Native Americans. One scientist has floated this same scenario—inadvertent biological warfare—to explain the disappearance of Neanderthal man. *Homo sapiens,* freshly arrived from Africa, carried some bug that proved deadly to its cousin *Homo neanderthalensis.*

These dynamics can also keep subspecies separate and on different evolutionary trajectories. African and Asian elephants, for example, each have a native herpes virus. Each tolerates its own with few symptoms. But one day zookeepers discovered that the African elephant's herpes was lethal to the Asian elephant, and probably vice versa. If the Indian subcontinent were to suddenly slam into Africa, the two elephant species couldn't remerge even if they wanted to. Their viral hangers-on keep them apart.

But the killing of potential rivals aside, what direct benefit do viral commensals—if they deserve the rubric—provide to their host?

The virologist Erik Barton addressed this question experimentally. He infected mice with two herpes viruses, mouse versions of the cytomegalovirus and Epstein-Barr virus. He suspected that the heightened antimicrobial defenses seen after EBV infection might protect against third-party pathogens. So he challenged his infected mice with opportunist bacteria. Mice already carrying the EBV stand-in were, indeed, unusually resistant. They better combated *Yersinia pestis,* the bacterium that causes the bubonic plague, and a bacterium called *Listeria monocytogenes,* which is often associated with food poisoning. A rough head count of the invaders allowed him to quantify the relative protection. Compared with their herpes-naïve counterparts, infected mice harbored 100 to 1,000 times fewer invading bacteria. The virus, however, improved its host's defenses only once it entered latency—precisely that phase when it appeared to be doing nothing. Meaning that, from the superorganism's point of view, it was doing something after all: it bolstered fortifications.

Viruses generally have a bad reputation. Many consider them little more than pernicious genetic parasites. Some say they fail, even, to qualify as alive. Their appearance doesn't help: images from electron microscopes show them to be spidery opportunists, weird geometrically shaped aliens that dock on innocent rotund cells and destroy them.

So when, in 2007, Barton published his results in *Nature,* they sparked some controversy. A colleague forwarded him an outraged e-mail from someone working on an EBV vaccine. The e-mailer charged that Barton was acting irresponsibly, that his work would inevitably feed the fires of antivaccine hysteria. But Barton had larger concerns: What if, thirty years hence, after vaccination against EBV became universal, we realized that the now-extinct herpes viruses provided some indispensable benefit? Then we'd berate ourselves for not exploring these nuances before eradicating the bug. "If it's biologically true, we should know about it," Barton told me.

Another group replicated Barton's results, although they found that the virus protected for just six months. That wasn't long enough, in their book, to qualify for mutualist status. For Barton, this interpretation overlooked the critical component of timing. Given the life span of a mouse, six months amounted to murine middle age. In other words, the virus defended mice well into their reproductive years. To his mind, protection early in life was perhaps the most important aspect of the mammal-virus mutualism.

In premodern humans, he explains, and especially among mothers who premasticated food for their children, the virus would arrive just at the point of weaning. That's when the passive protection conveyed by antibodies in breast milk wanes, and exposure to pathogens increases. This perfectly timed grand entrance is not, Barton thinks, accidental. The virus ends up enhancing immune defenses just when the child is most vulnerable—or as Barton puts it, the virus serves as a shot "of penicillin right when they need it most." Speculating further, Barton says it's probably no accident that other herpes viruses, all of them lifelong residents, hide out near regions—mucous membranes of the mouth and genitalia—where microbial invaders tend to gain purchase. Herpes viruses strengthen defenses in our most vulnerable areas, he says.

Although Barton's thinking borders on heresy in some circles, in others, such observations have become rote. Indeed, some look to these associations for guidance in treating certain infections. People carrying hepatitis G virus and HIV-1, for example, progress to full-blown AIDS more slowly than those without Hep. G. The hepatitis virus apparently boosts anti-HIV immunity. So should HIV treatment emulate the effects of Hep. G infection?

Prior infection with the human cytomegalovirus, another herpes virus, also suppresses replication of HIV-1. And hepatitis A may defend against infection by the much more debilitating, liver-scarring, cancer-causing hepatitis C.

To a certain breed of geneticist, these mutualistic-seeming relationships aren't too surprising. One of the major revelations to emerge from the decoding of the human genome, the first draft of which was published in 2001, was that 8 percent of our genetic material consists of viruses that have integrated themselves into our very DNA. Scientists initially considered these viral tidbits the equivalent of garbage left over from past intrusions, but more than a decade later, it has become increasingly apparent that these inserted viruses have enabled several important evolutionary leaps.

Take our adaptive immune system, for example—the white blood cells that learn what to recognize and whom to attack, or tolerate, over a lifetime. Central to this capacity is an ability to generate receptors that can recognize anything. How do we produce nearly endless possibilities from our finite genome? By reshuffling a limited set of instructions until a combination sticks. And this talent, scientists think, was imparted by a virus that slipped into our genome.

Or consider the placenta, the organ that allows mammals to gestate internally. The placenta does many things, including concealing the fetus from the mother's immune system. A viral integration also made this cloaking ability possible, scientists think. The integrated virus helps subvert Mom's immune response.

Interesting as they are, these examples don't have much bearing on the question at hand: If we coevolved with EBV, and it's more mutualist than pathogen, then why does it now incite multiple sclerosis? A tardy arrival probably changes the virus's impact on the host, but that alone can't explain the great variation in MS incidence around the world. As in previous cases, the context of the greater superorganism—what else is present or absent— may determine the virus's ultimate contribution to the greater whole.

In the past, EBV would arrive much earlier—at weaning, perhaps. (It still does in the developing world.) *Helicobacter pylori* might arrive at the same time. You'd also pick up a few worms beginning in childhood. Diet and constant exposure to microbe-enriched environs would have left you with a very different microbiota, one that we can imagine was less inflammatory. The signals coming from Mom while you were in the womb would also have differed. You'd almost certainly have more circulating regulatory T cells from the moment you arrived in the world. Now, the question: In that context, might the enhancement of the inflammatory response brought on by EBV confer all benefit and no cost—all protection against invasion and no autoimmunity?

I ask Barton this question and, like a good scientist, he replies that no one has addressed it experimentally. "Not even close," he says.

As it happens, the varying prevalence of other diseases associated with EBV suggests that coinfections are, in fact, important in determining the damage wrought by the virus. Where MS is rarest—in Africa and South Asia—Epstein-Barr is associated with horrific cancers enabled by secondary infections. Using them as a guide, we can devise some very rough equations: Malaria plus Epstein-Barr, for example, predisposes to a cancer of the white blood cells, and occasionally jaw, called Burkitt's lymphoma; Epstein-Barr plus HIV, another virus that imbalances the immune system, also increases the risk of lymphomas. With these EBV-associated malignancies demarking one extreme, and the EBV-related problems of high latitudes (mono and MS) the other, we can almost discern a "sweet spot," an optimal configuration of the superorganism. It includes the virus's early arrival, a microbiota that soothes, constant microbial pressure from environmental microbes, and a few parasites that engage regulatory circuits. And in that regulatory function, worms aren't alone.

CAN PARASITIC BACTERIA WARD OFF MULTIPLE SCLEROSIS?

As you'll recall from chapter 3, in the late 1990s, Italian scientists achieved extraordinary results by injecting a weakened mycobacterium native to cows—the bacillus Calmette-Guérin, usually used to vaccinate against tuberculosis—into MS patients. The trial was small, just twelve people. But as measured by MRI, disease progression slowed by 60 percent. Maybe because no one understood how this was possible, the study moldered. No one followed it up—at least not in humans.

But Zsuzsanna Fabry at the University of Wisconsin, Madison, set out to clarify the immunology. She'd collaborated with Joel Weinstock in the late 1990s, and studied helminths' power to stop MS in mice. Prior infection with *Mycobacterium bovis,* she found, was also 100 percent protective against MS in mice—but only in mice already carrying the bacterium when she tried to induce the autoimmune disease.

How did it work? Parasitic mycobacteria invade the immune cells that arrive to devour them, called macrophages. The bacteria turn these would-be heroes into safe houses for themselves. The immune system isn't entirely ignorant of their presence, of course, and a crowd of lymphoctyes gathers outside the hijacked macrophages, and beats its collective shields. This fraught state of affairs characterizes a latent mycobacterial infection. The walled-off invaders can't spread, but the immune system can't entirely eradicate them either. A long-term standoff ensues. And the stalemate, Fabry found, also weirdly purged self-reactive immune cells—lymphocytes that would otherwise cause autoimmune disease.

What did that mean?

Nearly everyone had latent tuberculosis 150 years ago—except, of course, those who had active tuberculosis. Scientists also know that some mycobacterial strains cause no disease; they're harmless. So if Fabry's findings in mice carried over to humans, the implication was that, in the past, the friendly fire that causes lesions in the central nervous system would have been nearly impossible to produce. With mycobacteria present, self-directed lymphocytes stood down.

These days, about one-third of the world remains infected with *M. tuberculosis,* mostly in places where MS is rare. The inverse relationship holds even in the developed world. One 1981 study found that if Danish children tested positive for TB before age seven, their risk for MS later was substantially reduced. If they tested positive for TB later in life, however, they weren't protected.

Which brings us back to the question at hand. Does the Epstein-Barr virus really cause MS, or does EBV plus some other factor, such as the wormless, mycobacteria-naïve postmodern immune system, lead to MS?

Much evidence suggests that vulnerability to MS is lost or gained even in adulthood. In the mid-twentieth century, people who moved from the north to the south as adults diminished their MS risk compared with the people they left behind. They'd already presumably acquired EBV. How did they lower their risk *after* its acquisition? Conversely, adults from the developing world who'd likely already acquired the EBV virus in their countries of origin could ever so slightly increase their risk of MS with adequate time spent in a high-risk area.

Consider the dramatic increase of MS in Martinique and Guadeloupe. Before the 1990s, multiple sclerosis was extremely rare on the Caribbean islands. But beginning in the late 1990s, the prevalence shot up from near zero to about 15 per 100,000 people. Scientists investigating the increase found that a stream of reverse migration—of émigrés returning after decades of living in urban France—drove the increase.

On average, returnees had double the risk compared with those who remained. But the odds weren't uniform among them. Those who left before age fifteen had quadruple the risk compared with those who emigrated later. Was this due to late EBV infection in France? It didn't quite square up. If the French West Indies were like other developing regions, the émigrés would have acquired the virus in infancy or childhood. They would have already had it when they left.

What's more, the MS risk increased with time spent in France even for adults. If you spent less than five years there, the risk of developing MS was only slightly elevated compared with people who remained at home. But if you spent more than a decade, your risk was more than triple that of people who'd remained on the islands. Maybe vitamin D deficiency explains it—lots of time spent at high latitudes with dark skin. But African (North) Americans, who had equally dark skin and who lived at less sunny latitudes, were also once considered immune to MS. In the mid-twentieth century, they developed the condition far less often than Caucasians—and this despite presumably suffering from greater vitamin D deficiency. Something else had to be at work.

In the 1950s, when the islanders first began leaving, parasites were ubiquitous on the two islands. The most common were the blood fluke schistosomes, which exert a powerful dampening effect on the immune system. As late as the 1970s, 70 percent of children between five and fifteen had at least one parasite. (Twenty years later, they were mostly gone.) The scientists were probably seeing, in other words, the gradual weakening of regulatory immune networks with time spent in France, rather like what happened to nonallergic immigrants to Sweden who, as the years passed, developed allergies for the first time in their new homelands. With the loss of stimuli by "old friends," immune dysregulation crept in.

The importance of ongoing stimuli in maintaining a balanced immune system is certainly one takeaway from the final phase of Jorge Correale's study in Buenos Aires.

A DEWORMING IN ARGENTINA

After observing his MS patients for more than five years, Jorge Correale ran his experiment in reverse. Four patients had begun to suffer from worm-related side effects—fever, diarrhea, and anemia, among others—and they asked to be dewormed. In a study like this, observation of a phenomenon that occurred by chance, one always wonders if the presumed cause (worms) is responsible for the observed effect (remission). Could some other unaccounted-for environmental factor—unclean water, housing conditions, even placebo—really be responsible? Now, like in the deworming studies in Gabon, Correale could quantify the worms' contribution by noting what happened when they disappeared.

After sixty-three months of hosting worms, four MS patients took antiparasite medication. The changes were quick and dramatic. A mere three months later, all the protective immune changes Correale had observed reversed. Pro-inflammatory excess rebounded; anti-inflammatory signals declined; and the disease resumed its terrible advance.

When he graphed these changes in immune function alongside those from healthy controls and uninfected MS patients, the pattern was telling. Following the parasites' eviction, the immune profile had switched from one that almost completely matched that of healthy controls to one approximating uninfected MS patients.

John Fleming's early results from five MS patients showed the same phenomenon. After three months on Weinstock's pig whipworm—2,500 eggs every two weeks—the number of new lesions appearing fell by more than two-thirds. Anti-inflammatory interleukin-10 increased. But soon after discontinuing the treatment, these protective changes reversed, and the pace at which new lesions appeared again quickened.

As I write this, another trial is under way at the University of Nottingham, testing hookworm on MS. The scientists are using twenty-five hookworms instead of the ten they tried with asthma. Assuming this works, parasites could prove an easy and cost-effective way to treat this degenerative disease. They won't cure it, of course. At best, they will bring it to a halt. But if prevention is the ultimate goal, assuming the link between MS and Epstein-Barr holds up, an EBV vaccine could quite literally eradicate MS, not to mention the associated cancers.

Of course, that approach totally ignores Erik Barton's concerns that the virus plays an indispensable role in the superorganism. As if to underscore his point, following his and Nilsson's studies on EBV, scientists in New York City have noted that infection before age eight with the varicella zoster virus—a member of the herpes family, as you'll recall, like Epstein-Barr—

protects against asthma and eczema later. Vaccination with the attenuated version of the virus, however, which has been available since 1995, does not ward off these disorders. This resembles Barton's worst-case scenario: The vaccine will probably soon drive the "wild" virus extinct—one more potential "old friend" gone from an ever-dwindling list.

In an ideal world—one that allowed nuance—we'd harness EBV's powers for good while heading off its deadlier aspects. Even in the real world, it's impossible to ignore that, absent an EBV vaccine, the easiest way to decrease the risk of MS may be deliberate inoculation with the virus at a young age. We're all destined to get it anyway. Intentional infection might preempt mono *and* MS. Acquiring the virus before age two may also curb allergic disease. And if Barton's correct, it could boost our defenses against pathogens. That's four birds with one stone.

"I personally would consider inoculating my children now if I knew any of them were EBV negative," Barton told me in an e-mail, "and if I had access to an EBV stock. (Neither is true.)"

The greater lessons of this research go something like this: The piecemeal disassembly of the superorganism changes how the remaining members interact. Remove worms (and mycobacteria and maybe *H. pylori*), delay commensal viruses, and strange new malfunctions arise. The relative contribution of each to the breakdown remains unclear. But Correale's, Fleming's, and Fabry's work highlights the stabilizing force of parasites that enhance regulatory circuits. They could hold the key not only to treating established MS, but to its prevention. At the very least, they offer a starting point for drug development—drugs, it's worth noting, with the unique distinction of having originated in parasites from our evolutionary past.

Correale's work also shows, as he noted, that the influence of gut-residing parasites extends throughout the organism, all the way to the central nervous system. And that reach is important for the next chapter on modernity's developmental disorder—autism.

CHAPTER 11

Modernity's Developmental Disorder: Autism and the Superorganism

> The immune system must be understood, not just as a defense against microbial invasion, but also as a sensory organ that informs the brain.
>
> —Betty Diamond

In the summer of 2005, Stewart Johnson, father to an autistic teenager, received the kind of phone call that filled him with dread. Someone from the camp where his son, Lawrence, spent a few weeks every summer was on the line. Johnson and his wife, Marjorie, loved their son dearly. But they looked forward to these few blessed weeks of peace when Lawrence was away. Their sanity, and to some degree their marriage, depended on the yearly respite. Immediately, worst-case scenarios came to Johnson's mind. Had Lawrence blown up in a violent fit? Was he smashing himself in the face? Were restraints required? Were administrators calling to tell Johnson to pick up his son early, because he'd become unmanageable?

None of the above, it turned out. Lawrence was fine—better than fine. He was participating in activities and he was sociable, the caller announced. He'd earned a pass to wander about camp unsupervised. In fact, they were calling to ask what new therapy Lawrence had begun, because it was working wonders.

"Are you sure you're not hallucinating?" Johnson remembers responding. There was no new therapy, he explained. He doubted the described changes were real. And after he hung up, he forgot about the call.

But a few days later, when he went to collect his son at the camp, he saw the transformation himself. Lawrence hugged his father, which was highly unusual. He gave Johnson a tour around the camp, calmly relating what they'd done—unheard-of. The car ride home was the most compelling evidence that something had changed. Trips were always difficult. Any deviation from the standard route—a different highway exit taken, a detour around roadwork—sent Lawrence into histrionics. On the way to camp, in fact, he'd thrown a prolonged tantrum. But now he sat unperturbed during the nearly three-hour ride from upstate New York back to Brooklyn.

Johnson had expected this newfound self-possession to dissipate at any moment. Now that it hadn't, he was intrigued. He decided to test it: He took Lawrence to a noisy barbecue joint in Brooklyn. At that point, Lawrence hadn't eaten out for years. And this particular restaurant had the sort of atmosphere—crowds milling, dishes clattering, babies crying—that had triggered meltdowns before. But now Lawrence sat unfazed by the hubbub, even as the waiter took his sweet time.

"The idea of me waiting for food in a restaurant for forty-five minutes with my son was so alien you might have been talking about something on another planet," says Johnson. "So my mind is totally blown at this point."

What had happened to his son? Johnson was at a loss. Only when he undressed Lawrence for bed did he noticed the bites. From the hem of his shorts to his sock line, hundreds of red, scratched-at, scabbed-over chigger bites covered Lawrence's legs.

Chiggers are a type of mite that passes through a parasitic phase. During hot and humid seasons, they await their victims atop blades of grass. When animals lumber by, the larvae, which are invisible to the naked eye, latch on and inject digestive enzymes into the host's skin. A cavity forms, the walls of which harden to create a minuscule tube of sorts. The larvae then feed on predigested skin cells by sucking them through this straw. Once sated, the parasite drops off to continue its life as a vegetarian. But for a long period during and after this feeding, the body mounts a strong immune reaction to the mess the chigger leaves behind. We experience the response as God-awful itching and swelling—the bites covering Lawrence's legs.

Johnson learned all this on the Internet after Lawrence went to bed. The strong immune response evident in the chigger bites had somehow improved Lawrence, he thought. The idea jibed with another observation: Johnson and his wife had noted that, whenever Lawrence got a fever, the worst of his autistic symptoms—the agitation and self-directed violence—abated. They'd half-joked about intentionally making him sick to cure him. (As it happened, many parents of autistic children had observed this. Some years later, scientists at Johns Hopkins would formally study the phenomenon.) And now, similar to what happened during fever, these bug bites had prompted a kind of remission. The immune response, Johnson realized, was the common thread.

Sure enough, as the bites faded over the next two weeks, Lawrence regressed. Watching his son descend back into a state of agitation, Johnson realized he'd chanced upon something important. If he could fake the chigger bites by inducing a similar immune activation, who knew how his son's life might improve?

From then on, Johnson, a money manager by day, spent his nights researching. He was familiar with the idea behind the hygiene hypothesis. He himself had an autoimmune disease called myasthenia gravis. His immune system attacked nerves where they connected to muscle. In his case, the end result was a slightly lazy left eye. So when he came across Joel Weinstock's work using worms to treat inflammatory bowel disease, and then discovered the company manufacturing the porcine whipworm eggs, he ordered the eggs for Lawrence.

After testing them on himself—no side effects—he started his son on 1,000 eggs every two weeks, less than half the dosage used in the University of Iowa studies. Nothing much happened. He increased the dosage to 2,500. Eight weeks later, Lawrence, who they'd once had to commit because he'd become so perturbed, was again transformed. He was calm. He answered questions directed to him. He smiled.

Johnson called Lawrence's neurologist of ten years, Eric Hollander at Mount Sinai Hospital in New York City. "You have to see this," he said. And when he brought Lawrence in, Hollander found himself marveling. Over the years, they'd tried everything—behavioral therapy, antipsychotics, antidepressants. One or two treatments seemed to work in the short term, but they always lost effectiveness over time. Hollander described Lawrence's autism as "refractory": it failed to respond meaningfully to treatment.

Which made the changes he now observed all the more extraordinary. If you think of autism as a collection of disparate symptoms piled into one—difficulty communicating, social withdrawal, agitation, obsessive-compulsive-like behavior—the symptom set that had most improved was also the most difficult to manage.

"He had almost one hundred percent improvement in his disruptive and repetitive behaviors," says Hollander, who's now at Albert Einstein College of Medicine in the Bronx. "Pretty remarkable."

THE BEGINNING OF AUTISM

These days, because of its (spurious, it seems) association with vaccines, autism has become a cultural flashpoint of sorts. The condition first appeared, however, well before vaccines were commonplace. In 1943, the Austrian-born psychiatrist Leo Kanner described eleven children "whose condition differs so markedly and uniquely from anything reported so far." These children, eight boys and three girls, had "excellent rote" memories, but were easily startled by "loud noises and moving objects." They engaged in monotonous repetition, and expressed an "anxiously obsessive desire for the maintenance of sameness."

Kanner, who was at Johns Hopkins in Baltimore, was well versed in psychiatric disorders such as schizophrenia and the various neuroses and hysterias. But this combination of symptoms appeared to be unique. He christened it "infantile autism," from the Greek *autos,* or "self." The children were imprisoned within themselves. And although he would later opine differently, he initially blamed overly cold parenting—parents he described as "refrigerated."

A year later, the psychiatrist Hans Asperger described a similar condition in Vienna. These four boys had no problems communicating verbally—in fact, they were happy to lecture at length on their favorite subjects—but they had difficulty comprehending emotion and making friends. Asperger, who may have suffered from what became known as Asperger's syndrome himself, called them "little professors."

The fourth *Diagnostic and Statistical Manual of Mental Disorders,* DSM-IV, includes five "autism spectrum disorders." The core symptoms include social impairment, obsessiveness, repetitive behaviors, and an early age of onset. Rett's syndrome is clearly genetic, the fault of a mutation in the X chromosome. The four other disorders, however, haven't been clearly linked to any single genetic mutation. And there's great variation in how they present. Some children diagnosed with autism have impairments evident from the day they're born. Others—maybe 40 percent—develop normally, only to regress between one and two years of age. For mysterious reasons, three of every four autistic children are boys.

Most alarming, autism diagnoses have shot up since Kanner first described the condition seventy years ago. During the 1970s, 3 in 10,000 children were diagnosed with the condition. By the early 2000s, 1 in 150 received the diagnosis. And in 2009, the CDC revised estimates to 1 in 110. As of this writing, early in 2012, the prevalence has again been amended to 1 in 88. The disorder is immensely costly to both the parents of autistic children and society as a whole—about $3.2 million for each autistic person over a lifetime, and $35 billion yearly in the U.S.

Scientists continue to argue over how much of that increase is real. The DSM began to include diagnostic criteria for autism only in 1980. And the criteria were updated repeatedly thereafter, both relaxed and then made more restrictive. (Revisions currently proposed for DSM-V, the forthcoming update, would further restrict the definition.) Some studies find that, going by the most recent criteria, many children who would have previously received a different diagnosis are now deemed autistic. But one careful study in California that controlled for these confounding factors still measured an increase between 1990 and 2006, although it was less than half the seven- to eightfold rise seen without controlling for these factors. And in conversation, scientists generally say that's about right: Half the increase is likely real, the other an artifact. Of course, a near quadrupling of any condition in just fifteen years, let alone one so debilitating, is startling.

What's going on?

THE VACCINE DEBACLE

In 1998, British scientists made a provocative argument, the consequences of which are still playing out. In the prestigious journal *Lancet,* the authors, led by Andrew Wakefield, argued that the measles, mumps, and rubella vaccine—called MMR—caused a kind of inflammatory bowel disease and autism. Authorities had introduced the three-in-one vaccine in the U.K. in 1988. So very generally speaking, the timing seemed to match. Wakefield also claimed to have found living measles virus in the guts of autistic children. He charged that the shots, usually given around eighteen months of age, just when parents received an autism diagnosis, triggered regressive autism.

The following year, the FDA further stoked fears around vaccines, but for a different reason. A vaccine preservative called thimerosal contained ethylmercury, a potential neurotoxin. The FDA announced that children who'd received their vaccinations according to schedule had possibly been exposed to more mercury than was deemed safe by the EPA. Authorities recommended that thimerosal be phased out.

The two announcements impugning vaccines—one focused on mercury, the other on an attenuated virus—caused some to panic. The percentage of vaccinated kids fell to 74 percent in London and other places, worryingly below the 95 percent needed for crowd immunity to function. Minimeasles epidemics began flaring. By 2006, new cases in Britain were at their highest in twenty years. Vaccination levels regained the 90 percent level only in 2011.

But in the meantime, scientists could neither replicate Andrew Wakefield's central finding that autistic children harbored viruses from the MMR vaccine, nor could epidemiologists show that the MMR vaccine—or any vaccination—correlated with autism onset among children. In California, where autism had nearly quadrupled between 1980 and 1994, the immunization rate had increased by only 14 percent. In Japan, where authorities ended the MMR vaccine in 1993 over worries about meningitis, and began administering each shot separately, the rates of autism continued to rise steadily. And finally, in Poland, where the MMR vaccine became mandatory only in 2004, both vaccinated and nonvaccinated children could be autistic. All throughout, even after drug companies phased out thimerosal from most vaccines, autism continued to rise.

Pressure built on Wakefield and his collaborators. Ten of Wakefield's coauthors from the original 1998 paper issued a "retraction of an interpretation." "We wish to make it clear that in this paper no causal link was established between MMR vaccine and autism as the data were insufficient," they stated in 2004. Britain's General Medical Council launched an investigation on ethical concerns. And in a 2010 verdict, the council branded Wakefield "callous," "unethical," and "dishonest." He'd administered painful and potentially dangerous lumbar punctures to autistic children without good cause, it charged. He'd paid children for blood at his son's birthday party. A later investigation by the *British Medical Journal* charged that he'd fudged data to better fit his hypothesis. The journal dubbed his work "an elaborate fraud."

The Lancet retracted that 1998 paper, and authorities stripped Wakefield of his license to practice medicine in the U.K. Wakefield, who years earlier had left his position at the Royal Free Hospital and School of Medicine in London to helm Thoughtful House Center for Children in Austin, Texas, resigned that position as well. He maintains that vaccine manufacturers have launched a campaign to discredit him. But on every level, the science does not support his claims. Vaccines are indeed associated with problems in some—high fevers, seizures, and very occasionally a demyelinating disorder that resembles multiple sclerosis—but considerable research by scientists around the world has failed to link vaccines to the autism epidemic.

And that brings us back to square one: What's behind the late-twentieth-century rise of autism?

AUTISM AND AUTOIMMUNITY:
A FORTY-YEAR CONNECTION

In 1971, the Johns Hopkins psychiatrist John Money described a peculiar case of infantile autism. After hospitalization for a severe mumps infection at two and a half years of age, a boy named Tommy had stopped showing emotion, become "very meticulous in play," and regressed developmentally. What struck Money as remarkable about Tommy's case was the greater familial context. Various autoimmune disorders plagued the family. Two older brothers had hypothyroidism, Addison's disease, alopecia totalis, and autoimmune diabetes. His mother had ulcerative colitis.

Many still thought autism was a psychological problem—you could blame your upbringing, in other words. And Money followed this line of thought, attributing Tommy's autism to "the constant threat of near-death" of his circumstances. But given the rampant autoimmunity in the immediate family, he also theorized that the autism could result "from the formation of autoantibodies affecting the central nervous system." That is, Tommy's immune system might have turned against his brain and spinal cord.

Thirty-odd years later, the Johns Hopkins scientist Anne Comi conducted a formal survey of families with autistic children. And nearly half, she found, had two or more family members with autoimmune disorders, compared with one-quarter of control families. The more autoimmune disorders present in a family, the greater the chances that someone would also have autism. If one family member had an autoimmune disorder, the chances were nearly double. But three cases of autoimmune disease, and the odds of the family also having an autistic member nearly sextupled. Most provocatively, Mom's autoimmune disease mattered most. If a mother had an autoimmune disorder, the chances of her child having autism increased nearly ninefold.

The study was small, involving just sixty-one autistic patients and forty-six controls, and it was conducted via questionnaire, an additional weakness, but it sparked interest in the connection between autism and autoimmune disease. Larger studies with better methodology kept finding links. The most authoritative study to date comes from Denmark. Scientists there parsed the records of all children born between 1993 and 2004—689,000 births in total. Doctors diagnosed more than 3,300 with autism spectrum disorder. Mothers of autistic children, the scientists found, tended to have more rheumatoid arthritis and celiac disease than controls. The former increased the odds of autism by over two-thirds; the latter tripled it. Dads weren't exempt. A father's type-1 diabetes increased a child's odds of autism by one-third.

None of this would have been too surprising to geneticists at Utah State University. Since the late 1980s, they'd noted both immune abnormalities in autistic children themselves and an enrichment of gene variants associated with autoimmune disease among autistic children and their immediate family members. Again, a mother's having the variants seems to matter most, an observation that drew attention to the maternal-fetal interface. Could autism result from a collapse of the immunological truce between mother and fetus? It wasn't necessarily that Mom's autoimmunity caused autism in her unborn child; rather, her autoimmunity (celiac disease, say) was likely evidence of deeper immune dysregulation, and this imbalance increased the odds that she'd mount an attack on the developing fetus. That "friendly fire" then caused autism.

Indeed, scientists identified peculiar antibodies in the mothers of autistic children, immunoglobulins that targeted proteins in the unborn child's nervous system. Researchers at the MIND Institute at the University of California, Davis, found that a mother's having these antibodies increased the risk of a child with regressive autism nearly sixfold. To show causality, the scientists injected those fetus-directed antibodies—immunoglobulins isolated from human mothers of autistic children—into four pregnant rhesus macaque monkeys. Control monkeys received antibodies from mothers of normal children. And sure enough, the babies of mothers who'd received "autistic" antibodies had behavioral problems. They were hyperactive; they engaged in repetitive OCD-like behaviors; they had trouble socializing with other monkeys in the troop.

HOW DOES THE AUTISTIC BRAIN LOOK?

As an infant develops, she prunes unnecessary neurons rather like a gardener trims branches off a plant, directing its growth. For a decade, scientists had noted that autistic children seemed to have larger-than-average heads early in life. Some argued that autism resulted from a failure to prune neurons. The brain ended up a chaotic, overgrown bush.

But after following autistic children, marking their growth trajectory, and comparing it with that of unaffected children, Eric Courchesne at the University of San Diego concluded that, if anything, the problem underlying autism wasn't a failure to prune, but a propensity to overgrow. Children who developed autism didn't have oversized brains at birth; rather, almost immediately after birth, they entered a period of frenzied cerebral expansion. Between two and four years of age, the period during which children seemed to regress, brain regions such as the cerebrum and cerebellum were 18 and 39 percent larger in autistic children compared with typically developing controls.

Soon after that peak, however, the accelerated growth slowed to a pace well below that of the typically developing brain. By age seven, the autistic brain matched the typical seven-year-old's. Thereafter, the normally developing brain outgrew the autistic brain. Some evidence suggested that the autistic brain began to shrink in adolescence. If cerebral development was a marathon, the autistic brain sprinted in the beginning and then collapsed halfway through.

Specialized regions of the brain were also malformed in autistic toddlers. The amygdala, which, in addition to initiating the fight-or-flight response, was critical in social interactions—if you're constantly terrified, you'll never make friends—was enlarged. And the degree to which it was oversized correlated directly with children's impairments.

Major brain abnormalities were turning out to underlie a disorder still defined by behavioral criteria. But one critical question remained unanswered: Were these peculiarities a cause or a symptom? As Harvard University's Martha Herbert put it, was autism "a brain disorder, or a disorder that affects the brain?"

THE INFLAMED BRAIN

On that front, Johns Hopkins scientists made a breakthrough find. Andrew Zimmerman, coauthor of the earlier studies looking at autoimmunity in families with autism, was convinced from his many years working as a pediatrician in and around Knoxville, Tennessee, that immune dysregulation contributed to autism. Allergies were rampant in the eastern Tennessee Valley, and pollen wasn't the only culprit. A haze emanated from a cluster of coal-fired power plants downwind. The Great Smoky Mountains to the east trapped the westward-drifting smog, producing some of the worst air quality in the country. He thought the high rates of allergic disease and the autism he'd seen must be connected, but he had no idea how immune activation, perhaps prompted by smog and pollen, might contribute to a behavioral disorder. So he consulted a colleague at Johns Hopkins who studied neurological disorders. His name was Carlos Pardo.

"If the immune system is doing something, it has to be in the brain," Pardo told Zimmerman. They needed to directly examine autistic brain tissue. The scientists located eleven samples from deceased subjects ranging in age from ten to fifty. Indeed, Pardo found that the autistic brain showed dramatic evidence of ongoing inflammation. This wasn't like the degraded myelin of multiple sclerosis, which resulted from a self-directed attack. It was more like a pan left on the burner for too long.

The brain has its own resident immune cells, called microglia and astrocytes. They hug the much larger neurons, functioning as maintenance men and keeping them in tip-top shape. In autistic brains, these handymen cells were visibly enlarged from long-term activation.

When the scientists checked spinal fluid from living autistic people, they also observed elevated markers of inflammation. No infection of the central nervous system was apparent, but the autistic immune system seemed engaged in chronic low-grade activation. Pardo cautions not to interpret the findings too literally. The inflammation might not cause the condition, but rather be indicative of an attempt to correct some ongoing malfunction—a healing response permanently switched "on." "Some of the inflammatory responses, instead of being bad, are good," he told me.

Yet evidence that inflammation was causal, not secondary—and that it began in the womb—kept piling up. The Johns Hopkins scientist Harvey Singer reproduced this brain inflammation experimentally. He injected pregnant mice with those antibodies from mothers of autistic children. The offspring of these treated mice lacked curiosity, were less sociable, and had a heightened startle response—rather like the changes seen in the monkeys that had received the same antibodies. And when he looked directly at their brains, he noted activated immune cells bathed in proinflammatory signaling molecules similar to those Pardo had observed in autistic human brain tissue.

At this point, let's take a breath and meditate. We know that mammalian reproductive success entirely depends on the mother's ability to tolerate her fetus. In the case of pregnant mothers with those "autistic" fetal brain-directed antibodies, it appears that maternal tolerance of the fetus has broken down. Why? The connection between autism and autoimmune disease suggests a central role for immune dysregulation—an inability to halt inflammatory processes—in both disorders. If you're having trouble tolerating yourself, perhaps you're also more likely to turn against your fetus. But it's worth remembering several things: First, those gene variants that predispose to autoimmune disease—and also apparently autism spectrum disorders—didn't always cause the diseases they do now. Second, we can reasonably assume that, in the past, they had a purpose, a job. And that job involved amping up defense against pathogens.

We saw this rule writ large in the reproductive success of those island-dwelling sheep in chapter 3. In a highly infectious, stressful environment, sheep with a tendency to create self-directed antibodies had a survival advantage. But even there, the talent had an evident cost—not autism, but diminished reproductive success. The sheep with "autoimmune" antibodies didn't have as many offspring. We also saw that people with inborn autoimmune tendencies—the Fulani in Burkina Faso—better repelled pathogens, such as malaria.

Let's remember the immunological context that prevailed during our evolution. In the past, the immune system's regulatory musculature was likely much stronger compared with today. Inflammation that wasn't directed at a pathogen, parasite, or other useful processes, was quickly subdued. So we'd do well to ask, if a mother's aberrant immune response to the developing fetus triggers autism—and ongoing inflammation maintains it—would mothers have been as prone to this vicious cycle in the past? Would a pregnant woman who was receiving "soothing signals" from environmental microbes, and whose regulatory circuitry was highly developed, make this mistake?

More broadly, can we fold autism into the allergy and autoimmunity epidemics? Are these three apparently disparate trends—a tendency to attack the self (autoimmunity), to recklessly pursue foreign proteins (allergy), and to turn on the developing fetus (autism)—all symptoms of a single problem, an inability to quash inappropriate inflammation?

The NIH geneticist Kevin Becker thinks so. In a 2007 paper sparked, in part, by what he sees as the misguided emphasis on genetics in the autism field, he pointed out that, very broadly speaking, autism followed the same epidemiological patterns as asthma. It was more prevalent in cities compared with the country. (That was true in countries where health care was free, meaning that everyone had access.)

For every autistic girl, there were four autistic boys. Few knew, but childhood asthma also disproportionately afflicted boys, who were about twice as likely to wheeze as girls. And while autoimmune diseases generally afflicted adult women about four times more than men, if the autoimmune disorder struck before puberty—before the onset of immune-suppressing testosterone—the preponderance was again male, not female.

There was, it turned out, an intriguing and pertinent explanation for the male predominance. Scientists studying fetal health observed that male fetuses were generally less resilient than female. If Mom was under stress, infectious or otherwise, boys were more likely to spontaneously abort, or to be born prematurely. The male fetal immune system was also more responsive to immune imprinting. If Mom was inflamed while pregnant, her sons were more likely to display inflammatory tendencies than her daughters. So if autism originated in prenatal immune dysregulation of some sort, these observations went a long way toward explaining why it disproportionately struck males. Boys were more sensitive to maternal immune imbalance than girls.

The epidemiology of autism mimicked other patterns of the hygiene hypothesis. Studies from the U.S., Australia, and the U.K. found that first-born children were more likely to have autism compared with later-born. And over the years, scientists had documented numerous immune abnormalities in autistic children themselves (more on this in a bit).

"Something environmental is causing the rise" of autism, Becker told me. "And it seems to parallel asthma and autoimmunity." The hygiene hypothesis—the notion that old friends educate the immune system, and without them, the immune system malfunctions—was, in his view, directly relevant to the autism epidemic. The consequences of immune imbalance were slightly more serious in this case—not sneezing or wheezing, but altered neural circuitry.

MEETING LAWRENCE JOHNSON

Stewart Johnson has frosted eyebrows and a quick, wry smile. He's a native New Yorker—grew up in the West Village—and he retains the energetic New York affect when he speaks. Behind his glasses, his left eye angles just slightly outward, the result of his autoimmune myasthenia gravis. The asymmetry lends his face an aura of hard-won wisdom, the contented fatigue of someone who's put in a long day's work.

One crisp, sunny spring day, I meet Johnson outside a coffee shop in Boerum Hill, Brooklyn, where he lives. The cherry trees lining the streets are in full, pink-blossomed glory. A breeze swirls their petals about in the air.

When I'd asked if I could meet his son Lawrence, Johnson had forewarned me that, now twenty years old, he was by no means cured. "He'll never be able to live alone," he said. But his most unmanageable symptoms—his agitation and self-directed aggression—had improved almost beyond Johnson's wildest hopes. This was important. Johnson had seen where out-of-control autistic kids ended up: institutionalized, restrained, and made to wear boxing gloves and football helmets to prevent their gouging out their own eyes and beating up their own faces. Just before that fateful summer of the chigger bites, in fact, Johnson and his wife had decided to send Lawrence to live at a residential school that specialized in handling autistic children. By then a teenager, he'd grown too big and strong for Johnson to restrain. And his self-destructive behavior aside, they worried that he might harm his younger sister. (He never has.)

They remained terribly distraught over the decision. On the one hand, they shuddered at the idea of sending their only son away. On the other, as Johnson tells it, "You reach a point where it's almost no longer a choice. It's simply a matter of survival. You cannot live like this."

When Johnson told me this, he added, his voice quavering ever so slightly, "If some omniscient being had come and said, 'That button on the wall, you press it and Lawrence not only doesn't exist, but he never existed.' I mean, I would have lunged for it." So it's difficult to overstate the impact Joel Weinstock's pig whipworm eggs have had on the Johnson family. Without the treatment, they wouldn't be together.

I watch Johnson and Lawrence approaching from down the street. Lawrence, who's dressed in a hoodie and blue jeans, follows close behind his father. His gait is shuffling and just a tad lopsided. He has brown hair and big boyish teeth. He gives me a hug upon our introduction. He seems excited to meet me.

After I introduce myself, he asks if I want to hear a joke.

"Yes," I say.

"Why did the chicken cross the road?"

"I don't know."

"To get to the other side," he says.

I laugh. Lawrence says, "You're not laughing."

I intuit what he means: I'm pretending to laugh. I'm going through the motions, but there's no hilarity in my laughter. Lawrence is full of somewhat unnerving insights like this. He's very attuned, Johnson tells me, to the emotional tenor of people's voices.

We head to a breakfast spot down the street, and sit outside in the rear patio. Lawrence orders French toast. Bells from a nearby church sound. Lawrence mimics the tone. He has perfect pitch, Johnson tells me. He always has. (I later read that Leo Kanner also noted great musicality in his original case studies.)

"What key are they in?" Johnson asks Lawrence.

"The sad key—yesterday sad," says Lawrence. Johnson translates: he's referring to the Beatles song "Yesterday," which is in a minor key.

The French toast arrives garnished with pecans.

"Lawrence used to have a terrible allergy to nuts," says Johnson. "Do you remember, Lawrence? Your face used to blow up."

Lawrence snorts in agreement. After the *Trichuris suis* ova, those nut allergies disappeared. The Johnsons noticed this improvement quite by accident: Lawrence ate a candy bar with nuts one day—and nothing happened.

Our conversation turns to meltdowns—the worst in the past. Lawrence remembers them all, and he also recalls what he was eating during each one. At the bowling alley below Union Square, he was eating nachos, which he threw on the ground. His father recounts that he also punched his own nose. Their lane ended up smeared with blood.

At the IKEA incident, where Lawrence was jumping in a ball pit when the tantrum struck, he was eating a cinnamon bun. At the Kennedy Krieger episode in Baltimore—he was getting evaluated there—he ate a Whopper, which he again tossed on the floor.

As we talk about these outbursts, I ask for his version of events. His answers are startlingly poetic. "I sobbed in the gutters," he says of the bowling alley. Another time, when they were driving through the Holland Tunnel and Lawrence became upset, he says, "I was crying in the tunnel lights."

"I was sobbing a river of tears," he says about a drive over the Verrazano Bridge.

"I was crying into the seat of the car," he adds.

"I had to drink the tears in the seat."

"I was underwater."

When I ask him why—does he remember why he was so upset—he ignores the question as if it were meaningless.

"I've been trying the 'why' question for ten years," Johnson says. "I've never got an answer."

The same could be said of autism spectrum disorders generally. For decades, anguished parents have been asking "why?" without a meaningful answer. For a time, vaccines served as a scapegoat, but the case never really stuck. Only as scientists have begun to understand the prenatal origins of disease has a plausible answer come to the fore. Modernity's developmental disorder begins with inflammation in the womb.

THE INFLAMED WOMB

Just a few years before a vaccine against rubella became available in 1969, the virus swept the U.S. In 1964, roughly 2 million expectant mothers contracted the disease. Many suffered minor symptoms—a fever, a rash (rubella means reddish), a few days in bed, but the pandemic had far-reaching consequences. Between 20,000 and 30,000 children were subsequently born with congenital rubella syndrome. The symptoms included some combination of deafness, cataracts, and mental retardation. Later analysis found that one of every ten of these children was also autistic. By one estimate, a mother having contracted rubella while pregnant increased her child's chances of presenting with autistic symptoms two-hundred-fold.

Rubella wasn't the sole culprit. Other maternal infections, including syphilis, chicken pox, and mumps, could cause developmental problems in unborn children. Even minor viral infections such as herpes were occasionally implicated. For years, no one really understood how or why maternal infection could cause problems. But the logical assumption was that the infectious agent directly interfered with fetal development.

That assumption held until Paul Patterson, a neurobiologist at the California Institute of Technology, found that even viruses that never went anywhere near the fetus could also interfere with fetal brain development. Pregnant mice infected with the influenza virus, which remains in the lungs, had offspring with behavioral problems. They didn't explore as readily; they avoided socializing; they disliked novelty. (These were also characteristics of schizophrenia.) In this case, timing made all the difference. Only mice infected in mid-gestation had offspring with altered behavior, not mice infected later.

Patterson thought that the mother's inflammatory response produced this outcome. To prove it, he repeated the experiment not with a living virus, but with viral RNA. Although the RNA was inert, the immune system would recognize the material as an invasion, and respond. Whatever changes occurred, he knew, they would result purely from inflammation. Again, the offspring of these mice displayed the same behavioral problems. Now, to clear any lingering doubts that this interference resulted purely from maternal immune activation, Patterson injected just interleukin-6, a pro-inflammatory signaling molecule, into the mice. This experiment gave the same results. Acute inflammation during pregnancy could interfere with brain development, and cause autistic-like behavior in offspring.

In the meantime, others were cementing the link between prenatal infection and another mental illness: schizophrenia. Alan Brown at Columbia University found that a mother's having contracted the flu during the first trimester increased the child's risk of later developing schizophrenia sevenfold. He came to this conclusion by analyzing preserved sera from pregnant women collected between 1959 and 1966. Brown calculated that one-third of all cases of schizophrenia were due to maternal infection, and therefore—in theory—preventable.

What does schizophrenia have to do with autism? They share certain symptoms—obsessive-compulsive behavior, agitation, and aggression among them—but they differ in age of onset. Schizophrenia usually shows up in adulthood, whereas autism first appears during toddlerhood or earlier. Nonetheless, the fact that both are linked to prenatal infection is important. The common thread is maternal inflammation.

And that brings us to the same issue we encountered when exploring the prenatal origins of allergy and asthma in chapter 7. Yes, prenatal infection can cause autism. But the incidence of prenatal infections has largely decreased in the same period—the latter twentieth century—that autism has increased. The epidemiology doesn't match. Of course, it's possible that, as populations urbanized, they were more often infected with certain viruses, like the flu, than ever before. There's probably a grain of truth to this. Different living conditions favor different pathogens. And we've never been this city-centric. More than half of humanity now lives in towns and cities, and urban birth is indeed a risk factor for both autism and schizophrenia.

But more generally, it's simply not accurate that mothers in the late twentieth century had more infections during pregnancy compared with the earlier twentieth century, or for that matter the nineteenth century. A glance at Jean-François Bach's famous graph in chapter 1 depicting the decline of infectious disease and the rise of allergic and autoimmune diseases quickly reminds us that the opposite is, in fact, true. Broadly speaking, we're more infection-free these days than probably ever before. The mid-twentieth-century rubella epidemic notwithstanding, infections are unlikely to be driving the autism epidemic. And anyway, as we've seen, modern populations have a tendency to become chronically inflamed. Moreover, the strong connection between autism and familial autoimmunity suggests that immune dysregulation underlies both disorders. Indeed, scientists in California have noted that a pregnant mother's being diagnosed with asthma or allergies during the second trimester more than doubles the unborn child's risk of autism. Her psoriasis triples it.

"The connection with autoimmune disease and allergies is telling us something important," says Patterson. Again, the message isn't necessarily that Mom's asthma or psoriasis causes autism. Rather, it's that inadequately controlled inflammatory responses in Mom can also predispose to autism in the child. Indeed, scientists find that a mother's having metabolic syndrome, one of whose symptoms is chronic, low grade inflammation, also increases an unborn child's risk of autism. Healthy fetal development depends on a balanced maternal immune system.

Two studies, one observational and the other experimental, demonstrate just how. Remember those fetal brain-directed "autistic" antibodies? A team of scientists in California asked what distinguished mothers who produced those antibodies from those who didn't. They looked at 365 mothers, 202 of whom had autistic children. And they identified a gene variant that these mothers possessed at a greater frequency. The gene was called MET, and the variant in question tended to reduce levels of the anti-inflammatory signaling molecule interleukin-10. Mothers who carried the gene were hardwired for a strong inflammatory response. They'd have a harder time halting inflammatory processes. Their immune systems were more prone to breakdowns of tolerance.

Work by the Swiss scientists Urs Meyer and Joram Feldon demonstrated just how important this peacekeeping, anti-inflammatory signal was during pregnancy. The scientists genetically programmed mice so that their macrophages (white blood cells that, as you'll recall, devour invaders) invariably responded to stimuli with anti-inflammatory IL-10. You could bully these cells tirelessly, but they always insisted on calm. Then the scientists repeated Paul Patterson's experiment: they injected the mice with viral RNA during early and mid-pregnancy.

Unlike standard lab mice, these pregnant mice squelched inflammation immediately. And when the mice pups were born, they developed normally—no autism- or schizophrenia-like symptoms. But if these genetically manipulated mice never encountered *any* pro-inflammatory stimuli at all—if only anti-inflammatory signals dominated during pregnancy—the offspring had other problems. The lesson: quashing inflammation protects the fetus, but the fetus requires both pro- and anti-inflammatory signaling for normal development. "The immunological balance during pregnancy is critical," says Patterson.

INFLAMMATION AND THE OVERGROWN BRAIN

Scientists at the University of Wisconsin, Madison, turned up another piece of the autism puzzle. Chris Coe and colleagues were replicating Patterson's studies exploring the effect of viral infections in rhesus monkeys. They'd confirmed that, as in rodents, a pregnant monkey infected with the flu virus had schizophrenic-seeming offspring.

Now, in an effort to better control precisely how much inflammation they incited—a variable that's difficult to manage with a self-replicating virus—they used the bacterial substance endotoxin as an inflammatory agent. Too much endotoxin could prompt a spontaneous abortion, they knew, so they started with the smallest of doses—nanograms compared with the micrograms or milligrams other experimenters used in rodents. To further diminish the risk of abortion, they gave the dose over two days rather than all at once.

To their surprise, this protocol produced not a schizophrenic-looking monkey but one that seemed autistic. The infant monkeys cried more. And between eight and nine months of age, they withdrew like regressive autistic children. When, at one year of age, the scientists took an MRI of their brains, they found that, rather than having the slightly shrunken cerebrums of the monkeys in the influenza experiments—a neural architecture that approximated observations made in human schizophrenics—these monkeys had enlarged brains. The scientists had accidentally produced a brain that, in the broadest sense, paralleled what Courchesne had observed in autistic children. It was overgrown. And acute inflammation didn't prompt the overgrowth; chronic low-grade inflammation did.

"It was paradoxical," says Coe. "It almost functioned as a stimulus." So here's the emerging picture: chronic low-grade inflammation predisposes to autism; sharper, acute inflammation like that accompanying an infection predisposes to schizophrenia. When the inflammation occurs during gestation also influences the outcome. This is Meyer and Feldon's argument, in fact. And the explanation affords a prediction. If infectious disease predisposes to schizophrenia, and immune dysregulation to autism, with the decline of infectious diseases in the developed world, you'd expect less schizophrenia and more autism.

And although we should take the following with a thimbleful of salt, some think that schizophrenia has indeed become less common. "Where have all the catatonics gone?" asked one scientist in 1981. "Is schizophrenia disappearing?" asked another in 1990. In a 1999 study, Finnish scientists made the connection explicitly. As polio cases declined after a vaccine arrived in 1954, diagnoses of schizophrenia also, they found, inched downward.

For our purpose, here's the major question: Is an immune system that's properly educated by "old friends" less prone to interfering with fetal development? Would mothers receiving those "soothing signals" from barnyard microbes, for example—signals that, as we saw, extend to the placenta—have autistic children less often?

Meyer's experiment suggests that the ability to control inflammation, to quash it, determines the vulnerability to autism. But where he used genetic manipulation to reinforce the anti-inflammatory response, we'd ideally want an experiment that addressed the question evolutionarily. Would a pregnant animal with a natural parasite load—a sewer rat, or a pig raised outside in the mud—produce the same outcome as that seen in either Patterson's or Coe's experiments? Would these immune systems, which scientists know work quite differently, turn off inflammation before it interfered with fetal brain development?

I ask Coe about this idea. "That sounds like a great experiment," he says. "Do you want to write a grant application?"

I ask Paul Patterson if, once scientists better understand the signs of impending trouble—elevated markers of inflammation during pregnancy, perhaps—they might head off autism with something as simple as a probiotic given to Mom. "It's a very interesting question," he says. "I don't think we know enough yet." Given the potential for worsening the outcome, however, he's in no rush to mess with pregnant moms. For now, he says, "It's important to know that immune disturbance starts early, and is ongoing."

FEVER QUELLS MADNESS

Back at Johns Hopkins, Andrew Zimmerman and the epidemiologist Laura Curran formally investigated those stories about fever improving autistic children. The literature, it turned out, had plenty to offer. In 1980, after a viral infection swept the ward housing autistic children at Bellevue Psychiatric Hospital in New York City, the children who contracted the virus improved, only to regress as their illness faded. One scientist estimated that a "moderate fever" of between 1.5 and 2.5 degrees C. (in the 102 degrees F. range) caused autistic children to "display dramatically more normal behavioural patterns." In 1999, the University of Tennessee psychologist Gary Brown wrote about an autistic boy whom he called "the sometimes son."

"Of course, all children quiet down when they are sick," he noted. "But the changes that occur in these autistic children are more dramatic—more like a metamorphosis in which the autistic child suddenly becomes almost normal."

Zimmerman and Curran distributed questionnaires to parents of thirty autistic children. The parents were to fill them out if, and when, their child became feverish. When the surveys came back, the pattern was clear: The most difficult symptoms—the irritability, hyperactivity, repetitive behavior, and lack of impulse control—all improved when body temperature went up. And the improvement was independent of fever-associated lethargy. After the fevers resolved, the children regressed.

The study suffered from the obvious weaknesses of a survey conducted by parents. On the other hand, the scientists hadn't informed the parents beforehand about their hypothesis. The parents weren't, therefore, biased toward viewing fever as curative. In combination with the earlier observations, the results hinted that these parents' secret hopes might actually be true: "It really tells you that the wiring is basically intact in a lot of kids," says Zimmerman. Buried beneath the dysfunction, there might be a normal child.

Fever, it turns out, has a long and storied history as a therapy for mental illness. In the late nineteenth century, the Viennese psychiatrist Julius Wagner-Jauregg noticed that psychotic and delusional patients improved when they contracted infections. He subsequently explored methods to intentionally induce fever. First he tried injecting proteins from tuberculosis bacteria, but settled on plasmodium-laden blood from malarial patients. And the approach worked—sort of. One patient of nine treated died from the fevers, but six were cured of the madness associated with advanced syphilis. Wagner-Jauregg earned the 1927 Nobel Prize for his work on "pyrotherapy." Others utilizing the remedy also noted benefits. After infection with malaria, formerly catatonic patients in the U.K. started reading, writing letters, and seeing relatives. But they invariably relapsed within months of their fevers resolving. The treatment served only a temporary stay.

Only now are scientists beginning to understand how it worked, and what it means that it worked. In the cases of syphilis-associated madness, the strong inflammatory response may have cleared out lingering infection. But in other cases, improvements likely stemmed from rebalancing an imbalanced immune system. It has become increasingly apparent that cognitive problems can arise directly from a dysfunctional immune system.

Research by Jonathan Kipnis, a neuroscientist at the University of Virginia, Charlottesville, highlights the somewhat unexpected overlap between the central nervous and immune systems. He's found that incapacitating mice's T cells messes with their heads. Without these white blood cells, the rodents can't navigate the usual mazes, and they perform poorly on a number of cognitive tests. When he puts these cells back, the mice regain their brainpower. When he knocks out just one immune-signaling molecule—interleukin-4—the mice display marked cognitive deficits. (Kipnis dreamed up these studies when, during a fever, he found he could think more clearly.)

Kipnis's findings may explain why, for example, HIV positivity so often leads to dementia. The virus depletes T cells—cells that help, according to his research, with memory formation and retrieval. In short, a dysfunctional immune system has major ramifications for brain function. And scientists at the MIND Institute have documented multiple immune abnormalities in autistic children. They have more TNF alpha floating about, a cytokine important in fighting infectious microbes. They have elevated levels of a hormone called leptin, which is involved in appetite regulation but also promotes inflammation. When stimulated with endotoxin, their white blood cells respond with greater force. They have less anti-inflammatory TGF beta, and the less they have, the worse their symptoms. And they also have fewer circulating regulatory T cells. With all this pro-inflammatory capacity, you'd think they'd be masters at self-defense. But the autistic immune system is in fact less effective at clearing infections.

Pardo's warning notwithstanding (he thinks inflammation in autism may arise from a deeper problem), standard treatments for autoimmune disease also work in autism. In small studies, scientists have reported good results treating autistic children with immune-suppressing steroids. But one can't use them indefinitely. Intravenous immunoglobulin treatment has also benefitted autistic children. IVIg, as it's called, consists of human antibodies extracted from donors. No one really understands why, but the antibodies both tamp down on brain inflammation and boost regulatory T cells. The major downside to the therapy is its prohibitive cost and limited supply, but that it works suggests—again, contrary to Pardo's argument—that ongoing inflammation contributes to autism, and that turning the inflammation off can turn off the disease.

Then there are the worms.

AN AUTISTIC BOY NORMALIZES

After a difficult labor, Shelley Schulz delivered her son via C-section in June 2003. He weighed 7 pounds, about average, had a slight case of jaundice, and from the beginning had trouble latching on to breast-feed. "He couldn't make that sucking motion," says Schulz. "To this day, he cannot blow out a candle without spitting."

The boy, whom I'll call Leo, learned to walk early—at nine months. And from the start, he focused obsessively. He never pointed, and never babbled as other babies did. He started speaking late, at age two. When he did finally talk, he had trouble stringing words into sentences. He usually uttered just one at a time. And he suffered from bowel problems. His stool seemed almost poisonous. To this day, Schulz says, her son has "burn marks" on his buttocks from that early "toxic poop."

As his developmental delays became more apparent, Schulz made the rounds seeking answers. He wasn't violent, but he had frequent meltdowns—maybe eight per day. When she finally received a diagnosis of autism, he was three and a half. Schulz, a former hedge-fund analyst, decided to try anything that might help her son. The Specific Carbohydrate Diet seemed to assuage his hyperactivity. (Many parents of autistic children report improvements with this diet, which eschews refined sugars and starches in an effort to mimic the imagined human diet during the Paleolithic—lean meat, nuts, vegetables, low-sugar fruits. But scientists aren't universally convinced that it's effective.) Later she found that prednisone, an immune-suppressant, greatly improved his irritability, but one couldn't take the hormone indefinitely.

And then, one day in the parking lot of a Mount Kisco health food store, she ran into an acquaintance named Judy Chinitz.

Chinitz, a nutritionist, also had an autistic son. He'd regressed around eighteen months, stopped talking, stopped playing, become terribly agitated. And he'd developed pediatric inflammatory bowel disease— explosive diarrhea, gut pain, and a distended stomach. Although clinicians insisted that he had gut problems because he was autistic, Chinitz suspected that his gut problems played into his autism—that they caused some of his autistic symptoms. So, seeking to treat his IBD, she'd tried the pig whipworms developed by Joel Weinstock. Not only did the IBD recede, his behavior improved. The agitation subsided. The tantrums came less frequently. A boy emerged from the "autistic fog," one who was by no means "neurotypical," but one who could communicate and socialize.

After hearing Chinitz's story, Schulz ordered whipworm eggs from Ovamed. And in the spring of 2008, she put Leo on the full 2,500-egg dose. His hyperactivity worsened. She lowered the dose to 2,000. He improved. "He out of the blue started reading," says Schulz. "He got to a level of cognition that was out of nowhere." But he was still hyperactive. By then, Chinitz had moved on to hookworm. The whipworm eggs were very expensive. A single dose of hookworm, on the other hand, and you had a few years covered. She'd traveled with her son to Tijuana for inoculation, and found that it worked just as well as, if not better than, *Trichuris suis* therapy. (Chinitz, who has an autoimmune disorder of the thyroid gland called Hashimoto's disease, also infected herself with hookworm. The nodules on her hand, a symptom of Hashimoto's, disappeared.)

Schulz followed Chinitz's lead. She infected her son with three hookworm larvae. Maybe five months later, Leo's meltdowns began to fade. Life was still a roller coaster of good days and bad, but the number of good days steadily increased until the bad days ceased entirely. "I found it hard to believe that these things were not in my head," says Schulz.

Others noticed the changes as well. The previous school year, Leo's first in the special-needs kindergarten class, had been difficult. Kristen Ragazzo, Leo's special education teacher, recalls that he was always "high-functioning." He had an excellent memory and good academic skills, but he had a hard time socializing. He avoided eye contact. He didn't make friends. And he exhibited what she calls "behavioral inflexibility," a classic symptom of autism. When something changed—the task at hand in class, or their assigned classroom—Leo would become distressed. Fire drills sent him into a foot-stomping, arm-flailing tantrum.

All that changed, however, between kindergarten and first grade— pre- and post-hookworm. It was gradual, not sudden, says Ragazzo—fits and starts that progressed in "an upward spiral." Changes in the routine perturbed him less and less. He developed a sense of humor. "Ms. Ragazzo, I'm being flexible," he'd announce when something arose unexpectedly. And most important, says Ragazzo, where before Leo kept to himself, he became interested in others. He made a friend.

"We were just excited to see he made a true friend in a child," she says. "The whole thing with autism is, it's a social disorder."

In her admittedly short six years of teaching special-needs children, Ragazzo has never seen a child like Leo normalize to this degree. Without quite knowing the many things Schulz has tried, and not discounting that maybe some of her own hard work contributed, Ragazzo credits Leo's mother for attempting everything until she found something that worked.

Next year, third grade for Leo, he'll leave the special-needs group and enter the regular class. He no longer requires special treatment.

"There is hope for these children," Ragazzo says. "Is he still autistic? Yes. Is he quirky? Yes. But there is hope. And it's just a matter of finding the right combination that works for that child."

THE GUT-BRAIN AXIS

When Leo Kanner described those first cases of autism nearly seventy years ago, he also noted, almost as an aside, that the children suffered from gastrointestinal problems. One boy had "large and ragged tonsils." Another vomited excessively. A third, a girl, stopped eating completely. To avoid starvation, doctors put her on a feeding tube for a time.

Ever since, pediatricians and parents have noted gut problems in autistic children. By one count, 40 percent of regressive autistic children have abnormal stool patterns. For those with a family history of autoimmunity, that number nearly doubles to 78 percent. Still others have noted a preponderance of food allergies and wheat intolerance among autistic children.

And yet controversy dogs the idea that gut problems are a reliable feature of autism. The disagreement may stem in part from the many conditions grouped under the "autism spectrum" umbrella. No one's quite talking about the same entity. Indeed, the association with gut problems seems strongest in the subset of regressive autism.

Another reason for argument: In that now-infamous 1998 paper, Andrew Wakefield claimed to have identified an autistic inflammatory bowel disease, or "enterocolitis." He may have tainted the idea. Long before Wakefield made his charges, of course, physicians observed gut problems in autistic children. Nonetheless, a 2010 consensus report on gut issues in autism reached the noncommittal conclusion that nothing was certain, and more research was required. Ever hedging, the twenty-seven authors, all specialists in the field, added that "problem behavior in patients with [autism spectrum disorders] may be the primary or sole symptom of the underlying medical condition, including some gastrointestinal disorders." In other words, they allowed that gut issues could drive behavioral problems.

Who cares?

The body has two brains, a bundle of neurons around the gut that takes care of activities such as peristalsis, and the larger one in our heads that thinks of itself as the Brain. Evolutionarily speaking, the former probably evolved first, but the two brains remain inextricably connected in ways that scientists still don't completely understand. They're literally linked by the vagus nerve, the biological equivalent of a fiber optic cable running from intestine to head, but they're also bound together by a much more fluid medium, the immune system. And scientists have repeatedly observed that gut issues can manifest as psychiatric disorders. Madness can start in the belly.

Case studies pepper the literature. In one, a fourteen-year-old boy presented with what looked like schizophrenia. Symptoms included insomnia and aggressive behavior. But he didn't respond to antipsychotics. A year later, he began vomiting and passing bloody stool. Doctors peered into his colon and found him to have an inflammatory bowel disease. After treatment with immune suppressants, the boy became "essentially normal." Ominously, the problems had come on, the physicians noticed, following a course of antibiotics. But ultimately, he wasn't mentally ill. Instead, his gut was inflamed.

In another case, a five-year-old boy with a diagnosis of autism was found to have celiac disease. When he went on a wheat-free diet, his autism improved. He regained lost verbal ability, showed affection to his mother, and basically normalized. This case study in particular raises a sticky issue. Presumably the gut inflammation was interfering with his cognition, and maybe his brain development. Catching and treating it at age five allowed the boy to resume a normal developmental trajectory, but if doctors had noticed the celiac disease at age twenty, it might have been too late. Chronic intestinal inflammation might have permanently altered the wiring of his brain.

In Italy, scientists continued to unabashedly link autism to gut problems. There, celiac disease in particular seemed more prevalent among autistic children. And they made another critical gut-related discovery: Autistic children had much more endotoxin, a bacterial product that incites inflammation, circulating in their bloodstream. The more they had, the greater their low-grade inflammation and the worse their autistic symptoms. Where did the endotoxin originate? It came from resident microbes in the gut. Why were microbial products leaking through? The autistic gut, it seemed, was dysfunctional.

We've seen three things that increase gut-barrier permeability, none of which are mutually exclusive: a junk-food diet; prenatal inflammation; and a deranged microbial community.

PROBING THE AUTISTIC MICROBIOTA

By the late 1990s, the Chicago pediatrician Richard Sandler had heard the same story from parents of autistic children so many times that he could no longer dismiss it as coincidental. The narrative went like this: Their toddler developed some minor infection, maybe of the ear. A doctor prescribed a broad-spectrum antimicrobial, which was standard practice. After completing the course, the child developed terrible diarrhea, which became semichronic. She then stopped speaking and playing. She lost interest in socializing with others. Eventually, after increasingly frantic parents had shuttled her from specialist to specialist seeking answers, she received a diagnosis of regressive autism.

Autism was, of course, considered a developmental disorder—that is, whatever defect or malfunction occurred, it happened in the brain. This sequence of events bore an eerie resemblance to another troubling infection, however—not of the brain, but of the gut. Antibiotic-resistant *Clostridium difficile* had become a scourge in hospitals around the country. It, too, struck after antibiotics had perturbed the microflora. It also caused debilitating diarrhea, even death. And—few knew this—it could trigger symptoms that on the surface appeared neurological, but really resulted from severe intestinal inflammation: delirium, hallucinations, even repetitive OCD-like behaviors.

Sandler wondered: Could regressive autism result from some opportunist invading the perturbed gut ecosystem after antibiotics?

To test the idea, he treated ten autistic children with the broad-spectrum antimicrobial vancomycin. The drug doesn't pass into the bloodstream when given orally, so any observed changes would stem from alterations to the microflora.

Almost miraculously, eight of the ten treated children improved. They began looking their parents in the eye, and speaking. However, the improvements were short-lived; when the treatment ceased after eight weeks, the children regressed. That, Sandler thought, was a clue. Whatever bug was behind autism—perhaps a neurotoxin-producing species such as *Clostridium tetani,* which causes lockjaw—it was apparently resistant to antimicrobials.

Or maybe there was no particular pathogen at all. Maybe the entire ecosystem was out of whack.

During the following decade, Sydney Finegold, a microbiologist at the Veterans Administration Medical Center in Los Angeles and a coauthor on this early study, stayed on the case. He found that regressive autistic children had a significantly different microflora compared with normal controls.

Most notable, perhaps, autistic children harbored an unusual bacterial species called desulfovibrio. The bacterium, which was common in oil fields, produced hydrogen sulfide as a by-product. In sufficient concentrations, hydrogen sulfide could corrode steel. Desulfovibrio was also extremely resistant to antimicrobials. If you treated a common ear infection, you'd clear plenty of good bacteria, but leave this one bad seed. And in the clear-cut environment, it could bloom.

With an eye to Finegold's work, the Canadian researcher Derrick Mac-Fabe developed a rodent model to show that resident bacteria could affect brain development. He was intrigued by anecdotal stories of autistic children who craved junk food, and whose symptoms grew worse after gorging on it. Were bacteria responsible? Some of those bacteroidetes that seemed abundant in autistic guts produced a by-product called propionic acid. In the right amounts, propionate was good for you. But too much, MacFabe suspected, could cause neurological problems. In the 1970s, an antiepilepsy drug that resembled propionic acid, called valproic acid, was found to induce autism in children when given to expectant mothers early in pregnancy. Might something similar be happening in people—not with a drug, but with an overgrowth of propionate-producing bacteria?

Injecting rats with the stuff certainly made them look autistic. They became hyperactive, had an out-of-control startle response, and impaired social behavior. Like humans, male rats seemed more sensitive to the treatment than female. And they showed an inflamed brain akin to what Carlos Pardo had observed in people.

MacFabe has yet to run the experiment from start to finish—from bacteria in the gut to behavioral abnormalities—but his model shows that, in excess, products from everyday bacteria can cause problems. He uses rabbits in Australia as an analogy. Rabbits live in North America as an integral part of the ecosystem. Introduce them to Australia, however, which lacks natural rabbit predators—something that happened in the nineteenth century—"and you get a mess," he says. The population multiplied exponentially, eventually reaching tens of millions. And they denuded the fragile, arid landscape. "It's not as simple as what kind of bug you have," says MacFabe. "It's how they interact in the microbiome."

Having some bugs, however, is essential. Remember those germ-free mice from chapter 9? They had smaller hearts and lungs, and an insufficiently developed immune system. In 2011, Swedish scientists announced that germ-free mice also had slightly malformed brains and altered behavior. Without microbial stimulation, genes that should have been on "low" were cranked to "high." Their synapses—the connections between neurons—lacked plasticity. They had trouble learning. They readily explored open areas and were less anxious in general, two potentially dangerous behaviors for animals that regularly become dinner for a number of predators.

When the scientists put the microbiota back, these deficits disappeared—but only if the microbes were reinstated early in life. If the mice matured beyond some threshold without microbial stimulation, the neural deficits prompted by a germ-free existence were irreversible. Plasticity was permanently impaired.

"It takes guts to grow a brain," wrote the scientist Betty Diamond about the groundbreaking study. (She wasn't involved.) Absent immune-system stimulation by the commensal microbiota, brain wiring ended up permanently altered. By extrapolation, an aberrant microbiota—one containing the wrong bugs, one overly simplified, or one with communities in the wrong ratio—could in theory rewire the brain. "The implications are legion," observed Diamond.

THE GRAND SYNTHESIS

How to synthesize these multiple paths to autism—the prenatal, the postnatal, the microbial? One answer: Don't bother. Perhaps there are different ways of producing the disorder. One may involve maternal immune dysregulation that interferes with brain development during pregnancy. The other may result from acquiring an aberrant microbiota that then affects brain development. Both involve inappropriate inflammation.

More important, they're not mutually exclusive. In fact, they're mutually reinforcing. The immune system does triple duty as defender of the organism, manager of the microbiota, and handyman in the brain. Your mother's immune functioning imprints yours in the womb, and this pre-programming inevitably sets the tone for your own interaction with the microbial world. Immune dysfunction that starts in the womb could, in theory, lead to recruitment and cultivation of a microbiota that promotes further dysfunction, which then contributes to abnormal brain development.

In chapter 9, we saw this feedback loop operate in experiments. Rodents with an immune defect cultivated a microbiota that became pathogenic. And that microbial community, when transferred to peers and offspring, was enough to cause disease in mice without the primary defect. The microbiota reflected the disease back on the host.

It's all guessing at this point. Yet so many lines of evidence—experimental, microbial, observational—pointing to immune malfunction in autism are difficult to ignore. Moreover, that helminths have helped in the admittedly small, totally uncontrolled, and overtly self-selected group of parents and autistic children whose stories I've told suggests that the gut and ongoing inflammation are, if not the cause of the brain disorder, major contributors. The temporary stay afforded by fever, antibiotic treatment, and direct immune-suppressant therapy only reinforces this conclusion. Inflammation seems to partly drive autism.

The immunology certainly fits. Those components of the immune system deficient in autoimmune disease—anti-inflammatory signaling molecules, and peacekeeping regulatory cells—also appear inadequate in autistic children and their mothers. It's no accident, in other words, that worm infections help in autism. As we've seen repeatedly, they strengthen this very same regulatory circuitry.

If this link holds up—and if autism really is just another inflammatory condition of many in modernity—then the stakes involved in correcting our postmodern immune system will have risen. The consequences of inappropriate inflammation will now include interference in brain development. And insofar as derangement of the human superorganism contributes, its restoration takes on greater urgency.

Based on Lawrence Johnson's case and a few others, a formal study testing pig whipworms on autism has begun at Mount Sinai in New York City. As with allergies and asthma, however, the most effective treatment will likely begin in the pregnant mother—not really treatment, but preemption based on genotyping and immune profiling. At this early stage, intervention may be as simple as a probiotic yogurt.

On the other hand, Paul Patterson's work suggests that giving vaccines that prompt an inflammatory response to pregnant mothers—such as flu shots—may not be a good idea. Of course, contracting the flu while pregnant is a worse proposition. One workaround floated by Patterson is to proactively vaccinate all women of reproductive age.

A bright point in autism research is scientists' having identified those "autistic" antibodies in mothers. Assuming the observation withstands further scrutiny, the presence of these fetal brain-directed antibodies provides a way to determine who's at risk of autism in time to intervene. Again, suitable interventions will likely involve rebalancing Mom's immune system, thus preempting interference in fetal brain development altogether. One scientist has proposed giving at-risk pregnant mothers nonsteroidal anti-inflammatories—in this case, drugs developed to treat inflammation in diabetics. That's an elegant allopathic solution. But what about an evolutionary approach? What about restoring the superorganism?

Stewart Johnson, who's spent years reading the scientific literature on autism and immunity, understands that pregnancy is probably the easiest and most effective point to intervene. "I'm near sixty. We're not having kids," he confided at one point. "But if I was, my wife would take TSO"—the pig whipworm eggs.

Maybe in ten or twenty years, we all will.

CHAPTER 12

Beyond Allergy and Autoimmunity: Inflammation and the Diseases of Civilization

In the wake of immunologically new humankind, I add one recommendation to the means in the Finnish Allergy Programme to obtain and maintain tolerance; let us take care of the butterflies! Their disappearance indicates loss of life. When the last tiger is killed in India, it may hit the headlines, however, shrinkage of biodiversity also takes place in the micro-world close to us, but without notice. Preserving biodiverse life might have a preventive effect on allergy and other diseases of modern civilization.

—Finnish allergist Tari Haahtela in *Allergy,* 2009

The examination room in Michael Gurven's clinic in Bolivia is dark, the equatorial sun blocked by heavy blankets slung over the windows. A middle-aged Tsimane woman lies on her side on a table, mostly covered with a sheet. Two days ago, the woman was in the jungle along one of the many rivers of Tsimanía, the piece of the Bolivian Amazon reserved for the tribe. Today, a gregarious Bolivian physician with braces moves an ultrasound scanner across the woman's ribs. The woman, her hair in two braids, watches the pulsating image of her own heart on a laptop-size screen. Light plays across her lined and surprisingly unperturbed face. The color red indicates oxygenated blood passing through the right side of the heart, explains Edhitt Cortez Linares, the technician. Blue signifies oxygen-depleted blood going through the left side. The squishy sound of her heartbeat, superimposed over the latest hits from the Spanish-speaking world emanating from a radio somewhere, brings to mind an octopus thrashing around underwater.

Drivers have been trucking in the Tsimane from their remote villages for days. A pair of cheerful, roly-poly women cook for them, and they pass the time watching a small TV mounted in one corner of the open-air dining room. They seem to enjoy Discovery Channel shows about the ocean, which probably none of them have seen. And one by one, they undergo a battery of tests—stool and blood samples, and for older individuals, measurements of cardiovascular health.

I ask Cortez what she usually sees. Some heart damage from infections, such as the leishmaniasis parasite, she says. She notes hernias in men, and prolapsed uteruses in women—the former from heavy lifting, the latter from the high birthrate. (The average Tsimane woman has nine children.) But she doesn't find cardiovascular disease of the sort she'd see in similarly aged adults in the U.S. She sees some arterial thickening, but no lesions and no plaque buildup. And that's strange: judging by measurements of inflammation, the Tsimane should be dropping dead from heart attacks.

Let me explain. Scientists have come to understand that inflammation plays an outsized role in cardiovascular disease. And so, on the basis of both experimental and observational data, some have predicted that populations exposed to lots of infections, which increase inflammation, should have an elevated risk of cardiovascular disease. One shorthand measure for inflammation is C-reactive protein, or CRP. In the U.S., high CRP in middle and old age correlates directly with the risk of heart disease and stroke. When CRP is low, the risk of these complications is also lower.

When Gurven used this measure on the Tsimane some years back, he found high levels of inflammation as measured by CRP—in line with predictions—but no signs of cardiovascular disease. They hardly developed high blood pressure or elevated cholesterol, and this despite often presenting with too little high-density lipoprotein, the "good" cholesterol, and somewhat elevated "bad" triglycerides. The observation challenged established ideas about heart disease.

Gurven bought an ultrasound machine and lugged it to Bolivia to confirm his initial findings, which were gleaned with less-direct methods. And the results so far uphold his earlier observation. He's seen some arterial thickening, but still no inflamed arterial plaques, the hallmark of cardiovascular disease.

How to explain it? Was the invulnerability to heart disease genetic? Probably not. North American Indians, a geographically distant but genetically related people, had plenty of cardiovascular problems.

Did the decidedly premodern Tsimane lifestyle contribute? Probably. They were physically active, and their food—meat from wild game and minimally processed fruits and grains—might promote cardiovascular health. But these factors couldn't explain away the absence of heart dis-

243

ease in the presence of inflammation. CRP at those levels predicted heart disease in the industrialized world. And yet there wasn't any. So Gurven's attention turned to the ubiquitous parasite infections.

Three-quarters of the Tsimane carry one or more intestinal parasite. Just outside the ultrasound room, in fact, I see my first living hookworm egg. A lab assistant collecting stool samples waves me over. I look through her microscope to observe a deceptively innocent-looking oval surrounded by fibrous detritus.

Worm infection, it turns out, may protect against heart disease the same way it precludes autoimmune disease: by skewing the immune system away from the microbe-hunting Th1 response, and by strengthening regulatory circuits that tamp down on inflammation. Helminths may also beneficially tweak the third axis of cardiovascular disease: cholesterol.

In Egypt, scientists have noted that patients infected with blood flukes have lower levels of fatty lipids in their bloodstream. And when British scientists infected mice that were genetically prone to develop cardiovascular disease with blood flukes, they saw their risk of heart disease drop by half. Amazingly, that decreased risk held even when these mice continued eating a high-fat, Western-style diet. The worms helped them cope with junk food. Importantly, schistosomes live in the veins of their host, not the gut. So whatever influence they wielded, it wasn't directly on the intestinal tract, or on incoming food. Instead, they somehow changed the systemic response to the diet.

In recent years, some have attempted to reframe cardiovascular disease as an autoimmune disorder. That's because the usually blamed risk factors—bad diet, lack of exercise, increased body mass index, and so on—explain only about half of the elevated prevalence of cardiovascular disease seen in developed countries. What can explain the other half? Self-perpetuating inflammation. The plaques that characterize atherosclerosis aren't just the circulatory system equivalent of grease clogging a drain. The closer scientists look at them, the more they resemble lesions caused not by infection but by a failure to terminate inflammatory processes. And as we've seen, weak immune regulation contributes to inflammation of all types. Worms, of course, strengthen immune regulation.

And so Gurven began considering the possibility that helminths were protecting the Tsimane from the heart disease that, according to current scientific understanding, should be rampant. Parasite infection might uncouple inflammation from the risk of stroke and heart attack.

At this point, I should point out that Gurven's findings fit a pattern we've seen repeatedly. Measurements that signify disease in the developed world repeatedly fail to mean the same in environments that resemble those of our evolutionary past. High levels of immunoglobulin-E—the "allergic" antibody—for example, signify allergies in New York. But in the Amazon, where Amerindians can have hundreds of times more IgE circulating in their blood—and where allergy is almost totally absent—high IgE means worm infection, and that's it. Elevated levels of rheumatoid factor may mean lupus or another autoimmune disease in Europe or North America, but in Africa, high quantities of these self-directed antibodies indicate malaria, and nothing more. And whereas elevated CRP increases the risk of heart attack in the U.S., in the Amazon it signifies immune mobilization against infection. Cardiovascular health remains unaffected.

WORLDWIDE PATTERNS OF HEART DISEASE

In the broadest sense, and with a little creative interpretation, the global epidemiology supports the idea that losing helminths might predispose to heart disease. In India, China, and Africa, heart disease increases with urbanization, even in the single generation that moves from the country to the city. Dietary changes and a sedentary lifestyle usually get the blame, and certainly they deserve it, although maybe not for the reasons we think. (More on diet and the microbiota in a moment.) But we can't discount changes in immune function brought on by a loss of parasites.

Take South Africa in the early 1970s. Apartheid divided the country along racial lines. Native black South Africans tended to be poorer, and to suffer from more infectious diseases than other groups. As a result, they had the lowest life expectancy at birth of the country's ethnicities. But beginning in middle age, they managed to outlive the more affluent Caucasian and Indian South Africans. How? They were invulnerable, it seemed, to the degenerative diseases that plagued these groups late in life. "It is highly relevant that among South African Negroes, coronary heart disease remains virtually absent, and total cancer incidence (age specific) is still much lower than that prevailing with the local Caucasian population," observed one scientist in 1974.

As we've seen, at that time, this population was also relatively free from multiple sclerosis, helicobacter-related cancer, inflammatory bowel disease, and allergic diseases in general. Now we learn that they aged better as well. Was it all a fluke—the result of an ascertainment bias—or did this population simply have a superior immune system? And if so, how much of that first-rate immune functioning stemmed from greater contact with "old friends"?

Looking at present-day populations, the anthropologists Thomas McDade and Christopher Kuzawa have compelling evidence that early-life microbial exposure confers lifelong salutary benefits. Scientists have followed a cohort of mothers and their children in the Filipino city of Cebu since the early 1980s. Analyzing data gathered from this cohort, McDade and Kuzawa determined that adults who as infants had more diarrheal episodes, more animal feces at home, and more contact with mud during the rainy season had demonstrably lower CRP as adults. (They also confirmed the "fetal origins" hypothesis: underweight newborns went on to have higher CRP as adults.)

McDade then surveyed young adult Filipinos in Cebu. Looking at two immune-system signaling molecules, the pro-inflammatory IL-6 and the anti-inflammatory IL-10, he found an altered ratio compared with U.S. populations. The Filipinos had more anti-inflammatory signals and less pro-inflammatory. In essence, a different history of exposures had delivered an immune system with a less-frenzied idle speed than in the U.S. As a result, these Filipinos had a lower vulnerability to who-knows-how-many inflammatory diseases of modernity.

It's not that these individuals never mount an inflammatory response; rather, they unleash inflammation more judiciously. The Filipino immune system, McDade says, increases inflammation when needed, but quickly turns it off when the job is complete. He suspects that as Gurven monitors the Tsimane over time, he'll observe a similar talent.

INFLAMED, FAT, AND DIABETIC

Jeffrey Gordon at Washington University in St. Louis made the seminal observation that germ-free mice didn't gain weight no matter how much they ate. In a series of pioneering experiments—studies that sparked much of the work on microbiota we explored in chapter 9—he shattered the notion that microbes were along just for the ride. They were in the driver's seat. Transplanting the microbiota from fat to lean mice could make the recipients fat, regardless of diet. And the microbial shifts observed in lean and obese mice had parallels in lean and obese human twin pairs.

As you recall, in Belgium, meanwhile, Nathalie Delzenne and Patrice Cani observed that the microbiota drove obesity, metabolic syndrome, and the development of type-2 diabetes. (Remember, germ-free mice that gorge on junk food don't develop the syndrome. Microbes are key.) Low-grade inflammation has long been noted as a central feature of metabolic syndrome, but until Delzenne and Cani's work, few suspected that inflammation drove the development of insulin resistance and the accumulation of body fat—and that the microbiota fueled that inflammation.

The cascade of events the scientists described goes like this: Eat junk food, and the microbiota shifts. Certain microbes bloom, others decline. Microbial products begin leaking through the gut. They prompt systemic inflammation. Normally, insulin binds with receptors on the body's cells, prompting their absorption of sugar, but the ongoing inflammation interferes with the hormonal signal. That's insulin resistance. The incoming calories get stored as fat, but since you're not sated, you eat more junk food. More bacterial endotoxin leaks through, prompting more inflammation, and so on in a vicious cycle. If this goes on for too long, the pancreas is likely to collapse from exhaustion. Now you're diabetic.

Simply adding a plant fiber that only certain bacteria could ferment prevented the entire inflammatory cascade. Why? It kept a population of bifidobacteria thriving and happy. Those bacteria kept the gut barrier nice and tight. Nothing leaked through. No inflammation took hold. Metabolic syndrome never developed.

Between the 1970s and the 1990s, the rising prevalence of obesity in the U.S. closely tracked the increasing consumption of high-fructose corn syrup, mostly in sodas. Now, some two-thirds of the adult population is considered overweight, and one-third obese, which increases the risk of both heart disease and certain cancers. You'd imagine that scientists would be content to explain this increase in body mass on diet, but as is the case with cardiovascular disease, they think that lifestyle factors, including diet and inactivity, explain only some of the trend. Essentially, fifty years ago, more Americans could eat burgers, sodas, and candy bars without putting on pounds.

Again, on the surface, this idea seems ridiculous, but we've just seen two experiments—one involving helminths, the other bifidobacteria—where secondary exposures changed how animals responded to a junk-food diet. In both cases, quelling inflammation was key. So you might imagine that any increased inflammation, or weakening of regulation, could increase the propensity to become obese.

Indeed, gene variants that boost production of TNF alpha, an inflammatory signaling molecule that's important in fighting off pathogens, also, it turns out, predispose to obesity—but only among those eating a high-fat diet. Scientists made this connection after surveying twenty-first-century urbanized women in South Africa, a population that, in contrast with its mothers and grandmothers half a century earlier, was at ever greater risk for obesity. If 30 percent of their dietary calories came as fat, the women who carried promoter variants grew obese. Genes that helped defend us against pathogens in the past were now, by heightening inflammatory potential, contributing to weight gain.

What about contact with old friends—could they change the tendency to become obese? Scientists at the University of California, San Francisco, discovered that worm infection could halt metabolic syndrome in mice, and this despite a high-fat, obesity-promoting diet. The worms didn't preclude fatness by stealing calories from the host; rather, they interfered directly with the inflammation central to the disorder.

Macrophages, a type of white blood cell, populate fatty tissue in metabolic syndrome. They contribute to the problem by fomenting the inflammation that leads to insulin resistance and type-2 diabetes. But macrophages also have a kinder, gentler side—one that promotes not inflammation but healing. Immunologists call these macrophages "alternatively activated."

The UCSF scientists found that worm infection prompted macrophages to switch into this "alternatively activated" state, to transform from inflamers to healers. Either the parasite changed them in its attempts to tamp down inflammation that might dislodge it, or the host responded to worms with healing because it knew, from millions of years of coevolution, that large parasites migrate through tissue; time to get to work cleaning up the mess they left behind.

The worms' effect was profound and long-lasting. Infecting the mice for just eight days, and then eradicating the helminths, protected against obesity and insulin resistance for more than a month. "Can the rising incidence of metabolic syndrome, like the prevalence of allergies and autoimmunity, be exacerbated by the absence of certain parasites with which we have evolved?" asked Rick Maizels, the scientist who conducted the early work on worms, allergy, and T-regs, in a commentary in *Science*.

Perhaps. Although I've mostly ignored this worm until now, one helminth species did persist well into the late twentieth century: the pinworm. Not so long ago, *Enterobius vermicularis* was a childhood rite of passage. The small worms crawl out of the anus at night, lay eggs, and secrete a substance that itches like mad. That itch is how the worm enlists your help in finding a new host. You scratch. The eggs get on your fingers, under your nails. Maybe you ingest a few, self-infecting. They also spread to walls, curtains, and other children. They're extremely contagious, and if one child was obviously infected in a classroom, many more probably were as well.

Between the early 1970s, when 20 percent of samples submitted in New York City tested positive, and the mid-1980s, pinworms largely disappeared. "Is pinworm a vanishing infection?" asked one scientist a little ruefully in 1988. Others have since considered the worm's recent disappearance, which parasitologists tend to overlook, as perhaps contributing not only to the allergy epidemic but also to the increase in autoimmune type-1 diabetes. Quite by accident, in the late 1990s, the University of Cambridge scientist Anne Cooke discovered that pinworms could prevent autoimmune diabetes in mice bred to develop the disease.

Studies from Taiwan associate pinworm infection with protection from asthma and hay fever, suggesting a strong immunomodulatory effect. And now we've got to wonder, did the disappearance of this final helminthic holdout—and the immune priming it provided—not only contribute to the rising incidence of asthma in the same period, but also make children more prone to obesity? In the UCSF experiments, the protective effects of onetime helminth infection were ongoing. In the past, were young children similarly primed against obesity later in life?

Supporting this idea, scientists find they can cure type-2 diabetes and reverse insulin resistance in mice by directly augmenting the number of regulatory T cells. Again, these peacekeeping cells halt the disease without the mice discontinuing the high-fat diet. The lesson seems to be that crappy food alone doesn't incite metabolic syndrome. Your inability to quash unnecessary inflammation also contributes. Maybe not so surprisingly, one inflammatory disease engenders another in the womb. Mothers who are overweight during pregnancy have children with an elevated risk of asthma. Why? Mom's low-grade inflammation may program the fetal immune system for the same. As you'll recall, these mothers also have a slightly elevated risk of having a child with autism.

MICROBES, PARASITES, AND CANCER

Compared with their Alaskan and Greenlander brethren, the Canadian Inuit remained relatively unacculturated after World War II. Even then, Westernization proceeded unevenly in the Canadian Arctic. Communities in the central and eastern Arctic retained their traditional ways longer than those from the western Arctic. The uneven pace allowed observers to note that the more the Inuit acculturated, the more they began suffering from Western diseases. First came a wave of appendicitis. Then teeth began to rot. The younger generation grew more quickly than the older, and ended up taller. Inuit adolescents got acne for the first time. Asthma appeared in the most Westernized communities. So did heart disease. Girls reached puberty at younger ages. And the types of cancers shifted. In times past, tumors of the nose, throat, and salivary glands predominated—cancers now thought to partly result from viral infections such as Epstein-Barr. These subsided while cancers of the cervix and colon became more common. The first cases of breast cancer appeared.

"Few—if any—population groups have ever experienced such rapid changes in their way of living as the Eskimos," wrote Otto Schaefer, a German doctor who tended to the Canadian Inuit during this transition. He attributed the shifting disease burden to changes in diet. When the once free-ranging, migratory Inuit moved to town and settled down, they had a predilection for junk food, he noted. They switched from a diet of mostly wild game to one of highly refined, processed, sweetened foodstuffs.

Much science has indeed linked the refined, high-fat, easy-calorie Western diet with increased risk of everything we've covered so far, including cancer. And if you look at mortality statistics from the U.K. half a century earlier, you see Britons going through the same transition—more colon and breast cancers, more cardiovascular disease—but in Britain these diseases are stratified by class. The upper classes begin suffering from the degenerative diseases of civilization before the lower, who, in the early twentieth century, continued to succumb to infectious diseases such as tuberculosis and stomach malignancies we now know are associated with *Helicobacter pylori* infection. The classic reading of these trends is that Westernization and affluence allow us to indulge our innate sweet tooth, drink excessively, and avoid exercise, fruits and vegetables. These habits increase the risk of Western cancers.

But there's another aspect to this story, one that relates to—you guessed it—inflammation. In the past decade or so, scientists have implicated the same chronic low-grade inflammation we've been discussing in two aspects of malignancy: the transformation of a cell line from a cooperating member of the greater organism to one that goes rogue, and the growth and expansion of the tumor itself. In the first case, constant irritation incites cells to divide too often and too quickly, increasing the chances of a mutation. And in the latter, the inflamed milieu around the nascent tumor helps it recruit new blood vessels, co-opt nutrients, and expand into new tissue.

Inflammation is one reason metabolic syndrome increases the risk of cancer. (The other reason relates to excess growth hormones. More on that below.) By one count, between 15 and 20 percent of all cancers arise from inflammation. Others suspect the percentage is higher still. The instinct has been to blame microbes, such as *H. pylori*, for inciting these mutagenic fires. But some scientists now argue precisely the opposite. As is the case with allergic and autoimmune disease, they contend that microbial exposure and a properly educated immune system fight cancer.

As it did with heart disease, the epidemiology bears out the "old friends" hypothesis for cancer. Around the world, breast cancer risk follows the incidence of infectious mononucleosis, a marker of a clean environment. And other cancers of civilization, such as prostate and colon, occur inversely to cancers associated with infection, such as stomach cancer (*H. pylori*). What's more, when immigrants move from the developing world to the developed, they often have less cancer than the natives in their new homes, but the immigrants' children have just as much. In the U.S., this disparity is sometimes called the Hispanic paradox. And it extends to cardiovascular disease and mental health. Immigrants from Latin America suffer less often from all these problems than native-born, including their own children. How might this work?

In a series of studies on female textile workers in Shanghai, Harvey Checkoway at the University of Washington, Seattle, found that regularly inhaling cotton dust lowers the risk of pancreatic, lung (even for smokers), breast, and ovarian cancers. Why? Each speck of cotton dust contains a microbial universe. Protection was linked to how many microbes the workers encountered. Workers at plants that used synthetic fibers, on the other hand, weren't similarly protected. And exposure to silica in mills actually increased the risk of ovarian cancer.

The scientist Giuseppe Mastrangelo observed the same pattern in Italy. There, not only did working in a cotton mill prevent lung cancer, so did working in a sewage treatment plant, and on a dairy farm. The longer one worked with cows, the less one's odds of developing lung cancer. This held true irrespective of smoking. In Scandinavia, meanwhile, any profession that brought people into contact with many microbes—farming, gardening, fishing, and lumberjacking—seemed to protect against cancer.

How might microbes protect against malignancies? By enhancing both regulatory circuits and antitumor immunity. Cancer is not only a problem of a onetime cooperative cell line mutating and striking out on its own, but also of immune surveillance. If the immune system is paying attention, it destroys these rebel lineages before they expand and overwhelm the organism. Tumor growth is also, therefore, a failure to police. The constant stimulation experienced by cotton mill workers or dairy farmers may keep antitumor immunity on high alert.

At least, that's one interpretation. Another explanation centers on strengthening those same regulatory networks that protect against allergic and autoimmune disease. Work by Susan Erdman at the Massachusetts Institute of Technology highlights the importance of regulatory T cells not only in preventing cancer, but in reversing malignancies once they have taken root.

Again, this theory contradicts much accepted wisdom in the field of oncology. While those in the allergy field looked to the regulatory T cell as a savior, oncologists have generally seen it as a Judas. Suppressor cells, they think, protect tumors from the immune system that would otherwise destroy them. In cancer, T-regs betray you by shielding malignancies from elimination.

However, Erdman found that a T-reg's behavior in a tumor depended on its previous education. She began with mice that, unable to produce anti-inflammatory IL-10, could not tolerate their resident microbes. They developed colitis, and then colon cancer. (In humans, having ulcerative colitis also increases the risk of colon cancer.) Transferring regulatory T cells from wild mice into these mutants stopped the inflammation. More amazingly, transferred T-regs could shrink already established tumors. They choked off the fire fueling tumor growth.

But only T-regs from mice that had previously seen infections could stamp out the cancer. T-regs from clean donor mice that had never encountered a pathogen not only failed to quash inflammation, they became turncoats. Once in the inflamed environment of the tumor, they joined the fray and began fomenting inflammation themselves. Only battle-hardened T-regs held their ground. They could also quash breast and prostate cancer, two other malignancies of civilization.

Aside from suggesting a radical new approach to cancer treatment— stop the inflammation—Erdman's work also indicated that our anticancer immunity depends on priming from microorganisms. Hygienic living may weaken our ability to extinguish the inflammation that fuels malignant growth. Maybe that explains why people who take nonsteroidal anti-inflammatory drugs such as aspirin regularly tend to develop cancer less often. The drug helps quell low-grade inflammation.

BUT WE ALREADY KNEW THIS: COLEY'S TOXINS

After losing a young female patient to bone cancer in the 1890s, the New York surgeon William Coley looked into an old rumor. He'd heard stories that when patients with tumors contracted infections, their malignancies sometimes shrank and disappeared. He unearthed many documented cases of the phenomenon in the literature. So he devised a way to deliberately induce infection by inoculating patients with a streptococcal bacterium. Some died from the treatment, but others saw their cancers disappear. He refined his bacteria-based concoction, and a company began manufacturing it as "Coley's toxin." Over the course of his career, he treated more than a thousand people. But with the ascendancy of radiation and chemotherapy—originally derived from the mustard gas used during World War I—"Coley's toxin" fell from favor. To this day, however, oncologists and surgeons continue to document spontaneous remission of cancer following infection and fever.

Indeed, one treatment that follows in Coley's footsteps has gone mainstream. An infusion of *Mycobacterium bovis,* the vaccine originally developed to prevent TB, effectively treats superficial bladder cancers. Introducing living parasitic mycobacteria prompts the immune system to remove the tumor. This seems to occur naturally as well. In general, people who've had more infections and fevers are less likely to develop melanoma, scientists find. And the greater the number of illnesses suffered, the more protection from melanoma one has. When first observed formally in 1999, scientists explained this relationship as a case of enhanced anti-tumor immunity brought on by fever. But some argue that bacteria, and especially mycobacteria, have a special relationship with melanoma.

As you'll recall, 8 percent of the human genome consists of viruses that have inserted themselves in the past. When the mechanisms keeping them quiet fail, these viruses can reactivate and promote mad replication—cancer. The BCG vaccine, which consists of attenuated *M. bovis,* reduces the future risk of melanoma by 40 percent. Why? Both the bacterium and the cancer, driven by that reawakened virus, share certain molecular patterns. Exposure to the former protects against a reawakening of the latter. In the melanoma-fever study above, those with latent TB had the lowest risk of all, one-sixth the average.

Not long ago, of course, we were covered from head to toe in bacteria related to BCG. We harbored them in our tissues as tuberculosis infection. We imbibed them in water and dirt. Those internalized viruses would have had far fewer opportunities to reactivate. In Graham Rook's parlance, it's as if we outsourced control of a latent enemy to mycobacteria. And when we lost contact with these bacteria, we lost control of the virus within.

Melanoma is among the fastest-increasing cancers in the industrialized world. The incidence tripled between 1970 and 2000. Sun exposure and tanning beds usually get the blame, but it's hard to ignore the connection with fevers and mycobacteria, and the fact that we now suffer from both much less often than even sixty years ago. The scientists Bernd Krone and John Grange argue that the BCG vaccine, intended for TB, should be given universally just for protection against melanoma. "[I]mmunization strategies of the future may have a two-fold aim," they say—"to compensate for the loss of natural encounters with 'old friends' and protection against potential enemies lurking within."

THE MYSTERIOUS CANCERS OF YOUTH

Two other cancers of youth closely track hygiene. Hodgkin's lymphoma, a cancer of lymph tissue, follows the same pattern as allergic disease.

Younger siblings get it less often than older. Day care protects. In twin pairs, the twin with more exposure to microbes has a diminished vulnerability. And then there's Epstein-Barr: having had infectious mononucleosis, an indicator of late exposure to the virus and, more generally, a microbe-depleted environment, quadruples the risk of young adults developing Hodgkin's lymphoma. Another childhood cancer, acute lymphoblastic leukemia, hews to the same epidemiology. Firstborn children have it more than later-born. Early day-care attendance protects.

While both these cancers have increased in the developed world, they remain relatively rare in the developing world. The pediatric oncologist Mel Greaves suspects that delayed colonization by a still-unidentified virus is at fault. Weak regulatory circuits may also play a role. Children who develop the leukemia have an elevated risk of allergic disease, it turns out, direct evidence of immune dysregulation.

And then there's colon cancer, the number-two cancer killer in the developed world. (Lung cancer ranks as number one.) The University of Pittsburgh gastroenterologist Stephen O'Keefe worked in South Africa for years and never saw colon cancer or polyps, but when he started practicing in the U.S., every second African American patient, it seemed, had polyps. The difference in risk was stark: 1 in 100,000 rural Africans got colon cancer, versus 1 of every 1,500 African Americans and 1 of every 2,000 whites. O'Keefe's attention was immediately drawn to diet—not the absence of fiber in the U.S., necessarily. The Africans he studied ate lots of corn, a low-fiber crop. Rather, he wondered about the relative abundance of meat in the U.S. diet. What kind of microbiota was it cultivating?

The colon houses the bulk of our resident microbes. So intimate is the relationship that some colonic cells derive their energy not from circulating blood but directly from by-products of microbial fermentation. O'Keefe noted that the rate at which these cells proliferate, a risk factor for colon cancer, was much lower in black South Africans compared with African Americans, who got much more colon cancer. The microbial communities they harbored differed as well. The black South African microbiota produced far more butyrate, which is both anti-inflammatory and fuel for those colonic cells. "The secret is that there is no secret about good nutrition whether it's for the heart or the colon," says O'Keefe. Fruits, nuts, and veggies, he says, not hamburgers, sodas, and fries.

As always, large parasites can't be ignored. Scientists have long debated the inverse relationship between allergic disease and cancer. Essentially, people with allergic disease develop several malignancies less often. Some have speculated that the allergic response is really a form of antitumor immunity. Sneezing, in other words, is an unfortunate consequence of having beefed-up anticancer surveillance. The actual relationship is probably more complex. It's likely that in the past, living parasite infection enhanced antitumor immunity without causing allergic disease.

Indeed, French scientists find that eosinophils, the cells that help expel worms from the gut, are also quite adept at inhibiting and killing colon cancer cells. And surveys indicate that colon cancer occurs in inverse proportion to how many eosinophils any given individual has in circulation. The more one has, the less the chance of colon cancer. There's certainly a pleasing symmetry to the idea that one cell type would handle two large multicellular problems: worms and tumors. Here's the broader point, however: Throughout human evolution, the person whose gut remained parasite-free for his or her entire life, and whose eosinophils were never called to action, would have been a rarity. Parasites would have continually primed antitumor immunity.

THE IMMUNE SYSTEM AND HORMONE BALANCE

Cancer is a complicated beast, and clearly factors other than inflammation contribute to its emergence and metastasis. One of them is hormone balance. Growth hormones, as their name suggests, help the body build muscle and bone. Sex hormones, which cause body and facial hair to grow on men, and breasts to emerge on women, prepare you for procreation. Elevated levels of all these hormones also increase the risk of several cancers.

Compared with hunter-gatherers, agriculturalists, and people from the developing world, Westernized populations have skyrocketing levels of both sex and growth hormones. Western men in the prime of life have testosterone levels that soar above those of men living in Congo, Nepal, and Paraguay. During menstrual cycles, Western women have far more circulating sex hormones compared with non-Westernized women. "Ovarian function in western women is not, as often considered by western biomedicine, the human 'norm,'" writes Tessa Pollard in *Western Diseases: An Evolutionary Perspective*. "The high levels of ovarian hormones seen in pre-menopausal western women are, in fact, evolutionarily novel." Why?

Energy balance partly determines the set point for hormone levels, and food availability is a huge determinant of energy balance. Obviously, we have more food than perhaps ever before. So it's tempting to blame those overblown hormone levels on the Western diet. But that is likely an oversimplification. There's also a hierarchy in how the body invests its energy. Hormones, which represent an investment in the future, rank toward the bottom of that list. So if your life is active, like a hunter-gatherer's, you'll spend energy on day-to-day activities rather than on hormones. Likewise, if you're constantly fighting off parasites and pathogens—immune activation is quite energetically costly—you'll spend energy on self-preservation now before investing in hormones for the future.

What's more, in a wormy, pathogen-filled environment, testosterone, which suppresses the immune system, has an immediate cost: parasites will overrun your body. (The comparable female hormone is progesterone.) As you'll recall, the dominant male in chimpanzee troops—he with the highest testosterone levels—also carries the heaviest worm burden. As a proof of principle, experimentally infecting rodents with parasites sends their sex hormones plummeting. So it's likely that during our evolutionary past, and in many existing people's present, there was a ceiling to how high sex and growth hormone levels could rise, and that ceiling was partly set by parasites.

Indeed, scientists commonly observe that children grow less hurriedly in parasite-ridden, infectious environments. They don't necessarily end up stunted, but they grow at a different pace. The human organism is probably spring-loaded to invest in growth and procreation when pressures that threaten immediate survival ease off—and the pressure has definitely eased off. Probably not coincidentally, in the West, the velocity of growth has increased steadily for a century. Puberty has arrived earlier and earlier. Both are risk factors for metabolic syndrome and certain cancers.

Bangladeshi women who immigrate to the U.K., meanwhile, maintain the progesterone levels of their homeland, which are much lower than those of native-born Britons. But Bangladeshis who migrate before puberty end up with progesterone levels similar to those of British-born women—and this despite their mostly sticking to a traditional diet. The major difference: immune challenge in Bangladesh, and the lack of such challenge in the U.K. The Western Hormonal exuberance may result in part from an unprecedented freedom from immune activation.

The instinct is often to discount the cancers of the West as the result of a longer life. We've got more elderly people around; cancer is an affliction of old age; that's why breast, prostate, and colon cancer seem so common in the U.S. and Europe. But if you compare the same age group in different regions of the world, you'll still see dramatic variation in the risk of these cancers. Your average sixty-five-year-old American woman has quintuple the risk of breast cancer compared with your average sixty-five-year-old Chinese woman, for example. For prostate cancer, the gap is wider still—a seventy-four-fold difference between China and North America. And the invulnerability is not genetic. The children of East Asians who immigrate to the U.S. quickly acquire the same risk of "American" cancers as native-born Americans, even as they stop developing cancers common back home, such as stomach cancer. Intriguingly, American-born women whose grandparents hailed from rural areas in East Asia have a lower risk of cancer than women whose grandparents came from urban areas. One interpretation is that the robust immune education their grandmothers received continues to reverberate epigenetically two generations later, on a different continent.

None of these observations absolve diet and inactivity, but they suggest that we ignore the immune system's contribution at our peril. And they remind us that immune dysfunction has far-reaching consequences. It may even influence our state of mind.

AND HOW DO YOU FEEL?

Back in the late 1990s, the neuroendocrinologist Christopher Lowry began injecting *Mycobacterium vaccae* into mice. *M. vaccae* was the bacterium isolated from Ugandan "hippo mud" we heard about in chapter 7 that beneficially stimulated the immune system. Lowry wanted to understand how the lung and the brain communicated, so he stimulated the former and monitored the latter. He was surprised to see that after *M. vaccae* injections in the body, special neurons in the brain began churning out serotonin.

Insufficient levels of this hormone contributed to depression, he knew, so it seemed that he'd just improved the rodents' mood—with a bacterium. He tested the treated mice. They were less sensitive to stress as measured by a forced swim test, a sink-or-swim scenario that gauges animals' tendency to become gloomy and despondent. Immune signals had boosted brain serotonin, and that translated to enhanced emotional resilience. Graham Rook, a coauthor, pointed out that these signals would have been constant in times past when we were liberally slathered with earth and mud. It seemed ridiculous, but they had to wonder: Could some of the increasing rates of clinical depression in modernity result from having lost this stimulus?

Over the same period, trials testing *M. vaccae* as immunotherapy to treat malignancies were in the works. The underwhelming results suggested failure; mycobacteria didn't help with cancer. But when John Stanford, the scientist who'd discovered the bacterium decades ago, reanalyzed the data, organizing the cancers by type and excluding participants who hadn't complied with the regimen, he observed two phenomena. First, *M. vaccae*–treated patients with adenocarcinoma of the lung had, in fact, lived longer. And second, everyone who received the injections reported improved well-being irrespective of increased survival. They felt better. Lowry's findings indicated how and why. The immunotherapy was increasing serotonin in the brain.

Meanwhile, the psychiatrist Charles Raison was looking at depression from a different angle: inflammation. He was exploring why up to half of patients undergoing therapy for hepatitis C, a potentially deadly viral infection of the liver, fell into a deep funk. The therapy involved injecting a pro-inflammatory signaling molecule called interferon alpha to amplify the immune response and help clear out the virus. By increasing inflammation, Raison realized, the medicine was inducing depression.

Raison expanded his inquiry. Chronic fatigue, he found, was also associated with low-grade inflammation. Others made similar observations in the context of cancer treatment. In this case, oncologists deliberately cranked up inflammation in an effort to provoke immune recognition of the tumor. Patients undergoing this therapy often sank into despair. Now scientists understood that it wasn't just their predicament; the amped-up inflammation depressed them.

Depression was a common "comorbidity" in heart disease, cancer, and obesity in general, which wasn't surprising. They're depressing maladies. Now scientists wondered if the low-grade inflammation common to these disorders might not be fueling the depression. Indeed, they found that inflammatory markers predicted low spirits better than obesity or heart disease. The classic immune-mediated diseases were also directly associated with mood disorders. Asthma in childhood increased the risk of depression in adulthood. Autoimmune diseases such as Crohn's and psoriasis often co-occurred with depression. Again, the obvious interpretation was that having a disease with no cure was distressing, thus the despondency, but now scientists began to suspect that the inflammation underlying the disease drove the melancholy directly.

Treatments for depression were, many noted, decidedly anti-inflammatory. Exercise, one of the most effective "natural" approaches to bettering one's mood, both increased serotonin and induced a range of anti-inflammatory immune responses. Omega-3 fatty acids, which are anti-inflammatory, improved mood in one study. Even serotonin-reuptake inhibitors, the gold standard of pharmacological treatment for depression, were, scientists found, anti-inflammatory.

Raison, Rook, and Lowry spelled out much of what I describe here in a 2010 *Archives of General Psychiatry* paper that, judging by the attention it received online, struck a chord. Here's their reading: Stressful events, like your boss yelling at you, clearly have a role in precipitating gloom. Acute stress incites a burst of inflammatory immune activity. Normally, that inflammation resolves, but in modernity, it was prone to entering a permanent feedback cycle. The signal got stuck in the "on" position. Chronic inflammation interfered with mood-regulating hormones. What should have been a passing low became a permanent funk.

They don't discount the weakening of social networks in modernity for some of this tendency—isolation is unhealthy for this social ape—but they also think that our inflammation-prone immune system, which results from a loss of contact with "old friends," can't be ignored. With the right microbial stimuli in childhood, we'd perhaps better handle the boss's tirades. The immune tolerance taught by microbes and parasites has direct bearing on emotional forbearance.

Raison, Rook, and Lowry's hypothesis lends credence to the baffling—and only half-believed—observation that well-housed, -clothed, and -fed North Americans suffer from about twice as much depression as Africans, who surely face more disease, privation, and other threats to their immediate well-being. Maybe there's a real biological difference underlying these observations. Maybe it's the immune system.

In Western Europe, where health care is mostly free, the prevalence of depression follows that of allergic disease: less in the country, more in the city. And anthropologists who study the Tsimane (not Michael Gurven) find them remarkably even-keeled and happy. When they measure Tsimane levels of cortisol, a stress hormone, they find them to rank among the lowest ever recorded for humans.

These observations suggest a new approach for treating depression that targets immune function. Immunotherapy is one possibility; another is to treat the microbiota. Fix the dysbiosis, the community derangement, and you fix the funk. Fecal transplants, presumably with microbes from a donor with a sunny disposition, may one day be standard treatment for clinical depression.

AGING GRACEFULLY

We've gone from immune dysfunction in youth (allergies and asthma) to immune derangement in adulthood (autoimmune disease) and chronic inflammation in the diseases of middle age (cancer and heart disease), have explored how we feel about the voyage (depression), and now we arrive at the twilight years: old age.

What's the secret to aging successfully? Obviously, reaching old age at all depends on avoiding the degenerative diseases mentioned above, depression included. Just as clearly, survival takes considerable inflammatory capacity. You must fight off the pathogens and parasites that would otherwise turn you into dinner. Unfortunately, aging is associated with something called immune senescence. The immune system loses its pizzazz. Long quiescent viruses, such as chicken pox, reactivate to cause shingles. Latent tuberculosis becomes active.

But there's another, somewhat paradoxical side to immune senescence. The weakening of the immune system leads to low-grade inflammation, and this inflammation, gerontologists think, drives the aging process: frailty, cognitive decline, Alzheimer's, and even Parkinson's. Scientists have coined a term to describe the connection: inflammaging.

So it's notable that the elite group of people who survive to very old age—ninety and beyond—often have a talent for quashing inflammation. Inflammatory potential helps you fight off pathogens, but promptly turning it off allows you to avoid wear and tear. Some studies find that, comparatively speaking, the superold produce bucketloads of anti-inflammatory cytokines such as IL-10 and TGF beta, and that they produce less pro-inflammatory IFN gamma. Gene variants that enhance this anti-inflammatory capacity are overrepresented among the elderly. In Turkey, people with a variant that leads to less IL-10 don't reach old age as often as those who naturally produce more. Among elderly Bulgarians, gene variants that churn out more pro-inflammatory TNF alpha are also relatively scarce. The same is true in Sicily. There, genes that augment pro-inflammatory responses are similarly underrepresented among centenarians. A gene variant that attenuates this inflammatory signal, meanwhile, is overrepresented.

Before we proclaim noninflammation as our savior, it's important to remember that survival depends on balance, and the optimum balance depends largely on context. That context is defined, in part, by coinfections. An elegant comparative study by Rudi Westendorp and colleagues at Leiden University illustrates this principle. Among Ghanaians living in a highly infectious environment, having genes that amped up the pro-inflammatory response increased the likelihood of surviving to old age—but only for those who regularly drank from a river or well. In an area near a borehole that provided clean, uncontaminated drinking water, those same gene variants were underrepresented among the elderly. There, presumably, too much inflammation was a liability. The people with pro-inflammatory genotypes had worn themselves ragged early.

Here was the twist: Westendorp and his colleagues couldn't be certain if having lost worms—clean drinking water meant fewer parasites—might have served as a tipping point. Eighty-five percent of Ghanaians also had malaria. If they lost helminths, but still hosted *Plasmodium falciparum,* they might become more vulnerable to complications from malaria. These pro-inflammatory genes, in other words, might have worked best in the context of worm infection. Absent the dampening effect of helminths, they overshot.

Any mention of malaria, of course, brings to mind Sardinia, land of centenarians. Some suspect that Sardinians live so long precisely because they have a genome honed by the malaria parasite. Surviving malaria requires an intermediate response—not so strong as to kill yourself, but not so weak as to allow the parasite to overwhelm you. That's also what's required for successful aging. And although proof is lacking that Sardinian longevity has anything to do with malaria, there is evidence from elsewhere in the Mediterranean that a balanced immune response is critical to successful aging.

Among Italian, Greek, and Tunisian elderly (over ninety), the nonfunctioning version of a gene called Chit-1 is relatively common. People with two functioning versions—two versions that promote attack—don't survive to old age as often as those with just one. They're so quick on the draw that they often shoot themselves in the foot. But people with two nonfunctioning versions don't make it either. They're not aggressive enough. Only those with a copy of each—balance—join the ranks of the superold.

Thankfully, scientists estimate that only 15 to 30 percent of longevity can be attributed to genetics. How well you age depends much more, in other words, on modifiable factors—diet, sun, exercise, happiness—than genes. One of those adjustable factors is the immune response. When it comes to aging, we may be able to tweak the immune system by targeting the microbiota. In mice, probiotics can correct the natural drift of the microbiota that occurs in the elderly, boost levels of bifidobacteria, and sharpen the immune response.

But there's another, less-addressable issue here. Today's centenarians had a very different immune education compared with tomorrow's. Those old Sardinians—and all elderly around the Mediterranean rim, in fact, where people often age well—matured while being challenged by malaria, worms, tuberculosis, and who knows what else. Doubtlessly, they also benefitted from the Mediterranean diet and lifestyle. But the unanswered question is how much did this favorable immune tuning early in life help?

The more dreadful prospect is that the current generation—one with a lousy immune education and a propensity for immune dysregulation evident in childhood—won't age nearly as well. The concern is warranted. In the U.S., with an eye on the obesity epidemic (an inflammatory disease), public-health types bemoan that the current generation may be the first since the sanitary reforms of the nineteenth century to have a shorter life expectancy than its parents.

And while diet and inactivity certainly play huge roles in this apparent backpedaling, so does immune malfunction. It's no coincidence that the suite of Western diseases—heart disease, certain cancers, metabolic syndrome, maybe dementia, allergy, autoimmune disease, depression, and even acne—share uncontrolled, self-damaging inflammation as a core feature. The modern immune system routinely fails to terminate inappropriate conflagrations.

Now, a foray into the community of people trying to address these problems on their own. You've met some of them—Dan M., who treated his MS in chapter 10, and Shelley Schulz and her autistic son, in chapter 11. There are hundreds more. They mostly suffer from the worst of these diseases. They see no good options in Western medicine. And they've taken matters into their own hands. They are the hookworm underground.

The Hookworm Underground: Crowdsourcing "The Cure" in an Age of Immune Dysfunction

The caution is understandable: helminths are living organisms, not chemicals that can be precisely dosed, and they can cause illness. But given that there are already hundreds of people experimenting with therapies, it is clear that caution has been overtaken by reality.

—2011 *New Scientist* editorial entitled "Let Them Eat Worms"

For Jasper Lawrence, trouble began with a sting in the late 1980s. He and his brother Caleb had a landscaping business in Santa Cruz, California. While they were clearing brush one day, a swarm of insects attacked. Jasper remembers the aggressors as bees, Caleb as yellow-jacket wasps. Whatever the case, they accosted Jasper, stinging him repeatedly. His face swelled. His throat began constricting. Caleb rushed Jasper to the hospital, where doctors treated his anaphylactic reaction with a shot of adrenaline.

Lawrence had suffered from hay fever since his youth, but just months after that fateful encounter, his seasonal allergies morphed into a case of adult-onset asthma. He lost his breath climbing stairs. Cats sent him into a tizzy. The usual bronchodilators didn't help. Only the steroid prednisone offered some relief. But it caused him to gain 40 pounds. And still he ended up in the emergency room several times a year. He describes himself during that period as "fat, pale and sallow, with dark circles under my eyes." Depression set in.

Then, while visiting family in England during the summer of 2004, he heard about David Pritchard's work, his observations about hookworm infection, the relative absence of asthma in Papua New Guinea, and his planned experiments using hookworm to treat asthma. To Lawrence, the theory made perfect, sublime sense.

That same day, Lawrence decided to acquire hookworm himself. The task proved harder than he'd imagined. The Centers for Disease Control had logged next to zero natural hookworm infections in the U.S. for years. He was surprised to find that no one sold the worms on the Internet. And so, after researching countries with high hookworm prevalence, Lawrence set off to acquire the parasite the old-fashioned way: by treading on infective larvae in the wild.

In 2005, he traveled to Cameroon, a country in West Africa where *Necator americanus,* the species he was after, was common. His strategy went like this: he sought remote villages, asked where the latrine area was—usually there was a designated outdoor area for defecation—removed his shoes, and then walked around barefoot in the waste. Children often ran screaming at the sight of a white man. Adults sometimes asked him aggressively what the hell he was doing. He talked his way out of it, or sometimes paid them off.

He carried on like this for two weeks.

Hookworm goes through the skin, hitches a ride on venous blood through the heart to the lungs, migrates over the pharynx, passes through the stomach, and latches on in the small intestine. Lawrence developed a cough five days after his first barefoot stroll, he says. After returning home to California, he isolated eggs from his stool. He'd acquired *Ancylostoma duodenale,* he thought, which lives between one or two years, not the longer-lived *Necator americanus* he originally sought. He cultivated larvae from his own eggs, and built up his worm colony. And one miraculous day as he was driving with the windows down, he realized his eyes weren't full of gunk, his nose was open and clear, and he was breathing freely.

He'd sent his asthma and allergies into remission.

And now he had an idea: He'd start selling the worm to others with autoimmune and allergic diseases. To do that, however, he needed the hookworm species he'd originally wanted. *A. duodenale* sucks more blood than *N. americanus,* and it doesn't live as long. It can go dormant in its host and infect nursing infants via breast milk. For a proper worm-supplying business, he'd need *N. americanus.*

In the meantime, he posted online a version of the story I have just told. A fellow named Garin Aglietti contacted him. In an effort to treat his psoriasis, Aglietti, as you already heard in chapter 1, had traveled to Kenya in search of tapeworm. He'd succeeded, but the experience of passing the mature tapeworm's egg sacks—called proglottids—had proven disturbing. He terminated that experiment, and decided to procure a smaller parasite.

After reading about Lawrence's experience online, Aglietti wrote to him. They chatted and bonded over their similar adventures. They decided to join forces.

Partly funded by Lawrence, Aglietti headed off to a remote area of Peru, microscope in tow, where *N. americanus* was common. He set about collecting stool samples. He slept in a rented hammock. He eventually isolated eggs, incubated them, and self-infected with a few larvae. But when he returned to the U.S., he found that the worms had apparently failed to take. Lawrence was rankled; Aglietti was disappointed. Nonetheless, the two planned a second trip, this time to a remote area of southern Belize near the border with Guatemala. According to their research, people there hosted *N. americanus* almost exclusively.

This was the first extensive amount of time they'd spent together. From the beginning, tension flared. Aglietti found Lawrence woefully unprepared. He showed up without a microscope; he didn't seem to understand the parasite's life cycle. Lawrence found Aglietti pretentious—a year of medical school hardly made him an expert. And Lawrence understood the life cycle just fine. Eventually, they isolated hookworm eggs from stool samples finagled from Mayan villagers and cultivated infective larvae. Depending on who tells the story, it was *he* who was responsible for successfully isolating and incubating the parasite.

As Aglietti recalls, Lawrence self-infected with one hundred and he with sixty. (Lawrence says he infected with two hundred.) They returned to the U.S., importing their future livelihood in their bowels. Back in California, Lawrence fell deathly ill and dewormed. Aglietti held on, but barely. Strange eruptions appeared on his skin. He felt weak.

To avoid trouble with U.S. regulatory authorities, they began collaborating with Jorge Llamas, a holistic doctor in Tijuana. But they couldn't resolve their differences. Lawrence took issue with Aglietti's use of tapeworm. (Later, Tyra Banks also lambasted him on her talk show for trumpeting what she called the "tapeworm diet.") Aglietti found Lawrence reckless and potentially dangerous.

Not long after returning from Belize, they parted ways. Lawrence sold worms from a website called autoimmunetherapies.com. "Harness nature to heal," the home page read. "Stop treating the symptoms, fix the problem." Aglietti eventually launched his own site, called wormtherapy.com. He posted before-and-after pictures of the psoriatic plaques on his elbows.

Lawrence also started a Yahoo discussion group dedicated to all things autoimmune, allergic, and helminth-related. He had experience in Internet promotion, and he bought a number of related domain names—asthmahookworm.com, helminthictherapy.com, even hygienehypothesis.

com—and linked them to his business site. People searching for the therapy were inevitably guided his way.

At first, he charged $3,900, and guaranteed three years of infection. He added human whipworm to his menu. (Later, when Aglietti charged less—$2,300—Lawrence reduced his price to $2,900.) He'd retained the relationship with Jorge Llamas, but over time, he traveled less often to Mexico to treat people. Sometimes he flew to meet clients in the U.S. And he also began shipping hookworm self-infection kits—a vial of larvae in fluid, a plastic pipette to remove them, and a bandage to place them on—directly from his home in Santa Cruz. Word was getting around. A CBS affiliate in San Francisco did a story on Lawrence and his "patients." ABC wrote about him on its website. Even the *New York Times* mentioned him in a 2008 story about David Pritchard.

And then near midday on November 3, 2009, the FDA showed up at Lawrence's house in Santa Cruz. The local office had caught wind of his operation from one of the news stories. At first, the agent seemed more curious than anything else, as Lawrence recalls, eager to ensure that Lawrence was complying with safety and quality protocols. But the tenor of their interaction quickly changed. A second inspector appeared horrified by his in-home lab—which included sand, sugar, buckets, and heaters, and of course involved stool. Lawrence had heard rumors about people who'd unsuccessfully tangled with the FDA. The stories involved fines and jail time. He was frightened. His wife, Michelle Dellerba, who helped him with the operation, was trembling with fear. Less than a week after the FDA agent showed up at his door, Lawrence and Dellerba dropped everything and left for Mexico. A few months later, they were back in the U.K., where both had family connections and citizenship.

JASPER LAWRENCE IN PERSON

And that's where I first meet Jasper Lawrence. I find him at his brother-in-law's house on the outskirts of Lancaster, a picturesque town close to the northwestern coast of England. He has blue eyes, a slightly pockmarked complexion, and bears a passing resemblance to the British pop singer Phil Collins.

I had planned to begin my worm-hosting experiment there. A few minutes after I sit down at a worn wooden coffee table in the living room, Michelle Dellerba approaches with a small plastic vial presumably containing hookworm larvae. How many do I want? she asks.

Lawrence inquires how I'm planning to pay.

"Hold on," I say. Before coming, I'd asked that he get tested for dangerous viruses such as HIV and hepatitis—tests that the scientists at the University of Nottingham studying hookworm had conducted as well. Does he have his results? I ask.

No. The British health-care system isn't like the American, says Lawrence. It's impossible to walk in and get blood work done.

I'll wait until you get the tests, I say.

Michelle shrugs, sucks the worms out with a pipette, puts them on gauze, and applies them to her own arm. "That's how you do it," she says. Both she and Lawrence have discolorations on their left forearms, presumably from constant reinfection. We retire to a lawn in the back. Lawrence and I talk. The way he tells it, he had a rough, unhappy childhood in the U.S. For a society that supposedly values individualism, the U.S., he thinks, is brutishly conformist. The class-conscious U.K. is much less insistent on conformity. He was a punk during adolescence, he says. He stopped reading books after Richard Dawkins's *The Selfish Gene*. "I didn't want to pollute my mind with other ideas," he says.

A Radiolab piece in the U.S. that featured Lawrence had just brought him nationwide attention, the second in as many years. And tomorrow another favorable, feature-length article on Lawrence will appear in *The Guardian*'s Sunday newspaper *The Observer*. He's getting the kind of free advertising that most businesses could only dream of, but he rails against journalists: They get everything wrong. They don't care about the science. All they want to know, he says, is "what it was like putting your foot in a pile of shit."

"Well, how was it?"

"Fucking terrible."

He chain-smokes hand-rolled cigarettes throughout. At one point, he thanks David Pritchard. "Pritchard allowed me to start smoking again," he says. He got into the hookworm-selling business because he wanted to have an impact. "I understood I was gifted and I wanted to make my mark," he says, "to start the very first human parasite clinic in the world." A few moments later, he adds that he'd always wanted to coin a word. "*Helminthic therapy* is largely my word," he says. He corrects himself. "'Invented'—I wouldn't say that. But popularized." He's keen on me realizing, however, that he's not in this for the money. "There are easier ways to make money," he says.

I had first spoken to Lawrence late in 2007, after I'd found his story online. He'd come across as passionate and articulate. His half-British accent only added to this impression. He was just the sort of visionary, slightly crazy character who, because his outrageous-seeming assertions were probably correct, appeals to anyone in the business of telling stories. He also had a sardonic sense of humor. "I'm extraordinarily average," he'd said about himself, after he'd recounted his story of walking through fields of excrement in Cameroon. He offered up fantastic sound bites: "I feel like the guy in a bad horror movie, the only one who knows the truth, but who everyone thinks is crazy," he'd said.

I'd subsequently watched as more and more people tried the therapy, and read their almost unbelievable testimonials in the Yahoo group. I'd heard about the FDA's arrival, about Lawrence's fleeing. And I'd observed that some interactions on the Yahoo group took on a cultlike tenor. Admittedly, Internet chat rooms don't bring out the best in people, but ridicule seemed to fly often. Sometimes it bordered on bullying. Any opinion besides the party line—that helminths cured allergy and autoimmune disease—was derided and dismissed.

Now that I'd met Lawrence, I could see that he was indeed emotional and a little megalomaniacal—rather like how he came across in cyberspace. He tended to confuse his understanding of parasites, which was certainly extensive, with actual training in parasitology and immunology. He displayed a baffling conviction in his own expertise, ranging from his assessment of the consequences of hookworm's migration through lung (none, he insisted, although scientists cite pneumonia as a possible complication), to the chances of its transmitting a virus (none, although the Nottingham scientists addressed the issue in their human hookworm studies).

These qualities didn't strike me as intrinsically problematic. They could pertain to your average CEO. As long as Lawrence was helping people, the good seemed to outweigh the bad. And I'd spoken to many people whom Lawrence had helped.

But then several things came to my attention. First, I discovered that Lawrence had posed as a patient online to drum up business. He'd explained his symptoms (asthma and irritable bowel syndrome) on a site called curezone.com, announced he was planning to visit a clinic in Mexico he'd heard about (his own) to try this new treatment (worms), tracked his progress after his fictitious visit, and then announced remission. "This is not the placebo effect, it is not wishful thinking. It worked," he wrote in early 2008. He added a link to his own operation.

Then Aglietti forwarded me an e-mail from Lawrence dated to January 2007. Lawrence had sent it after Algietti's trip to Peru. He'd succeeded in enrolling in one of the University of Nottingham trials testing hookworm on asthma. In the e-mail, he said that he'd filched a vial he thought contained the worms. He applied the solution to his arm. He said he felt the itch. "It was pretty funny and scary, and makes a decent story," he wrote to

Aglietti. The vial must not have contained living larvae, however. He and Aglietti were still one foray to Central America away from acquiring a living parasite. (Johanna Feary, who led the study, says patient confidentiality prevents her from commenting, but that, as a matter of course, study participants weren't left alone near viable larvae. The trial results, published in 2009, say two people dropped out, one because of abdominal pain, and another, from the placebo group, due to "psychological problems.")

In short, Lawrence had posed as a patient to encourage business. And he'd tried to steal parasites from the scientists whose work was the only thing lending his otherwise questionable operation even a veneer of legitimacy. The downside of the Lawrence phenomenon suddenly seemed to overshadow the upside.

When I asked Lawrence in a phone conversation about the attempted theft—was it true?—he said he didn't know what I was talking about. When I read him the e-mail he'd allegedly sent Aglietti, and suggested that perhaps it was a forgery, he relented. "I'm embarrassed and appalled by my behavior," he said. "Things were not going well for me . . . And I really, really wanted hookworm." Didn't he worry about interfering with important science? No, he said. No way he could have.

How did he know?

"How on earth could that interfere with their science?" he responded.

When I asked about his posing as a patient, he again proclaimed his shame, and then rationalized it. What he'd done was no different from what any pharmaceutical company ad does, he said. They get actors to pose as patients and tell stories.

But the viewer knows it's an ad, I said.

That didn't matter, he replied, because his story was fundamentally true anyway. He took hookworm, and he saw improvement.

"I wasn't setting out to deceive," he said. "And what I said wasn't a lie."

Our subsequent interactions spiraled downward. He asked that I pass along FDA documents on his case I'd acquired via Freedom of Information requests. I declined, but directed him to online instructions on how to file his own FOI request. By that time, I'd started my own hookworm experiment with Aglietti. Lawrence was particularly incensed by my having used Aglietti's services. (He'd spent considerable breath during our first meeting haranguing Aglietti, whom I hadn't yet met at that point.) Hadn't I crossed some journalistic line? Hadn't I lost my objectivity? I explained that one journalistic approach was, in fact, to test the goods.

He exploded in a final e-mail. After years of watching him write in to the Yahoo group, it came off as vintage Lawrence.

"Fuck you, you dilettante," he wrote. "Have you ever walked in someone else's shit in the tropics . . . ? You have not even begun to grasp the subject you have chosen to write about, and your naiveté is obvious in almost everything you say. You have not grasped how important this is to any one of us who was, or is, truly sick. . . . You are asking the wrong questions. How about: why would a person go to Africa and risk his life, why would he steal, why would he mortgage his house, why would he sacrifice his entire life, to obtain for himself this therapy, and then make it available to others believing it would likely result in his becoming a fugitive or a prisoner?"

What still failed to register with Lawrence was the importance of credibility. Few would knowingly choose a pharmacist or doctor who stole the medicine they hawked (worm larvae) and then posed as a patient using it. And given that the hookworm underground operated beyond any oversight, his credibility was more important still. "I was afraid to take larvae from him," one former client confided. "God knows what he is giving me." Indeed, this client, who shared his story anonymously, and others told me that learning about Lawrence's posing as a patient invariably prompted them to question his entire story. A digital picture of Lawrence's stamped passport shows that he did go to Cameroon (assuming it's not faked), but did he really acquire hookworm—or any worm—there?

Aglietti, for one, doesn't think so. He maintains that, in Belize, Lawrence clearly didn't understand the parasite's life cycle, and that he seemed surprised by the hookworm itch—evidence, in Aglietti's mind, that he'd never dealt with it before.

There's clearly no love lost between the former partners, so Aglietti's assessment should be treated with care. But Lawrence's story presents other inconsistencies. After Cameroon, he claims to have cultivated larvae from the eggs he himself shed and built up a colony. But in our final conversation, he divulged that he didn't really know what he'd gotten in Cameroon—maybe it was *Necator americanus* after all.

Moreover, before he had a replacement worm in hand, Lawrence says he dewormed. Aglietti had just landed in Peru when Lawrence notified him that he'd purged his African worms. Lawrence told me he was certain Aglietti would return with the right species—thus the premature deworming. But the explanation seems forced. You travel halfway around the world, send your agonizing disease into remission by walking through latrines, spend weeks cultivating larvae, and then casually rid yourself of the curative agent before you have a replacement? And, to begin with, you may have had the correct species all along?

None of these observations are indictments. Only Lawrence knows if his foundational myth is fabricated. In one important respect, it doesn't really matter. He has a worm now. And his self-important bluster aside, he does get one thing right. People are suffering. They have terrible diseases. They may even be at death's door. Modern medicine may present options that range from weak to terrible to none at all. Assuming you're getting what's advertised, hookworm acquired from underground sources may be the most rational choice. And Lawrence and Aglietti's worms have clearly worked wonders for some.

HERBERT SMITH SAVES HIS GUT

Meet Herbert Smith. He's slightly built, has a broad forehead, glasses, and brown hair. He was diagnosed with Crohn's disease at age sixteen. Now he's in his early thirties. When an ulcer perforated in 1999, he had a section of his bowel removed. Two and a half years later, he required surgery again, this time for an obstruction. Two and a half years after that, another surgery for an abscess. "It was literally like clockwork," Herbert told me in a downtown Manhattan café one evening. Each time, surgeons removed a few inches of intestine, and the Crohn's would cease. Then, after a year, it would return. During severe episodes, he passed black, half-digested blood.

He went on an immune suppressant called Humira, which helped. But an acquaintance also taking Humira died from lymphoma, a cancer of the immune system. The drug increases the risk of this malignancy. He used prednisone, but the steroid gave him mood swings and insomnia.

He heard about Weinstock's work, saved money, and bought pig whipworm eggs from Ovamed, at the time nearly $5,500 for a three-month dosage. That helped, but the treatment was too expensive to maintain long term. He found Lawrence's story online. He joined the Yahoo group. After three years of watching, he "took the plunge." In the spring of 2010, while abroad on business, he infected himself with thirty-five hookworm larvae. No side effects. No pain. He eased off the Humira. Everything was good. No flares. Five months later, he added whipworm. After a time, he realized his many food allergies—to cherries, peaches, plums, and apricots—had faded. One day, he went to the market and bought an avocado, another food that used to cause his throat to swell, and ate it. Nothing happened. He'd weaned himself entirely off Humira.

"I was fine," he says. "Incredible."

Herbert asked that I not use his real name. He worried that if people knew he carried hookworm, it might interfere with his dating life, or his job. He'd been in total remission for a year when I met him. His doctor, Moshe Rubin, the director of gastroenterology at New York Hospital in Queens, New York, confirmed it. He couldn't recommend the treatment, he said, and he also couldn't discount spontaneous remission. But "I did scope him and his disease was quiescent," Rubin said. "It looked like remission."

JOSH COOLS THE MAGMA

Josh's first psoriasis plaque showed up at age eleven. It resolved, but around age twenty, a new plaque appeared at his hairline, and another on his chest. This time, although they continued to ebb and flow, the overall trajectory was toward increasingly severe psoriatic disease. He tried topical steroid ointments, but they weren't powerful enough. He was willing to take stronger drugs, but his insurance company, which considered psoriasis a cosmetic condition, wouldn't cover expensive new drugs such as Enbrel. Deep cracks formed in his skin. Moving became painful for the father of two. For a few years, ultraviolet light therapy helped, but it's well known to lose effectiveness over time, and it did with Josh as well. By then, angry scaly plaques covered much of his torso and arms—"a disgusting, ugly mess," he says. The constant pain interfered with his sleeping. At night, he couldn't roll over in bed. For the first time in his life, he began considering—even if fleetingly—suicide. "I was not functioning as a normal person," he recalls.

He heard Jasper Lawrence's story on Radiolab. He jumped on the Internet and came across Garin Aglietti's "worm therapy" operation out of Tijuana. They talked on the phone. "I was expecting him to say, 'We've done this with several people with psoriasis, and it's worked,'" he says. But instead he heard, "I've done it with myself. But other than that, you'd be the first one."

Aglietti's straightforwardness sold him. He flew to San Diego, met Aglietti, crossed the border, got in a car, and half expected to awake with a kidney missing. Instead, in October 2010, he received twenty-five larvae. Almost immediately, his skin "seemed a little bit less angry." An odd feeling of euphoria overtook him on the flight home. "I wanted to get up and do cartwheels," he says. (Others have also described a "hookworm high." Could it relate to a cessation of inflammation-mediated depression?) After some ups and downs, by April, his psoriasis had receded dramatically. He was wearing shorts for the first time in years. "My wife is just completely blown away," he says.

Josh's general practitioner, Hanno Muellner in Williamsburg, Massachusetts, corroborates the story, although with the usual caveats. Diseases like this wax and wane, he says, and he can't recommend the treatment. Nonetheless, this "severe" case "was clearly 90 percent better," he says.

"Very exciting." In photos, the difference between before and after is so dramatic that I suspected they might be fakes when I first saw them. In the before picture, Josh's torso is an angry red sea dotted throughout with islands of irregularly shaped white plaques. In the latter, just a few, small-ish plaques remain. The sea of magma has completely receded.

I have heard more amazing stories like this. A fellow in Oregon named Karl sent his Sjogren's syndrome, a painful autoimmune disease wherein the immune system attacks tear ducts and salivary glands, into remission. (Like Josh, he wanted only his first name used.) A man named Michael L., from Virginia, cured his hay fever. You've already heard some of the success stories of autistic children. There are more cases like those, I'm told, but given how fraught autism is as an issue, those parents wouldn't talk.

Others, meanwhile, are documenting their journeys with helminths publicly on blogs, more than ten of which have sprung into existence in recent years.

For some, however, things haven't turned out so rosily. Nikolai K., an undergraduate at the University of California, Berkeley, tried hookworms to treat his Crohn's. He wanted an alternative to the immune suppressant Tysabri. He saw the choice as "mild anemia over the risk of brain infection." He had several friends with IBD who were benefitting from helminths. He self-infected with maybe ten worms, got the hookworm high for a week, but then worsened—horrible headaches, fatigue, stomach distension. He'd been forewarned of possible symptoms, so he tried to barrel through. Within three weeks, however, he'd lost 15 pounds of lean muscle. He was suffering from severe anemia. When his doctor saw him, he ordered an emergency blood transfusion.

When I spoke with him, he wasn't planning to try worms again anytime soon.

Scott R. has a similar story. He translates websites in San Francisco for a living, and he tried Aglietti's worms for a case of Crohn's that extended "from throat to rectum." At first, the parasites seemed to work wonders. "I woke up one day, and it was completely night and day," he recalled. "Things had changed with me." His mood improved. He found it easier to concentrate. He was relieved not to have fourteen bowel movements a day.

But the improvement was short-lived. Within two or three months, the gains reversed. He discovered he'd lost his worms. He acquired more. And this time, his symptoms worsened. His weight fell from 140 to 110 pounds. He grew so thin and anemic that, one day, he admitted himself to the hospital. Doctors rehydrated him intravenously. Even given this bad experience, he says, he'd try hookworm again in a heartbeat. That period of improvement was magical.

His doctor, Jonathan Terdiman at the University of California, San Francisco, also confirms the improvement, and the subsequent decline. He worries about quality control of the drug. Is everyone getting the same thing? Is it always viable? "I think it's cool, interesting, but I advise you not to do it pending working out these details," he told me. He's caught wind of Aglietti and Lawrence's feud, the mutual bad-mouthing carried out periodically across the Internet. "When I hear that kind of stuff, the whole thing is not mature," he says. In all, he had six patients who tried worms. After initially improving, five then worsened. Only one, when I spoke to him in 2010, remained infected with worms.

Her name was Debora Wade.

I meet Wade one cloudless, sun-drenched California day at her home in Santa Cruz. She has a freckled face, auburn hair, and a nose that descends straight from her brow in classical Greek fashion. We sit in her lush backyard. A large garden occupies a lot to the side. Chickens cluck in a pen.

Wade received a diagnosis of Crohn's disease at age sixteen. In her early thirties, she had her descending colon surgically removed. By 2007, she'd tried everything, including Humira. Nothing worked. She'd lost 20 pounds. She was having bloody bowel movements daily. She'd heard about Weinstock's research, but, unable to work, couldn't afford the porcine whipworm eggs. She came across Jasper Lawrence's operation. Of course, she thought, when she found out where he lived. Of all places, he lived in Santa Cruz.

In December 2007, she went to Mexico, where she'd never traveled for fear of catching parasites, and received ten worms. Her symptoms worsened at first. Her ankles swelled. Her joints stiffened. But six weeks later, her condition improved. Her appetite returned. Four months on, the pain had receded and her bowel urgency had greatly diminished. She gained back those 20 pounds. And her doctor (Jonathan Terdiman) said that her levels of C-reactive protein, a marker of inflammation, fell to zero. By spring of 2008, she was in remission. She felt better than she had in years.

Wade began adding more worms, two and three at a time. That may have been a mistake. She entered a turbulent phase. By the following winter, she'd regressed to square one, and a stool test indicated she'd lost all her helminths. She started over: she infected with ten worms, and this time let well enough alone. Within a month, she was back in remission. She taught herself how to do egg counts—a method for tallying worm burden. She found she could tolerate foods that previously were troublesome, like pizza. Wade, who's almost six feet tall and thin, reached a healthy 165 pounds.

And then she started to regress. To make matters worse, she and Lawrence, who had become quite close, had a falling out. She had pressed him to prove that his worm was the correct species of hookworm. He never did, but he accused her of breaking their contractual agreement. That's when the FDA showed up at Lawrence's house, and he fled the country. Without access to more parasites, Wade again descended into the hell of Crohn's disease.

Months after I first met her, I watched her tell this story, dressed in high heels and a maroon-colored dress, at a meeting of the International Biotherapy Society in Los Angeles. Things were not going well. She'd lost weight since our first meeting. She'd tried porcine whipworm, but it worsened her condition. Now she was on prednisone. "I should not have to be traveling to Tijuana to get infected with hookworms. I should not be doing my own egg counts," she said at the end of her talk. "I should not be worried about my legal rights if I wish to self-infect. I should not be paying thousands of dollars for some larvae. I should not have to wait years for this research on the depleted microbiome theory to be proven. I should not have to wait for a pharmaceutically derived worm product, when the worms themselves are available now."

The audience, mostly scientists, seemed shocked by her presentation and its implicit plea. The crowd included researchers who used maggots to clean wounds and leeches to control swellings. So her using worms couldn't have seemed too far-fetched. Notably, however, Wade was the only patient who gave a talk, the only person who recounted life with an incurable disease.

When she finished, a Korean scientist who studied apitherapy—bee stings to treat rheumatoid arthritis—approached and explained, not without emotion, that the measures of C-reactive protein she'd cited were scientifically meaningless, and that regulatory authorities had to be placated, no matter how onerous their requirements. When he left some minutes later, Wade's eyes welled with tears.

She just wanted to be well, she said. Aglietti, who was also attending the conference and had taken her on as a client, tried to comfort her. "We'll figure it out," he said. "Think what this disease has made you. It's made you the person you are."

"I would rather be a shallow, one-dimensional person than to have suffered what I've suffered," she countered. She'd wanted to study abroad in Italy, she said. She'd wanted to go to medical school. "But I couldn't because I was too sick."

THE MEANING
OF MEANINGLESS MALADIES

When the gilded classes began suffering from hay fever in the late nine-teenth century, some interpreted their sneezing not only as evidence of intrinsic refinement, but of a civilization that had somehow gone wrong. This new coal-powered world had become too mechanized, they argued; it moved too quickly. As a result, they, the most sensitive individuals, were suffering. The solution was a return to purity, to nature—a retreat from cities to hay fever resorts in the mountains and along the coasts.

As Mark Jackson points out in his book *Allergy: The History of a Modern Malady*, a critique of modernity has coursed through our understanding of allergy, and the White Plague of tuberculosis before it, for more than a century. At times, the hygiene hypothesis (or the "old friends" hypothesis) appears to jump eagerly on that bandwagon. During several years of observing the burgeoning hookworm underground, I could see that this evolutionary-minded framework offered a psychological salve to people suffering from allergic and autoimmune diseases. It contextualized disorders that were not only largely untreatable, but also otherwise inexplicable.

David Elliott, the scientist who worked with Joel Weinstock on developing TSO, put it best. "It's amazing how quickly people understand this concept," he said when I asked if he had trouble finding recruits to try his worms. "We're saying this illness is . . . a feature of the world changing, not of you changing—that you are perfectly fine."

What a relief to hear. And how alleviating to know that there's a reason for your malady—that's it not your fault, and that it's not random. How terrific that the scientists exploring this treatment can explain these diseases in a way that the creators of immune suppressants and asthma inhalers can't.

The science we've explored suggests that the implicit critique of modernity is biologically valid—that in fact we *have* messed with the human superorganism in a way that causes immune dysfunction. However, these selfsame holistic overtones allow people in the hookworm underground to gloss over critical aspects of the science. Chief among them: The theory predicts that these disorders will arise less often when these organisms are present. It doesn't necessarily say that introducing them will fix already established dysfunction. Yes, it may, and has in small studies. But that critical application remains to be proven.

Meanwhile, the lack of true expertise or scientific rigor imbues the movement with the same instability that bedevils any endeavor crowdsourced on the Internet. Unverifiable anecdotes guide major decisions. People choose evidence that best supports preconceived notions. Risks aren't assessed clearly. The simple fact, for example, that hookworms are eating you, and therefore damaging your intestinal mucosa, is mostly ignored. And no one wonders what happens to people in the developed world—people with a measurably different intestinal flora and altered immune function—when they're suddenly colonized by a crowd of parasites in adulthood. That, too, is an evolutionary novelty. No one yet knows the long-term costs.

Furthermore, safety precautions and guarantees of quality are completely self-imposed by "worm providers." There's no liability, and no account- ability if something goes wrong. No one can be sure—me included—that they're receiving the desired helminth species. Only certain parasites pass through the skin, which nicely eliminates many unpleasant possibilities, but that doesn't guarantee that you're getting *N. americanus*. (Michael Cappello, a hookworm expert at Yale, confirmed that I harbored *Necator Americanus* using a DNA test.)

All that said, given the explanatory power of the science, there's a certain inevitability to the emergence of the hookworm underground. Whether it was Lawrence, Aglietti, or anyone else, someone was going to get into the business of selling worms. The diseases are terrible. The science is promis- ing. The explanation is appealing. And there's the potential fix, residing in the bowels of one-sixth of humanity.

Indeed, not long after Lawrence and Aglietti began peddling parasites, an operation opened in Spain. For a few months, someone was selling hookworm out of Australia. (Both have since gone off-line.) Lawrence, meanwhile, clearly considers himself a Promethean character. He's taken a cure-all from scientists who were never going to share it, and delivered it to suffering mortals. Some of his clients agree, and sing his praises accord- ingly. In private, others express a weary resignation over his erratic behav- ior, volatility, and willingness to fudge the facts. Only someone who's slightly crazy would do what he did, they say. So some zaniness comes with the territory. Yet while they're glad someone made worms available, a little more professionalism would be nice.

I've come to view Lawrence as a parasite in his own right. For legitimacy, both he and Aglietti depend on research carried out by actual scientists published in peer-reviewed journals. But so far they haven't contributed to that body of work. It remains to be seen whether, like the parasites discussed in this book, Lawrence and Aglietti confer any benefit to the science, their host. They may, simply by fomenting awareness. Now doc- tors hear the question "What about hookworm therapy?" They've clearly helped a few people. A few people have also clearly not benefitted from the treatment. At this point, there's no way of knowing what the ratio is.

Despite these many uncertainties, the underground has flourished. Judging from Lawrence's and Aglietti's estimates—and the number of people participating in discussion boards—hundreds of people have tried it. Amazing stories of remission keep emerging. There's talk of circumventing Aglietti and Lawrence by freely exchanging egg-containing stool, a true do-it-yourself hookworm movement. A helminthic therapy wiki page that's not associated with either Aglietti or Lawrence serves as a reference library. It has scores of scientific articles on parasites and immune function, and more recently, instructions on how to incubate larvae at home. Partly to escape the perceived bias of the Yahoo discussion board, Herbert Smith, who's become one of the therapy's greatest proponents, launched a Facebook discussion group dedicated to it.

"I think it stands as a fascinating example of the power of the Internet and new ways of 'doing' medicine," wrote a graduate student in media studies who, when he learned I was writing this book, contacted me on Facebook. He'd sent his own ulcerative colitis into remission with whipworms. "[I] am fascinated having watched the emergence of the grassroots, seemingly underground and DIY 'worm movement' with sick patients taking their health into their own hands, sometimes even clashing with the medical establishment," he added.

From my standpoint, the hookworm "movement" looked like a Greek tragedy in slow motion. The real scientists had to proceed with their slow-moving experiments. Otherwise, how would we know that the therapy really worked? Regulatory authorities had to stop people from selling unapproved, potentially unsafe drugs. If not, chaos would reign. And people afflicted with incurable ailments had to seek help, even if that involved parasites procured from questionable sources. All parties were right to pursue their interests, but doing so inevitably led to conflict.

For this last group's sake, however—the people who matter most— one could only wish for greater harmony. And while bona fide scientists generally grumble about the hookworm underground—Joel Weinstock describes it as "bootlegged" science—some think that bringing it into the fold may be best. "I reckon regulatory authorities should recognize that it's happening. They shouldn't be forcing people to go out there [to Mexico], pay for worms, and then take them and not have a monitor," said John Croese, an Australian gastroenterologist who's experimenting with celiac disease and hookworm.

Debora Wade also has a point: People shouldn't have to wait decades, especially when they might not have decades to spare, to try a "drug" that not long ago the majority of humanity carried unawares.

* * *

After the biotherapy conference, I worried about Debora Wade. She had acquired a few more hookworms from Aglietti, but failed to improve. And even as her doctors recommended surgery, drugs, and antibiotics, she kept insisting on worms. I wondered if she might have misplaced her faith. Worms clearly weren't working, and yet her belief in them was unwavering. One day I mentioned in a Facebook chat that the whole theory could be utterly wrong. "Too many positives to be all wrong. Too many drug free remissions," she replied.

Later, I asked how she continued to believe in the approach with so much evidence to the contrary. Two and a half "years of total remission is hard to ignore," she wrote.

She exhausted other non-hard-drug options—a fecal transfusion, bacteriophages (viruses that target bad bacteria)—before finally acquiescing to antibiotic treatment. Soon thereafter, she underwent a colostomy—a surgically created bypass around her rectum. And she agreed to take a new immune-suppressant drug called Tysabri, which terrified her.

But by September 2011, the Crohn's had begun attacking the opening, called a stoma, that surgeons had created in her side for effluence. The Tysabri seemed to have worsened things. She had now tried and exhausted every option available both in allopathic medicine and the hookworm underground. There was just one more possibility. Aglietti was experimenting with a new worm species. She planned to try it, she told me. Even her doctor, who understood no other options remained, wasn't saying no, she said. Now that I better comprehended her predicament, I couldn't view her commitment to the idea of worm therapy as delusional. It seemed more like indomitable optimism. She'd run out of options, but she kept trying treatments she thought might work.

And then a miracle of sorts occurred. Wade's doctor convinced her insurance company to cover infusions of intravenous immunoglobulin, which was otherwise prohibitively expensive. IVIg consists of antibodies extracted from human donors. No one's sure why, but flooding the body with these antibodies enhances immune regulation and helps with many autoimmune diseases. Wade improved on the drug, but not 100 percent. And in December 2011, she traveled to Tijuana to acquire ten hookworms, and to test the new mystery worm species. Three months later, she'd regained lost weight and was feeling much better. "Yee haw," she wrote in the Facebook discussion group. "Yeah worms!"

By then, I was more than a year into my own experiment.

CHAPTER 14

Life with the American Murderer, and the Body's Largest Organ

Life is naturally tattered, infested, bitten off, bitten into. The stem with a broken leaf, like an animal with lesions on its internal organs or less-than-glossy feathers, is more normal than its unscarred counterpart. An unblemished animal— or person—is idealized and fictional, like the advertisements showing a solitary traveler at the Eiffel Tower. It doesn't really exist except in our imaginations. Disease is part and parcel of how we are supposed to look, of how we are supposed to live.

—Evolutionary biologist Marlene Zuk
in *Riddled with Life*

By the time I self-infected in November 2010, I had no illusions that worms could cure established autoimmune disease; at best, I could expect remission. If they helped, I'd likely have to remain infected indefinitely. The time when they might have prevented immune-system malfunction was far in my past. That was disappointing.

I also had misgivings about how benign the parasites really were. People I'd interviewed had reported few if any symptoms, but they'd often taken immune suppressants recently. I hadn't. I also suspected that unbridled optimism and relief clouded some people's honest assessment of the symptoms associated with infection. Curiosity aside, one reason I went forward with the experiment against much of my own better judgment, in fact, was to witness with my own eyes—or gut—how hellish it was. If I was going to argue that the immune system required contact with helminths, I wanted to know firsthand how a parasite infestation made you feel.

And almost immediately, I felt remarkably crappy. A metallic-flavored headache began shortly after they started burrowing through my skin in Tijuana. I didn't experience the "high" I'd heard others describe. I felt, rather, like I had altitude sickness, was sleep-deprived, and had a mild hangover all at once.

A week after inoculation, a slightly bloated feeling settled in. A week later, the mosquito-bite-like larval entry points on my arm, which had mostly faded, suddenly swelled again into itchy bumps. Cramps began around week three. I suffered occasional bouts of mild vertigo. Around the month mark—by then the immature hookworms had presumably reached my small intestine—the cramps intensified.

And that's about when the diarrhea began. I'll spare you the details, except to say that it wasn't the worst I've had, but it wasn't pleasant either. Scientists refer to diarrhea as the "weep and sweep" method of expulsion. My gut was trying its damnedest to shake the parasites loose. Millions of years of coevolution had left their mark. I knew something without ever having learned it in my lifetime.

That's also when I noticed the first benefit. My nose became extremely clear. Several mornings when I awoke, I lay in bed listening to the chambers in my sinuses popping open.

But the stomach malaise worsened. I'd sworn not to take any medication. On the Yahoo group, worm infectors discussed managing symptoms by taking prednisone. That seemed like cheating. I wanted to arrive at a natural balance between parasite and host, and I wanted to know what it felt like to get there. But the intestinal upset was bad enough that I broke down and quaffed some bismuth. I dreamed up metaphors for my predicament: Was I like the biblical Jacob, trying to wrest a blessing from an angel? Was I a cowboy trying to tame a wild mustang? And was I strong enough to "break" the horse? Or would I be bucked and broken by it?

My sinuses remained amazingly clear, but the headaches kept up. So did the urgent bowel movements and the sick feeling. I began to berate myself for ever thinking this might be a good idea. As someone who lived through plenty of suffocating asthma attacks, I should have remembered that without health, life loses its luster. And in this case, I was the agent of my own undoing. I should have left well enough alone!

By mid-January, another change: Alopecia often co-occurs with pitting of the nails. The protein in nails and hair, called keratin, is the same, and in this disease, the immune system pursues it in both locations. As a result, we alopeciacs have a "printout" of disease severity written in pits and grooves on our nails. By mid-January, some nails were growing in clearer. That was interesting. Another benefit, more a feeling than anything else, was so diffuse I almost didn't think it real: It felt like my skin was softer.

Since adolescence, I'd had a mild case of folliculitis, little bumps on my scalp. They unexpectedly subsided, not completely, but noticeably. More notably, several patches of eczema on my fingers faded away.

By early February, the gastrointestinal problems had greatly resolved. Minor headaches lingered on, mostly in the mornings. And very fine hairs began sprouting on the inner corner of my right eyebrow, the mole on my head, and eventually my left eyebrow. Let me be clear: These hairs were fine to the point of invisibility. I could see them only with my face pressed against the bathroom mirror. If 0 represents total hairlessness and 10 a full head of hair, I had perhaps advanced to 0.05, and even that's a stretch.

By then, the pollen count had started inching up. My wife began to sneeze. My nose remained open. I monitored her closely. She began to have middle-of-the-night allergic attacks, spasms of tears and snot. For the first time in years, I didn't. I gloated. My wife began to take the project more seriously. Maybe I wasn't crazy. Maybe I was on to something.

Spring progressed. The headache lingered. The pollen count rose further. The pollen types shifted. My sinuses remained joyously clear and free of obstruction. I rode the subway with the sneezing hordes and, for the first time in my life, didn't rank among the worst of them.

And then, one early May day, my wife and I took a walk through the cherry blossoms of the Brooklyn Botanical Garden. A few days earlier, I'd noticed a passing gritty sensation in my eyes. And now, among the cherry trees, my allergies returned to full intensity. In the subsequent week, all the benefits I'd seen were reversed. My eczema returned. My folliculitis flared back up. I sneezed along with everyone else. What had happened? Had I lost my worms?

Aglietti sent me plastic vials for a stool test. I now faced a challenge for which I was totally unprepared: How to get my own stool into a container slightly larger than a roll of quarters, and without making a mess? Aglietti outlined two techniques: Put plastic wrap across the toilet bowl, evacuate, and then collect. Or just evacuate directly into the bowl, and scoop some out before flushing. The sample didn't exactly need to be clean.

I decided on the latter approach, and shipped the sample in a special "biological substance" pack to his apartment just south of San Diego. No eggs. (Identifying eggs is a standard method for determining parasite infection.) Aglietti sent a second set of vials. He told me to collect samples over two days this time. Still no eggs. As far as he could tell, I was worm-free. Did I want to try again? Hell no! I thought. But I told him I'd think about it.

Given that I had apparently expelled the worms on my own, I didn't see the point of repeating the experiment. I was disappointed. But broadly speaking, the results fit in with one aspect of evolutionary theory. A person with allergic propensities was also preprogrammed to expel parasites. I was exhibit A on that front, apparently. I just didn't expect to be that good at it.

Work by John Croese, the Australian gastroenterologist who'd spent years studying hookworm, and who'd seen it confer some benefits on both Crohn's and celiac disease, allowed further interpretation. If you want to study hookworm, and you're in the developed world, you've got to deliberately infect someone just to have a source of the parasites. This particular human-adapted species doesn't do well in lab animals.

So, after acquiring infective larvae from David Pritchard in the U.K.— worms that, years earlier, originated right next door in Papua New Guinea— Croese and his colleague infected themselves. They each had very different reactions. Croese had violent pain and urgent bowel movements, but his partner, Rick Speare, scarcely felt the infection. They conducted capsule endoscopies on themselves, a new method whereby you swallow a miniature, pill-shaped digital camera that beams out photos of your insides.

By directly examining live infection, the scientists made two important finds: First, they noted the cause of all Croese's pain. The areas where the young worms attached literally turned to mush. The gut liquefied its own tissue to shake them free. Here was allergy in its proper functional context—the expulsion of worms. Worms that couldn't keep their grip were swept toward the exit.

The scientists also observed that no matter how many larvae they introduced into their bodies—fifty or one hundred—each one of them always ended up with the same quantity of adult worms. For Croese, who had the more severe reaction, that number ranged between six and nine. For Speare, who had a less severe reaction, the colony stabilized around sixteen. Here were two slightly different worm-fighting strategies. One person (Croese) paid a large price up front, whittled the worm colony down early on, and presumably had fewer resources siphoned off later. The other (Speare) avoided that high initial outlay of resources, but probably had to dole out more in the long term. These strategies were, it seemed, genetically predetermined.

Most important, although Speare's colony was ultimately larger than Croese's, both settled into harmony with their parasites. Both stopped warring with the worms once they reached their genetically predetermined "ideal" load.

So maybe my ideal load was zero. Also, unlike in times past, the well-nourished modern human organism has nearly limitless resources to throw at interlopers. Maybe that played a role. I was too well fed to tolerate worms.

I never took the deworming drug. And weirdly, despite seemingly having apparently lost my parasites, the pitting of my nails continued to fade. Other hookworm symptoms—an occasional faint headache and a mild bloated feeling at times—also lingered on. And in late September, by which time I'd given up on the experiment, the reversal reversed. My sinuses again became extraordinarily clear. My skin again felt soft and supple. The patches of eczema on my hand again disappeared. The folliculitis faded. Fine hairs began to sprout with renewed vigor. Ten months after having received the hookworm, and roughly four months after the worms seemingly vanished, they were back. And this time, eggs showed up in a stool test.

One factor bears mentioning here: A consensus of sorts has emerged that transient worm infections don't really modulate the host immune system. To the contrary: Light, transitory infections may prime allergic disease, a point that's generally overlooked in the hookworm underground, but which also explains the occasionally conflicting finds on helminths and allergy. Some studies show that worms exacerbate wheezing, and that deworming improves allergic symptoms. Only chronic infections with sufficient numbers—and with the right species—have the power to change how your immune system works.

So it's possible that in the first few months, my immune system was still wrestling with the new arrivals. Judging from the severity of my symptoms—and extrapolating from John Croese's work—maybe I successfully whittled the population of immature worms down to a few. The worms were struggling to survive and unable to procreate. That explains the absence of eggs. At some point, however, the law of diminishing returns kicked in. The cost of expelling the remaining survivors was too great compared with the benefit. My immune system flung its hands in the air and accepted the remaining adults as part of the new order. And that's when the immune modulation began again.

I'm still food-allergic, although some aspects of the anaphylactic reaction—such as the throat constriction—have become less ferocious. In contrast with others' experiences, like Jasper Lawrence's, my asthma has remained utterly unimpressed by the parasites, neither better nor worse, but my sinuses have achieved remarkable clarity even as allergy season begins anew, and my wife again starts sneezing. My nails continue to grow in with fewer pits and grooves. Fine hairs continue to sprout, although at a density so far below anything that would qualify as remission that I hesitate even to mention them. It is extraordinary and heartening, however, that after decades of repression by a wayward immune system, hair follicles can return to life.

And my eczema is simply gone.

So I've learned several things. First, if the ultimate goal is to prevent autoimmune and allergic disease by proactive infection in childhood, and if my experience is generalizable, I could never recommend hookworms. Given my symptoms, I would never deliberately administer them to my daughter, for example. On the other hand, if I knew with any certainty that she was destined to develop Crohn's disease and debilitating allergies, I might think differently. But I don't know that. I don't even know with certainty that worm infections early in life prevent these diseases. A pile of epidemiological studies suggests that this is indeed the case, and so do experimental studies in rodents. But we've got to be honest: Definitive proof in humans is lacking. Again, only science can settle the question.

On the other hand, as a layperson, it's simply miraculous that a creature residing in my gut can pry open my sinuses and soothe my eczema, all without the stupefying symptoms of an allergy medication. The worms evidently understand my immune system with a precision that allopathic medicine has yet to match.

I think the changes in my skin may comprise the most important find of the experiment. Skin dysfunction, it's turning out, is that fork in the road where minor immune dysregulation veers into major allergic disease.

HOW TO MAKE A FOOD ALLERGY

Anyone hoping to articulate an official policy in response to the rise of food allergies in recent decades must address the following questions: Do allergens themselves cause allergy? Or does some deeper problem lead to allergic disease?

In 1998, the U.K.'s Department of Health decided that allergens themselves were the problem, and it recommended that infants with allergies in the family avoid known allergenic foods such as peanuts. The recommendations stemmed from two findings: Mice that had never been exposed to a given protein, and thus never sensitized to it, failed to mount an allergic reaction when they did encounter it; and a study wherein scientists kept children away from major allergens for two years—peanuts, eggs, fish, among others—found these kids to develop fewer allergies than controls. In 2000, the American Academy of Pediatrics issued similar guidelines.

But the recommendations never sat well with Gideon Lack, an allergy researcher at Imperial College London. Clearly, if you had an allergy to peanuts, you should avoid them. However, avoidance failed to address the more fundamental question of how you became allergic to begin with. The huge variation in food allergy around the world strongly suggested that environmental factors played an important role, but it didn't indicate that exposure to allergens was foremost among those factors. If anything, he suspected that in places where children ate more peanuts, they tended to be less allergic to them.

In the years following the official recommendations, Lack unearthed hard evidence of this fact. He compared the prevalence of food allergies among Jews in Israel and London, and found them to be more common in the London group—nearly ten times as common for peanuts, and five times as high for sesame. What was the difference in exposure? In Israel, infants worked out teething pains by gnawing on a Cheetos-like peanut snack called Bamba. In the U.K., however, they tended to avoid peanuts altogether. Early oral exposure to allergens—not avoidance—seemed to prevent allergy.

Now Lack wondered how allergic British children, whose parents endeavored so diligently to avoid allergens, were becoming sensitized. From time spent in the U.S. during the 1990s, he had an inkling that rodents could develop allergies through their skin. When scientists lathered allergy-prone mice with egg protein, for example, the mice became sensitized. And if these now sensitized mice inhaled the allergen they'd previously merely touched, they developed asthmalike symptoms. What presented as an allergic lung disease had begun as proteins on skin. Could a similar process be occurring in young children? *Maybe oral ingestion wasn't required to produce food allergies.*

Lack surveyed British children with peanut allergies and their parents. He could discount one theory right away, he found. Sensitization wasn't occurring prenatally. No peanut-specific antibodies showed up in blood extracted from these children's umbilical cords. He did note, however, a strong association with environmental exposure to the allergens—not orally, but through the dermis.

Mothers didn't know it, but some popular infant creams meant to soothe diaper rash, eczema, and dry skin contained peanut oil. Mothers who used these ointments had children with a nearly sevenfold increased risk of peanut allergy. What's more, certain soy proteins, it turned out, resembled proteins in peanuts. Both belonged to the legume family. Some ointments contained soy products as well. Mothers using these creams could be cross-sensitizing their children to peanuts without, necessarily, sensitizing them to soy. Horribly, children with the most inflamed skin likely received the most ointment. Parents of the most allergy-prone kids were the most likely to inadvertently sensitize them.

The skin is the body's largest organ. For terrestrial animals especially, it's important for keeping moisture in. The epidermis contains waterproof fatty lipids and a hard scaly outer layer called the stratum corneum. The skin also serves as the first line of defense against a range of parasites, from ticks, fleas, mosquitoes, lice, chiggers, mites, and anything else that bites but remains mostly outside the body, to the helminths that burrow through it, like hookworms and blood flukes, to seek our insides. Our immune system is probably inclined, therefore, to treat eukaryote proteins it first encounters in the epidermis as belonging to parasites of one sort or another, and to counter with an antiparasite response—the allergic response.

Eating proteins, on the other hand, usually leads to tolerance. That's how oral immunotherapy—the process of deliberately training the immune system to tolerate peanuts, say—works. Unless otherwise interfered with, we are preprogrammed to treat proteins coming down our throats as food. Many pathogens and parasites also approach via the oral route, but the gut immune system—the most complex in the body—has ways of differentiating. An approach via the skin, however, is much less ambiguous. It signifies invasion.

So one interpretation of Lack's finding went like this: children developed food allergies because they encountered food proteins through their skin first, which prompted an antiparasite response thereafter. The problem he'd uncovered was partially one of sequence. The route of first contact mattered. These children weren't necessarily exposed to allergens too early; they were exposed via the wrong organ. If anything, they needed oral contact earlier, before they encountered the protein via their skin. Worst of all, keeping children away from allergenic food—preventing them from developing oral tolerance—might exacerbate the problem.

Indeed, after officials recommended steering clear of allergens, scientists found that food avoidance failed to curb the increase in food allergies. They continued to increase in both the U.K. and U.S. In 2008, the American Academy of Pediatricians backtracked, revising its guidelines. It now advised breast-feeding exclusively for four months, but after that, it no longer recommended delaying the introduction of any foods.

Parallel to Lack's work, geneticists were also zeroing in on skin dysfunction in allergic disease, but from a slightly different angle.

THE MUTANT PROTEIN
THAT SPARKS THE ALLERGIC MARCH

In 2006, a team of scientists in Irwin McLean's laboratory at the University of Dundee, Scotland, decoded two important gene variants involved in the skin's structure and integrity. The gene they pinpointed encoded a protein called filaggrin, an important component of the outer layer of skin, the epidermis. The variants the scientists had identified hobbled production of this protein. People who carried two copies of these "null" genes produced almost no filaggrin. As a result, they developed ichthyosis vulgaris, a condition characterized by chronically irritated flaky skin.

That fact was interesting in and of itself. But an observation by an Irish collaborator, Alan Irvine, elevated the significance of their discovery. In reviewing patient records, Irvine found that people with ichthyosis vulgaris also tended to have lots of eczema. Having insufficient filaggrin apparently predisposed to allergic disease. More important, whereas it took two mutant versions to produce severe ichthyosis vulgaris, just one copy sufficed to increase the risk of eczema.

A flood of studies on "null" filaggrin mutations followed, all of them strengthening the association with allergic disease in general. People with the nonfunctional gene were less likely to grow out of their eczema, and more likely to have asthma as well. They were twice as likely to have allergies to dust mites and cats. In Denmark, the gene predicted the so-called atopic march. Carriers suffered more severe eczema early in life, as well as more asthma and allergic sensitization later. Filaggrin mutations even increased the severity of alopecia areata.

Japanese populations had a different mutation in the same gene, but it was also a "null" version, and it also increased the risk of eczema. Overall, carriers of these genes had between a three- and fivefold increased risk of eczema, and increased odds of asthma (with eczema) of between 50 and 80 percent. The variants also elevated the risk of peanut allergies between two- and fivefold.

In the meantime, experiments firmly established the link between eczema, inflamed skin, and food sensitization. Without anyone realizing it, a laboratory mouse had spontaneously developed a mutation in the filaggrin gene in the late 1950s. Scientists had dubbed the mouse "flaky tail." Just applying egg protein to flaky tail mouse skin caused allergic sensitization. Others observed that, filaggrin mutations aside, after "stripping" rodent skin with adhesive tape, which removes the outermost protective layer, they could induce allergic sensitization through the dermis. If they carried out this experiment with peanut proteins, even mice that had previously eaten peanuts without problems lost their tolerance.

By extrapolation, it was possible to develop an allergy to food you already tolerated by encountering it through inflamed, irritated skin.

ALLERGY:
A PROBLEM THAT'S ONLY SKIN DEEP?

When it came to allergy, the different lines of evidence—genetic, environmental, and experimental—all now pointed toward the skin. The long-observed sequence of falling dominoes in the atopic march—eczema proceeding to hay fever, wheezing, food allergy, and possibly worse—wasn't just coincidence. Sensitization through the epidermis was a critical enabling step. We had a problem with the largest organ in our body, the waterproof sack that held us together.

Which brings us to a by-now-familiar question. The "null" fillagrin genes aren't new. They aren't exactly rare either. About 9 percent of Europeans carry them. That so many nonfunctional versions of a single gene exist—a lot of different approaches to failing in the same way, essentially—suggests that whatever purpose they serve, it was important in our evolutionary past.

So what were these genes for? Is there any advantage to having less filaggrin in the skin? McLean and Irvine hypothesize that increased permeability—and we're talking porous at the level of bacteria—may have been adaptive in the past by allowing the immune system to directly sample microbes in the external environment. By sneaking little morsels in, the immune system could encounter dangerous pathogens like the plague, the flu, or tuberculosis in small enough amounts that it wasn't overwhelmed, but large enough that it could remember them thereafter. When you saw them next—when they showed up as an invading horde at the gates of your mucous membranes, for example—you'd already be immune. Carriers of the null variants were, the thinking went, better able to prepare.

Why are these genes causing problems now? As it is, only half of those carrying "null" filaggrin mutations develop allergic disease. Environmental influences are therefore important. Rather than guarantee a collapse of the skin barrier, mutant filaggrin genes seem to lower the threshold beyond which other factors trigger a meltdown. Healthy skin is slightly acidic, for example, but these mutations make it less so. And lowered acidity may enhance the ability of certain commensal bacteria to adhere, and to inflame. That's one possibility. Others have pointed out that modern-day soaps aren't doing us any favors: Our skin is naturally acidic, but most soaps tend toward the alkaline. Yes, it's possible we're literally scrubbing ourselves to allergy.

But is the deciding factor really coming from without—acidic soaps, indoor heating, and so on—or is it emanating from within? One study found that, irrespective of genotype, mice with ongoing allergic inflammation degraded filaggrin on their own. Another observed that, in people with perfectly normal filaggrin genes—no mutations—allergic inflammation nonetheless decreased amounts of the protein in the skin. Now imagine that you have a genetically predetermined weakness in filaggrin production. Essentially, your skin barrier collapses sooner when you're inflamed. Which brings us to a chicken-or-egg-type question: What comes first in these allergic diseases, skin dysfunction that sets off the allergic march, and leads to so much more allergic inflammation? Or a primary tendency to become inflamed that's amplified by an already fragile skin barrier?

WHAT HAPPENS WITHIN REFLECTS ON OUR SKIN

In the developed world, eczema follows the same urban-rural gradient—more in the city, less in the country—of other allergic diseases. Farming protects against eczema, as does drinking unpasteurized milk and attending day care early in life. One study found that German schoolchildren who were positive for *Helicobacter pylori* were about one-third as likely to have eczema as their helicobacter-free classmates. Bengt Björkstén has also observed that a less-diverse microbiota at a younger age predisposes to eczema later. Likewise, lots of microbe-depleting antibiotics early in life increase the risk of eczema.

As with other allergic diseases, the vulnerability to eczema appears to begin in the womb. The more animals a farming mother encounters, the less her unborn child's chances of developing the rash in toddlerhood. Mothers who regularly tend to three or more farm animals while pregnant have children with less than half as much eczema compared with those whose mothers didn't. And we mustn't forget the worms. As we saw in chapter 7, Mom's helminth infections provide similar protection. Children of Ugandan mothers who were dewormed during pregnancy have more eczema than children whose mothers weren't.

In 2010, the Nottingham University scientist Carsten Flohr published results of the largest placebo-controlled deworming trial to date—1,566 rural Vietnamese children whom he'd treated repeatedly over the course of a year. They ranged in age between six and seventeen. Most had hookworm. Some had giant roundworm. After deworming, Flohr noted an immediate increase in the children's vulnerability to sensitization via the skin. They never crossed the mysterious threshold to overt allergic disease, which remained rare in the community, but just removing the worms inched them slightly closer to the precipice. Taken together, these studies suggest that systemic immune function plays an outsize role in how our skin works, and greatly determines whether filaggrin mutations contribute to skin-barrier collapse, or serve some other, more useful function.

In a 2011 review in the *New England Journal of Medicine,* Irwin McLean, who's a geneticist, mentions that genetically boosting filaggrin production could possibly treat eczema and the allergic march. That's certainly true. And once developed, such methods would doubtlessly be useful. But given the bulk of evidence suggesting that, while skin barrier problems amplify allergic disease, the cascade of events leading to failure begins deeper in the immune system—in prenatal immune imprinting and contact with old friends—wouldn't it be wiser to address the root problem?

My experiment was essentially that—an attempt to nudge my systemic immune function by introducing hookworm to the small intestine. That's why I find the changes in my skin so interesting. I can't say with certainty how the worms changed the immune response in my skin, but it was real and noticeable. And I'll conjecture that if I'd had immune input like this early in life, maybe even beginning in the womb, I'd have avoided the bevy of allergic diseases that followed and, who knows, maybe my autoimmune disorder as well.

CHAPTER 15

The Collapse of the Superorganism, and What to Do About It

When we try to pick out anything by itself, we find it hitched
to everything else in the Universe
—John Muir

There is a certain irony in the fact that we must now search
for new ways to reproduce the infectious diseases against
which we have been fighting with great success.
—Jean-François Bach

For the entirety of human evolution, the accretion of the human super-organism followed a predictable pattern, both in the cast of characters that aggregated and the timing of their arrival. That aggregation went like this: As you were born, your mother's vaginal and fecal microbiota coated you from head to toe, a first microbial inoculum. Her breast milk further seeded you with bacteria. Special sugars it carried fostered a specific community of microbes in your gut. And the milk itself carried signals on how to respond immunologically. It augmented certain aspects of immunity against pathogens, but it also conveyed a message of tolerance.

At weaning, your mother and father prechewed food for you. You acquired Epstein-Barr virus and *Helicobacter pylori*. Maybe *Mycobacterium tuberculosis* also showed up, established infection, and—in most people—went quiet. The large, close-knit family unit—lots of brothers and sisters, other children, aunts, and uncles—guaranteed a bonanza of other microbial exposures. As a result, the microbial community you harbored was complex and diverse at an early age. As an ecosystem, it was more stable, less vulnerable to invasion compared with today's. As a microbial organ, it provided a basal level of immune stimulation that was more soothing than in postmodernity.

In early childhood, you acquired a worm colony, not necessarily many, but always a few. With time, your immune system grew better at expelling these parasites, but they never went away entirely. You'd probably have a few worms for the rest of your life.

All the while, you drank surface water teeming with saprophytes and other soil-dwelling bacteria, microbes that accumulated as the water trickled through grasslands and forests, down mountainsides, and through a living world. Your immune system noticed them, registered their presence, but held back.

You subsisted on food that, despite its being cooked or fermented, was fibrous, rough, and, compared with modern fare, difficult to digest. Microbes acquired from your mother, from your peers, from the environment, from other animals perhaps—maybe even from raw food itself—helped you squeeze every last possible calorie from that diet.

You got plenty of diarrhea. Roughly 25 percent of you died before reaching your first year, sometimes more, sometimes less depending on group size, the climate, the abundance of food, conflict, and so on. Another 4 percent expired by age five. But if you survived until age fifteen, you could reasonably expect to reach your sixties. You grew at a different pace. The resources you invested in your immune system early in life demanded a more gradual growth trajectory than today's. You reached puberty later. You were probably shorter, but not necessarily. In the late Stone Age, heights were comparable to today's. And in preindustrial Europe, even as city-dwellers remained stunted, country-dwellers were tall.

When you were pregnant, the living world I just described communicated, via your immune system, with the developing fetus. Your immune system prepared the unborn child for arrival into a world teeming with microbes and parasites. That preparation involved enhancing machinery for sensing microbes. From the get-go, your white blood cells bristled with microbial sensors. But they also had a knack for quashing unnecessary inflammation.

One thing happened much less often: Amid all that activity, including the immune stimulation that began before birth, developing an allergic or autoimmune disease was, if not absolutely impossible, much more difficult. Contact with microbes and parasites didn't necessarily cure allergy and autoimmunity; rather, constant engagement with them precluded disease development. And gene variants that in today's relatively parasite- and microbe-free environment predispose to immune dysfunction were, in that context, protective. They helped defend against real pathogens and parasites, not imaginary ones. They may even have enhanced certain aspects of tissue repair.

For millions of years, this was how the superorganism came together. Even after people settled down to farm, the pattern remained roughly the same, although with an added veneer of crowd diseases.

And then came the Industrial Revolution.

In retrospect, the first sign of the superorganism's coming apart probably had nothing to do with allergic or autoimmune disease. In the late eighteenth and early nineteenth centuries, a wave of tuberculosis swept across Europe. The sudden spike, and then its decline beginning in the late nineteenth century, has always mystified historians. Genetic analysis indicates that *Mycobacterium tuberculosis* has been with humankind since before we dispersed from Africa. Archaeologists have identified traces of it in nine-thousand-year-old bones in the Near East. The Ancient Greeks were familiar with the disease it caused.

Yet the rate at which Europeans started falling ill with consumption in the late eighteenth century suggests a newly arrived infection. Robert Koch, who ultimately identified the bacterium responsible for the white plague, estimated that one of every seven deaths in mid-nineteenth-century Berlin was due to TB. Some think a new, more virulent version emerged, and genetic analysis does suggest that one strain swept to the fore in recent history. However, John Grange and his colleagues at University College London also think a subtler shift contributed to the wave of TB at the dawn of modernity.

As Europe urbanized, Europeans lost contact with environmental mycobacteria in mud and dirt. Those microbes naturally boosted immunity to TB. Rural and town-dwelling folk also probably drank milk from cows infected with TB's parasitic cousin, *Mycobacterium bovis*. That bacterium eventually became the basis of the BCG vaccine. *M. bovis* can cause illness in humans, but once exposed, you're resistant to TB. The tuberculosis epidemics, Grange and his colleagues argue, resulted from changing patterns of exposure to these other mycobacteria. These same bacteria, Graham Rook will tell you, enhance regulatory circuits and prevent allergic disease.

Other factors we've explored may have contributed to the white plague. If *Helicobacter pylori* helps keep mycobacterial infections latent, then changing patterns of colonization with *H. pylori* may have decreased resistance to tubercular disease. Remember, the generation born in the late eighteenth and early nineteenth centuries first saw an increased gastric cancer risk, evidence perhaps of later colonization with *H. pylori*. Or the rising incidence of stomach cancer might reflect the beginning of the exodus of worms. Helminths may make the bacterium less carcinogenic. This is a highly speculative reading of the past, of course, but the first real sign of population-wide immune dysfunction begins soon thereafter, in the late nineteenth century.

The emergence of hay fever and inflammatory bowel disease among the gilded classes suggests a loss of several or all of these organisms. The emergence of MS in the same social strata hints at delayed colonization with Epstein-Barr. The rise of all three diseases serves as evidence of weakening immune regulation. And a tangential phenomenon definitively links improved hygiene to the changing diseasescape.

The polio epidemics that began in northern Europe during the late nineteenth century suggest that, generally speaking, people had begun to ingest much less of one another's feces. Infection with the poliovirus was coming later and later, causing more and more paralytic poliomyelitis. Public hygiene was improving.

That change didn't occur evenly. Huge swaths of what we'd now call the developed world stayed wormy and feces-ridden until the mid-twentieth century. The southeastern U.S. and southern Europe, for example, regions that remained inured to immune-mediated diseases until much later, retained their parasites for decades.

Norman Stoll gave his famous "This Wormy World" lecture in 1947, decades after the Rockefeller Foundation funded the deworming of the American south. Sardinia eradicated malaria after World War II. And even as TB, measles, hepatitis A, and the other infectious diseases declined thereafter, some helminths, like pinworm, lingered on until the 1980s.

All throughout, changing diets and increasing affluence altered the microbiota. Widely available after World War II, antibiotics began pushing our microbial communities into new formations. Urbanization, smaller families, less crowded conditions, cleaner water, and improved hygiene overall depleted the pool of potential colonizers, and greatly attenuated what Bengt Björkstén, the microbiologist who compared the microbiota of Swedes and Estonians, calls "microbial pressure." A burgeoning consumer society that, at least since the 1920s, fetishized cleanliness and purity to a near pathological degree, also helped limit contact with our old friends.

Roughly around mid-century, the lower classes in developed nations passed through the epidemiological transition that the upper classes had initiated in the previous century. A decade or two later, with the many layers of the superorganism peeled back and its core—the teeming community inhabiting our gut—shifting into evolutionarily novel territory, the allergy epidemic began in earnest. And the incidence of IBD, MS, and type-1 diabetes rose in tandem.

That's the story of the collapse of the human superorganism in a nutshell. And the consequences are more far-reaching than just higher rates of allergy and autoimmunity. The intensified inflammatory tone of modernity contributes to cardiovascular disease, certain cancers, obesity, metabolic syndrome, depression, and maybe developmental disorders. The skittish modern immune system has become a major tormentor.

BUT WHAT ABOUT ALL THOSE CHEMICALS?

One reason to trace the beginnings of allergic and autoimmune disease back to the nineteenth century is to understand that immune dysfunction began before the man-made substances that now course through our veins and collect in our tissues became widespread. But that understanding doesn't completely absolve them. There are many ways to produce the same problems.

A few deserve mention: Several studies have associated the antimicrobial triclosan, in use since 1972, with allergies. These days, it's in detergents, hand sanitizers, some toothpastes, deodorants, and certain children's toys. The more people use it, scientists find, the greater their chance of allergies. Is this reverse causation? Are allergic people simply clean freaks? Possibly. But besides depleting our "good" microbes, and cracking our skin, triclosan degrades into dioxin. Dioxin binds to estrogen receptors—sensor cells used to receive important signals. Triggering these receptors haphazardly may distort growth and development. In rats, for example, triclosan has altered thyroid function. And it lasts for decades in the environment. In 2009, the Canadian Medical Association called for a ban on the stuff in household products.

Another widely used molecule, called bisphenol-A, also binds to the body's estrogen receptors. BPA leaches from the plastic lining of canned foods, from white dental fillings, and from some plastic baby bottles, among other sources. It has been linked to asthma and breast cancer experimentally in rodents, as well as observationally in humans. Children exposed in utero to excess BPA have a greater chance of wheezing. Mice prenatally exposed develop asthma more often. Female rats develop inflamed guts after exposure as newborns. In 2008, Canadian authorities banned its use in baby bottles. Two years later, they deemed it a toxic substance.

And finally, scientists have repeatedly noted an alarming association between the popular painkiller acetaminophen and allergic disease. The drug (known as paracetamol in Europe) is also sold under the brand name Tylenol in the U.S. Nothing is firm yet, but at this point, more than twenty studies have associated the over-the-counter analgesic with either allergic sensitization or wheezing. Prenatal exposure seems most important. And mothers and children with gene variants that "turn down" production of a natural antioxidant have the greatest risk. The antioxidant is called glutathione, and acetaminophen interferes with its production.

Reverse causation hasn't yet been ruled out. Children who go on to develop allergic disease, who are sicklier early in life, may simply take more acetaminophen. But that explanation seems less and less likely, especially given that certain genotypes appear more sensitive to the drug than others. Also, children who use other painkillers, such as aspirin and ibuprofen, don't have an increased risk of allergic disease. (In some cases, use of these nonsteroidal anti-inflammatories slightly decreases the relative risk of asthma.) And the relationship between acetaminophen and asthma is dose-dependant. The more you use it, the greater your chances of asthma.

Others suspect acetaminophen in autism, although with less evidence to support the link. But the hypothesized mechanism in both cases relates to the drug's interference in the natural detoxification processes, which then increases inflammation. The jury is still out—no animal studies yet, and no experiments demonstrating causation—but the repeated association is troublesome, especially because doctors routinely recommend this painkiller for both children and pregnant mothers.

The larger point is, just because we've weakened our immune regulation via biological means, that doesn't mean substances we encounter in modern life couldn't have contributed to the allergy epidemic. That goes for pollution, for flame retardants, for pesticides, and for tobacco smoke, to name other potential malfeasants. No one should read this book and conclude, "See? My daughter wheezes because the worms are gone. It's fine to leave the triclosan in the toothpaste, the BPA in the baby bottles, and smokestacks without scrubbers."

THE INNER-CITY QUESTION

As I wrote this book, whenever I mentioned its subject matter to people, the first question was almost invariably, "Why do people living in the inner city have so much asthma?" The short answer is no one's quite sure. But Matthew Perzanowski and colleagues at Columbia University are on the case. First, the inner city is not necessarily "dirty" in a way that protects against allergic disease, like a Bavarian cowshed. Inner-city apartments in New York City have no more bacteria than does the average suburban house.

That said, Perzanowski and others have found that allergic disease does occur inversely with environmental bacterial load even in the inner city.

The more bacteria in people's houses, the less their chances of allergies. The hygiene hypothesis applies.

What about the astronomical rate of allergies in general, a near tripling of wheezing among children in the few blocks separating the Upper East Side from East Harlem, for example? (That's 6.4 percent prevalence in the former versus 18.5 in the latter.) The answer is partly ecological. Perzanowski has found that buildings in low-income neighborhoods—old tenements and towering public housing projects—favor a proliferation of rodents and cockroaches. Children in the inner city are exposed to a lot more dander from pests than those living in better-off neighborhoods, but not more microbes.

How does this finding jibe with the research explored in this book? One interpretation: Kids in the inner city encounter lots of proteins that can be mistaken for parasites, but, like everyone else, they don't receive any calming signals either from environmental microbes or real parasites. They experience the same lack of a good immune education while also encountering more aggravation. And if they do encounter parasites, the timing and intensity of infection may not only fail to prevent allergy, but may make it worse. Transient infestations that begin late in childhood can, some think, exacerbate allergic disease.

All that said, an uncomfortable but nonetheless important question lurks in the inner-city asthma phenomenon: Are some U.S. minority populations more prone to asthma and allergy? Again, Perzanowski is teasing out genetic contributions from environmental ones. He's comparing African Americans living in both low-income and higher-income neighborhoods. So far, he finds that African Americans in higher-income neighborhoods, exposed to fewer pests, have less allergic disease. Vermin seem to be an important determinant. The high burden of asthma in these communities could be theoretically prevented. So, slumlords, this book doesn't provide an excuse *not* to clean up your buildings.

However, some suspect that even if perfect economic parity existed among different ethnicities, certain groups would still develop allergy more often than others. Perzanowski doesn't quite buy this. But we've seen plenty of evidence that gene variants that predispose to both autoimmune and allergic disease are enriched in the very populations most exposed to pathogens and parasites in the past. Why? Because in their proper context, those genes augment defense. What does this mean? In the modern environment, these populations, most of which hail from the tropics, will likely have a greater risk of asthma and allergy when stimuli from parasites and microbes goes missing.

The surging allergy prevalence in some parts of the developing world suggests that this is happening already. Costa Rica, which, rather than funding an army, famously invested in education, health care, and economic development—and which also sponsored major deworming campaigns—is alone among Central American countries in its extremely elevated prevalence of allergies and asthma. One quarter of Costa Rican teenagers wheeze. In Peru, 26 percent of teens wheeze. The percentage of asthmatic teens in Brazil is roughly the same—one of every four.

South Africa, the richest nation in sub-Saharan Africa, also has a comparatively elevated prevalence of allergic disease. And other parts of Africa where, forty years ago, allergic disease was so freakishly absent have since seen an uptick. (For that matter, younger generations in former Eastern Bloc countries are also more allergic than older ones.)

These numbers are very preliminary, and they're mostly from urban centers. They don't necessarily reflect countrywide patterns. Furthermore, they're based on surveys, which have many weaknesses. But they fit with both historical and contemporary patterns from the industrialized world. Cities tend to have more allergic people than rural areas, and allergy as a phenomenon begins in urban centers.

IN HUMAN ENVIRONS,
EVEN ANIMALS GET WESTERN DISEASES

One day in the late 1990s, it occurred to Ajit Varki, a scientist at the University of California, San Diego, to compare human T cells with chimpanzee T cells. The exercise might illuminate our similarities and our differences, he thought.

How right he was. He noted that, compared with chimp T cells, human T cells displayed a gross lack of certain receptors that acted as brake pedals. Human T cells were quicker on the draw, but they also had a harder time calming back down.

Why did we have hyperactive T cells? He hypothesized that in order to escape some pathogen that was hounding us in the past, we devolved the receptors. Others interpreted the loss more literally, surmising that since our ancestors parted ways with chimps, we'd become so filthy that surviving required jumpy T cells. Our own pestilence became a force in our evolution.

More important than why these changes occurred were the consequences of their occurrence: a propensity to overreach. Varki thought that, because of our comparatively berserker T cells, humans were constitutionally prone to inflammatory diseases such as asthma, rheumatoid arthritis, and type-1 diabetes, as well as self-destruction when fighting infections, like HIV-associated dementia or cerebral malaria.

I accepted Varki's argument at face value. But I had a nagging sense that, if true, the coevolutionary relationships explored here—dependence on microbes, and the requirement of contact with parasites—would apply to other mammals as well. Everyone was in the Red Queen's Race, not just *Homo sapiens*.

Soon enough, I learned that our closest animal companions suffered from diseases quite similar to our own. Dogs got inflammatory bowel disease and eczema. Cats got asthma and colitis. Horses could also get inflammatory bowel disease and allergies, which struck me as particularly absurd. Here's an animal evolved to roam the plains, an ungulate that has spent millions of years with its face in the grass, and it can't tolerate pollen? A wild allergic horse could literally starve from hay fever.

Of course, immune dysfunction in domestic animals could always result from inbreeding, but certain patterns were eerily similar to those we've seen in people. One scientist compared the IgE levels of Scandinavian gray wolves, zoo-living wolves, and their dog cousins. Wild wolves had double the IgE of allergy-prone domestic pooches. We don't know how often they got allergies, unfortunately, but we can extrapolate from human studies that it's less often than Fido. Parasitized human populations with skyrocketing IgE levels—the allergic antibody—tend not to suffer from allergic disease, while Westernized populations with low IgE levels are plagued by runny noses and wheezing.

Veterinarians, meanwhile, observed the same alterations in bowel flora seen in human IBD in dogs with colitis—more *E. coli*, and a paucity of those all-important clostridial bacteria that induce T-regs.

Horses even had their own immigrant-allergy syndrome. Equines raised in Iceland had lots of intestinal parasites. When imported to continental Europe, they were dewormed, and they often developed a vicious allergy to native biting insects. If they were first exposed to the bugs right after deworming, on the other hand, a period during which their regulatory mechanisms were still in high gear, they didn't develop the allergy. They tolerated the insects like native-born horses.

Then I learned about the primates. Captive primates did, in fact, develop humanlike inflammatory diseases, in some cases more than humans. In the 1980s, UC Davis scientists treated an elderly captive chimpanzee for allergic asthma, which came on seasonally. She was, they found, allergic to grass, weed, and tree pollen. In another case, zookeepers noted that a female gorilla developed a case of eczema. They blamed in-group conflict for the eruptions. This was a common fallback explanation for allergic and autoimmune disease in our great ape cousins. And why not? Until recently, many invoked the same psychosocial factors to explain human IBD, asthma, and even autism.

In another case, a captive gorilla developed a severe case of ulcerative colitis and died, leaving her handlers baffled. They'd treated her with antibiotics. Why hadn't she improved? Upon necropsy, they found no pathogens in her intestine, just commensal bacteria inciting severe inflammation.

A little more research, and I discovered that spontaneous colitis was a significant problem among certain types of captive primate. Large numbers of baboons and rhesus monkeys literally dropped dead from inflammatory bowel disease. Mustached tamarins, small South American monkeys, tended to develop a condition that approximated human Crohn's disease. One study of aged rhesus monkeys who'd lived their life in captivity found that they suffered from precancerous lesions and colon cancer frighteningly often.

In the late 1990s, scientists went so far as to survey wild primates to determine the incidence of these problems in a natural setting—not macaques in this case, but cotton-top tamarins in South America. They found the conditions to be almost entirely absent. More than two-thirds of sixty-nine captive monkeys had severe colitis. Twelve had colon cancer. In eighty-eight wild monkeys examined, however, the scientists couldn't find a single case of severe colitis or colon cancer. "The observation suggests that colitis and cancer in the tamarin model are linked to environmental factors," the authors observed dryly.

But which ones? In this case, the scientists noted threadlike worms in the wild tamarins. Wild capuchin monkeys, a different species, were even more grossly infested with helminths. Weinstock, you'll recall, thought worms protected from inflammatory disease by skewing the immune response and enhancing regulation. That may also protect against cancer. In chapter 12, we also saw that eosinophils were strongly tumoricidal. Parasites induce eosinophils.

Captive primates, who spent their lives indoors eating monkey chow (processed food) supplemented with fruits and veggies, also had a different microflora. The altered microbial communities didn't always associate with any particular disease. But in some cases, they did. Among captive macaques, for example, individuals with colitis had a measurably different microbiota compared with their healthy counterparts.

Japanese macaques in one captive colony even spontaneously developed a multiple sclerosis–like condition. As in humans, it was associated with a herpes virus infection. Notably, earlier generations—this particular colony was founded in 1965—never developed the disorder. Now it afflicts some 2 percent.

Primates also developed skin problems. Autoimmune-mediated hair loss, it turned out, was common enough. Two captive chimpanzees at zoos had alopecia universalis. One, in India, was named Guru. The other, in St. Louis, was named Cinder on account of her gray skin. (Cinder passed away, unfortunately.) Without their hair, both were impressively muscular. Why had they become hairless?

Again, explanations tended toward the stress-triggers-autoimmune-disease model. But read *A Primate's Memoir* by the neuroscientist Robert Sapolsky and you'll note that a wild primate's life is far from stress-free—not because of lions and hyenas, necessarily, but because of fellow primates. Moreover, despite the imagined stress of captivity—one likely characterized by mind-numbing boredom—captive primates tend to live longer than their wild counterparts. Meanwhile, in the many man-hours humans have spent observing primates in the wild—years, maybe decades collectively—no one has written a case study on the hairless chimp or gorilla.

I concede that absence of proof does not proof of absence make. But in 2010, scientists from Harvard Medical School, where captive macaque were prone to developing alopecia and eczema-like conditions, conducted a comparative study that again highlighted the primacy of a hygienic environment in immune dysfunction. They contrasted a group housed exclusively indoors, which was kept free of pathogens for research purposes, with one that lived outside. Regardless of where they ended up, animals born outside had about half as much alopecia, they found, and less-inflamed skin in general. What was the sole measurable difference? Unlike the inside group, the outside group hosted parasitic lung mites.

Captive and domestic animals are dewormed and treated with antibiotics just like modern humans. Indoor animals are also exposed to an indoor microbiota that likely differs from the one their organism "expects." They often eat sterile, processed food. And in these conditions, they're afflicted with humanlike autoimmune and allergic diseases.

Varki's argument that humans have a comparatively elevated tendency to develop immune-mediated diseases compared with chimps may be true. But by how much? If you compare New Yorkers with chimps, the answer is probably a lot. But if you compare chimps with, say, people living in the Amazon jungle, the increased relative risk would probably narrow significantly. The rules governing immune function and the microbiome appear to be universal. Whether you're a dog, a horse, or one of our primate relatives, the rule is: Extricate yourself from the web of life in which you evolved, and the immune system loses its bearings.

ENOUGH! WHAT TO DO ABOUT IT?

As of early 2012, trials testing helminths on multiple sclerosis, autism, peanut allergies, and finally a large trial using Weinstock's whipworm eggs on IBD, are in the works. In the coming decades, it's probably not overly optimistic to anticipate the development of immune-modulating drugs based on helminths and bacteria. However, it's worth remembering that a pill may not fully mimic the effect of a living parasite or bacterium. Helminths constantly subvert the host immune system. They adjust to the host's adjustments. Bacteria also respond, not only to you, but to other microbial residents. A pill can't easily replicate these constant maneuverings. We may be stuck with the real thing.

Stool transplants are already on the verge of going mainstream, at least for treating *Clostridium difficile* infection. But the subtler applications, such as correcting immune and metabolic dysfunction, still require development. Scientists don't call the microbiome "the forgotten organ" for nothing. They only just noticed it was there, and they're still in the early phases of learning how it works.

The work on farms is, in some ways, the most heartening. We can correct our postmodern immune system with a mix of microbes, and these microbes already exist somewhere in the proper formation. We just have to get them in a bottle.

Unlike with large parasites, however, no one's really sure how exposure to cowsheds, chicken coops, and pigpens protects against allergies. Do the microbes colonize your gut? Do they prevent worse bacteria from gaining purchase? Do they stimulate immunity so that, when asthma-causing viruses show up, you bat them away without a second thought? Or does the sheer abundance of microbes encountered from an early age simply keep your immune system healthily engaged? As the allergy researcher James Gern notes, "Resolving these questions is of critical importance if we are to bring the substantial health benefits of being raised on a farm to those who are not."

There's also burgeoning interest in probiotics, prebiotics—food for those indigenous beneficial bacteria—and a combination of the two called synbiotics. They clearly help prevent *C. difficile* infections, and they would seem to hold immense promise as a tool for immune modulation. So far, however, studies looking at that application have delivered conflicting results. And there's the nagging issue of diversity: If exposure to a diverse microbiota is critical in avoiding immune dysfunction—and much research suggests this is a cardinal rule—how will current probiotics, which comprise just a few species, really help?

The only sound recommendation at this point is dietary. Eat plenty of fruits and veggies—food for beneficial bacteria—lots of anti-inflammatory omega-3s, and avoid easy calories and processed foods. It certainly won't hurt, especially during pregnancy, and it may help. Don't deceive yourself, however. If you have asthma already, eating a Mediterranean-style diet is unlikely to fix it.

Although the lack of concrete recommendations seems like a fatal weakness, it's important to recognize something I've left mostly unsaid: The research you just toured, comprising work by thousands of scientists around the world, represents an ongoing paradigm shift in the understanding of our own biology.

To begin with, the body is a country where considerable effort goes toward cooperation with other organisms. The gut, perhaps our preeminent immune organ, serves as the primary command center for this endeavor—a reactor chamber that sets the tone for immune and metabolic functioning far and wide. The makeup and dynamism of the living community it houses, therefore, have far-reaching consequences. Perhaps most essential, for the immune system, *peacekeeping is an active process, not an absence of process*. Equilibrium is not necessarily the default setting, but a talent that's developed.

In this context, allergic disease is seen as a parasite control mechanism that, absent real parasites and key microbes, has spun out of control. Autoimmune disorders are viewed as tissue defense and maintenance processes that, because of weak regulatory oversight, have become self-destructive. Crucially, it's the microbes and parasites once considered mortal enemies that are teaching us these lessons—how our immune system really works. So while we don't yet know how we'll get there, we can already make out the destination.

First, future doctors will emphasize disease prevention, and that will begin in pregnancy. The doctor will genotype you and take a history of diseases in the family. She'll know your genetic predisposition to autoimmune and allergic diseases. She'll look at measures of inflammation and regulation. Accordingly, she'll begin nudging your immune function during—or likely before—pregnancy, not necessarily for your sake, but for your unborn child's. That will include dietary changes, probiotics, prebiotics, maybe a stool transplant. She'll endeavor to have the fetus develop in a favorable immunological environment.

When your child is born, the doctor will ensure colonization by the right microbes in the right sequence—that her microbiota is diverse, healthy, and stable as she grows. She may apply artificial microbial pressure, innocuous saprophytes, say, that your immune system recognizes from past lives, or barn dust. Maybe she'll introduce Epstein-Barr virus at a young age, maybe *Helicobacter pylori*. Perhaps she'll recommend a helminth or two during toddlerhood or later. These last three scenarios seem like a hard sell, but perhaps drugs that mimic these agents, without the potential harm, will be available. Note, however, that all of them take up long-term residence. Successful preventive treatment will have to be long term as well. The greater point here is that future doctors will guide your immune system along a course of development that's informed by the original assembly pattern of the human superorganism.

She'll also steer you clear of known pitfalls. When she prescribes antibiotics, say, she'll give you a probiotic. She'll keep a backup supply of your own unique microbiota in all its complexity, or maybe a premanufactured "ideal" microflora, to repopulate your intestine. As you age, she'll ensure your microflora stays in good shape, that it doesn't slide into a pathological state. This maintenance will help you avoid the degenerative diseases of middle and old age.

WHAT ABOUT CHANGING EVERYTHING ELSE?

The fantasy I just laid out is really a barely disguised ecosystem restoration project. The problem it addresses is really a biodiversity crisis. And it's hard to ignore that our internal extinction crisis coincides with a planetwide external one. For millennia, we've overhunted, burned, deforested, plowed, overfished, and generally thinned out the biosphere. More recently, we've begun altering the very chemistry of ocean, earth, and atmosphere. Scientists estimate that the extinction spasm now occurring will be the severest since a meteor slammed into earth 65 million years ago.

These are ecosystems nested within ecosystems. They're not separate, and neither are the crises occurring simultaneously on multiple levels. That's not to say that curing IBD requires solving the problem of global warming, or that reversing the worldwide decline in global reefs will fix asthma, but it does say something about the heavy-handed approach that got us here. The common theme is a casual indifference to the web of life, our place in it, and its place in us.

Moreover, as an ecosystem restoration project, the rules that guide the conservation of external biodiversity apply. Central among them is that you can't really save the tiger without saving the tiger's jungle and everything in it, from soil microbes, to ants, to trees. They come as a single, integrated unit. So if you want to keep the right microbes around, you've got to foster the correct ecosystem. That doesn't mean a return to Dickensian filth, or even the worm-ridden, lice-bitten splendor of the Paleolithic. There's plenty of middle ground, ample room to use intelligence and finesse. And there are important reasons why an ecosystem approach may prove most effective.

Throughout the book, I've implied that modern homes are relatively barren of microbes, which is true. But they're not totally barren. Nature abhors a vacuum. And while we like to think we've successfully engineered microbe-free zones, some evidence suggests that modern dwellings actively cultivate microbes that make us sick.

That's essentially what the Helsinki University microbiologist Mirja Salkinoja-Salonen has found. (She compared microbes in Finnish and Russian Karelia.) When occupants complained about "sick building syndrome"—people working or living in a building who feel constantly ill—she invariably isolated toxic bacteria from its nooks and crannies. She found these poisonous microbes in day-care centers and schools as well.

Dried and stored powdered formulas, and prepackaged convenience foods, often contained other types of noxious bacteria, Salkinoja-Salonen found. The high-sugar, high-fat, low-moisture environments worked like the Sahara Desert for most microbes, but not for these. They loved these conditions, and they thrived. Ditto with electronics. She's isolated toxic microbial by-products from office computers as well. Scientists at the University of Oregon's Biology and the Built Environment lab have made similar finds. Buildings with central air, such as hospitals, paradoxically select for pathogenic bacteria. Opening the window and letting in outdoor microbes literally makes a room healthier.

Critically, while these toxic bacteria thrive in modern human dwellings, they're totally absent in other environments. Salkinoja-Salonen fails to locate them in Finnish barns, for example. Like weedy species anywhere, they seem unable to compete in complex, highly evolved communities. They flourish only in man-made wastelands. These observations add a layer of complexity to the "old friends" hypothesis. Modern human environments not only exclude health-promoting bacteria, they may actively cultivate damaging microbes.

Given both the positive and negative impact of the environmental microbiota on our health—and the fact that we spend so much time inside—in the future, we'd ideally engineer buildings that naturally cultivated health-promoting microbes and precluded toxic and inflammatory ones. The easiest, most cost-effective way to "engineer" said dwellings may be to move farm animals into your apartment. One imagines, however, that some middle ground exists here as well.

The microbes that seem to benefit us most come from living soil and animals. Maybe vertical farms like those advocated by Columbia University's Dickson Despommier—food-producing ecosystems in skyscrapers—could also supply urban environments with this microbiota. It's titillating to imagine tubes connecting tower farms to nearby apartment buildings, vents through which living microbes waft into living rooms and bedrooms.

Other seemingly nonrelated currents are pushing in the same direction—new urbanism, foodie culture, green spaces, locavorism, community farms, and so on. Why not incorporate the fostering of, and exposure to, "good" microbes, and the exclusion of bad? Urban children could work in community farms while growing up. So could pregnant mothers. Assuming you tested your local dairy cow for pathogens, you could even regularly drink unpasteurized milk. Food manufactured locally might also allow the purchase of fermented food with live cultures—not the "probiotic" bacteria added to commercial food after the original cultures have been killed off, but the complex microbial communities that actually do the fermenting.

These are all admittedly fantastical scenarios. But what's not completely far-fetched—what seems more necessity than fantasy, actually—is a change in how we approach our day-to-day ecology, both inner and outer. More than twenty years ago, George Williams and Randolph Nesse argued that to truly comprehend the causes of illness, doctors had to understand human evolution. Otherwise, they'd be stuck treating symptoms and not root causes. They called this idea "Darwinian medicine."

We can expand the appeal: to treat modernity's ailments, consider not only the human genome, but also our metagenome—the other 99 percent of instructions necessary to smoothly operate the human superorganism. And rather than clear-cut, actively cultivate. Germ theory has brought us a long way. But in order to continue maximizing our health and well-being, and that of our loved ones, we've got to eschew the brute-force tactics of the exterminator, adopt the gardener's finesse, and grow the superorganism.

Acknowledgments

THANKS TO THE FOLLOWING SCIENTISTS
AND PHYSICIANS FOR GRANTING INTERVIEWS

Charles Akle, Emma Allen-Vercoe, Sidney Baker, Erik Barton, Kevin Becker, Bengt Björkstén, Martin Blaser, Janette Bradley, John Britton, Geoff Butcher, Harvey Checkoway, Yu Chen, Ilseung Cho, Chris Coe, Cris Constantinescu, Anne Cooke, Philip Cooper, Mark Corrales, Jorge Correale, John Croese, Mario Milco D'Elios, Les Dethlefsen, Maria Dominguez-Bello, David Elliott, Stanley Falkow, Alesso Fasano, Denise Faustman, Johanna Feary, Martin Feelisch, Caleb Finch, Sydney Finegold, Colin Fitzsimmons, Carsten Flohr, Detlev Goj, David Graham, Mel Greaves, Brian Greenwood, Richard Grencis, Michael Gurven, Martha Herbert, Mark Holbreich, Otto Holst, Anthony Horner, Kwang Jeon, Marie-Helene Jouvin, Jonathan Kipnis, Daniel Littman, Jorge Llamas, P'ng Loke, Christopher Lowry, Derrick MacFabe, Rick Maizels, Fernando Martinez, Sarkis Mazmanian, Thomas McDade, Edward Mitre, Anne Müller, Cathryn Nagler, Stephen O'Keefe, Carlos Pardo, Paul Patterson, Richard Peek, Gretel Pelto, Matthew Perzanowski, Sonja Praprotnik (via e-mail), David Pritchard (despite repeated attempts, I never interviewed Pritchard for this book; I learned most of what I know of his story during a 2008 interview for an article), Charles Raison, Eyal Raz, Karl Reinhard, Harald Renz, Meret Ricklin, Eleanor Riley, Karen Robinson, Graham Rook, Noel Rose, Moshe Rubin, Mirja Salkinoja-Salonen, Yolanda Sanz, Bianca Schaub (via e-mail), Leonardo Sechi, Scott Sicherer, Vijendra Singh, Manuela Sironi, Justin Sonnenburg, Stefano Sotgiu, Jonathan Terdiman, Meri Tulic, Peter Turnbaugh, Judy van de Water, Erika von Mutius, Joel Weinstock, Rudi Westendorp, David Whitlock, Maria Yazdanbakhsh, and Andrew Zimmerman.

THANKS TO THE FOLLOWING "PARASITE USERS"
FOR SHARING THEIR STORIES

Garin Aglietti, Judy Chinitz, Stewart and Lawrence Johnson, Josh, Nikolai K., Karl, Lisa, Michael L., Jasper Lawrence, Dan M., Scott R., Shelley

Schulz, and Debora and Karsten Wade. And thanks to those who shared stories anonymously.

Thanks to Raj Patel for so generously connecting me with his agent, and for advice throughout. My agent Kris Dahl for liking the book idea, and selling it. Also at ICM, Laura Neely for her diligence. My editor Colin Harrison for his interest, calming presence, and insightful advice during the writing process. Also at Scribner, Susan Moldow, Nan Graham, Kelsey Smith, Rex Bonomelli, Kate Lloyd, and Laura Wise for making this happen. Martin Blaser, Rick Maizels, Karl Reinhard, Karen Robinson, Justin and Erica Sonnenburg, Erika von Mutius, and Joel Weinstock for reading and commenting on sections of the manuscript. Graham Rook for reading and commenting on the whole behemoth. Your insights were invaluable; your enthusiasm was infectious. Janette Bradley and again Dr. Rook for inviting me to Nottingham. Nick Coronges for hosting me in London. My cousin Michael Molina for hosting me in Miami. Maria Yazdanbakhsh for so graciously spending an afternoon fielding questions. Stefano Sotgiu for a day among the ruins of Sardinia, and for a tasty Sardinian dinner with his delightful family. Michael Gurven for agreeing to my visit to Bolivia, and for letting me tag along in Tsimanía. P'ng Loke and Mahesh Gundra for indispensable technical assistance. Michael Cappello at Yale for taking my call and agreeing to help. In the Cappello Lab, Josephine Quagraine and Lisa Harrison for the swift and definitive identification of my parasites. Kaitlin Yapchaian for beautiful photographs taken on a rainy day.

Marguerite Holloway for the yearlong adventure at Columbia, and her unparalleled attention to detail as an editor. Jonathan Weiner for the example of his excellent books. Owen Thomas for giving me free rein, every feature writer's dream. The Columbia University libraries for alumni access. The Metcalf Institute for a yearlong writing fellowship at a great paper, an essential learning experience.

Mark Harris for his counsel on book writing in general, and for listening patiently to my anxieties. Tony Kushner for his unconditional support since the beginning, and for inspiration to go big. My uncle Robert Manoff for his intelligence, and for always championing my pursuit of journalism. My sisters, Una and Hannah, and my brothers, Isaac and Ashi, for being lovable characters. My father, Gregg, for his encouragement, advice, and support, and for his unfailing emphasis on the primacy and importance of books. He's partly responsible for this. My stepmother, Summer, for her enthusiasm and energy, and for being a great mother. My grandfather Richard Manoff for his example of discipline and purpose. My *abuela* Joaquina Pérez for her kindness and humor. My mother, Carmen, for her irrepressible sense of mischief, uncompromising goodness, and insistence on something else, something more, something beyond. You are sorely missed. The extended Manoff, Deutscher, and Velazquez clans for a lifetime of warmth and generosity.

The Park family for bringing me so easily and unquestioningly into the fold. My daughter, Selah, simply for being—for her luminescence. And most of all, my dear wife, Carol Park, for her herculean support and forbearance as I wrote and rewrote this book. Without you, this couldn't have happened.

One more thing: During research, I amassed a library of more than 8,500 papers and abstracts. They represent the work of tens of thousands of scientists around the world. For obvious reasons, I couldn't mention even a healthy minority of these contributors in the text. In fact, in the interests of simplicity, I tended to omit the names of people and institutions. But science is a collaborative effort, and the meat of this book represents a lot of grueling, painstaking work by lots of researchers. So to the tens of thousands I didn't mention, but whose research directly or indirectly informs this book, thank you. Your work is critical to our health and well-being, and to our understanding of ourselves.

Notes

For more sources and additional reading, visit epidemicofabsence.com.

Chapter 1: Meet Your Parasites

1 *Mother, it is no gain:* R. Tagore and S. K. Das, *The English Writings of Rabindranath Tagore: Poems* (New Delhi: Sahitya Akademi, 2004), 45.

2 *"Well-nourished persons often harbor helminths":* David E. Elliott and Joel V. Weinstock, "Inflammatory Bowel Disease and the Hygiene Hypothesis: An Argument for the Role of Helminths," in *The Hygiene Hypothesis and Darwinian Medicine,* ed. Graham Rook.

3 *Only 1 to 2 percent of the population got alopecia areata:* K. H. Safavi et al., "Incidence of Alopecia Areata in Olmsted County, Minnesota, 1975 Through 1989," *Mayo Clinic Proceedings* 70, no. 7 (1995).

4 *The first genome-wide association study of alopecia:* Lynn Petukhova et al., "Genome-wide Association Study in Alopecia Areata Implicates Both Innate and Adaptive Immunity," *Nature* 466, no. 7302 (2010).

5 *one of every ten children passing:* S. T. Holgate, "The Epidemic of Allergy and Asthma," *Nature* 402, no. 6760 Suppl. (1999). Asthma stats from the CDC: www.cdc.gov/vitalsigns/ Asthma/index.html. (Accessed February 15, 2012.) Hay fever stats from the American Academy of Allergy, Asthma & Immunology.

5 *one in twenty passersby had one of eighty:* The stats vary on autoimmune disease prevalence. Between 3 and 8 percent of the developed world suffers from one of more than eighty autoimmune or inflammatory conditions. Pierre Youinou et al., "Geo-epidemiology and Autoimmunity," *Journal of Autoimmunity* 34, no. 3 (2010).

5 *One of every 250 people:* N. A. Molodecky et al., "Increasing Incidence and Prevalence of the Inflammatory Bowel Diseases with Time, Based on Systematic Review," *Gastroenterology* 142, no. 1 (2012).

6 *I'd note glucose monitors:* Edwin A. M. Gale, "The Rise of Childhood Type 1 Diabetes in the 20th Century," *Diabetes* 51, no. 12 (2002).

7 *A 2009 study found:* Alberto Rubio-Tapia et al., "Increased Prevalence and Mortality in Undiagnosed Celiac Disease," *Gastroenterology* 137, no. 1 (2009).

7 *In 2002, the French scientist Jean-François Bach:* Jean-Francois Bach, "The Effect of Infections on Susceptibility to Autoimmune and Allergic Diseases," *New England Journal of Medicine* 347, no. 12 (2002).

8 *Vanishingly few children in Albania:* "Worldwide Variation in Prevalence of Symptoms of Asthma, Allergic Rhinoconjunctivitis, and Atopic Eczema: ISAAC. The International Study of Asthma and Allergies in Childhood (ISAAC) Steering Committee," *Lancet* 351, no. 9111 (1998).

8 *The incidence of type-1 diabetes:* DIAMOND Project Group, "Incidence and Trends of Childhood Type 1 Diabetes Worldwide 1990–1999," *Diabetic Medicine* 23, no. 8 (2006). Christopher C. Patterson and EURODIAB Study Group, "Incidence Trends for Childhood Type 1 Diabetes in Europe During 1989–2003 and Predicted New Cases 2005–20: A Multicentre Prospective Registration Study," *Lancet* 373, no. 9680 (2009).

8 *scientist calls "animals, faeces and mud":* G. A. W. Rook, "The Gut, Immunoregulation and Micro-organisms from Man's Evolutionary Past," *Nutrition Bulletin* (2010).

10 *In the 37,000 examinations:* Complete stats of immune-mediated disease in Tsimanía come directly from Michael Gurven. As of October 26, 2011, they are: 8 cases of allergic

rhinitis; none of asthma. Prevalence of allergies: 6.6/10,000; 15 autoimmune diseases (11 vitiligo, 1 rheumatoid arthritis, 1 lupus, 1 Graves disease, and 1 proliferative membranous glomerulonephritis). Prevalence of autoimmune disease: 12.3/10,000. The medical census population: 12,116 people. They've seen some patients up to ten times. They've done more than 37,000 person-examinations in total. That's probably more than 95 percent of the Tsimane population.

10 *Others studying unacculturated Amerindians:* A. Magdalena Hurtado and Kim Hill have spent years studying the Hiwi of the Venezuelan lowlands, and the Aché of the Paraguayan Amazon. They have no asthma, no high blood pressure, no diabetes. A. Magdalena Hurtado, Kim Hill, and I. Arenas De Hurtado Selva Rodriguez, "The Evolutionary Context of Chronic Allergic Conditions: The Hiwi of Venezuela," *Human Nature* 8, no. 1 (1997).

12 *Life here is not easy:* Gurven, Michael, et al., "Mortality Experience of Tsimane Amerindians of Bolivia: Regional Variation and Temporal Trends," *American Journal of Human Biology* 19, no. 3 (2007).

13 *Your parents, meanwhile, maybe had seasonal hay fever:* Mark Jackson makes this point in his *Allergy: The History of a Modern Malady.*

13 *"We as immunologists are now faced":* William Parker, "Reconstituting the Depleted Biome to Prevent Immune Disorders," *Evolution & Medicine Review* (2010): http://evmedreview.com/?p=457. (Accessed February 12, 2012.)

14 *In 1982, scientists:* Berkley R. Powell, Neil R. M. Buist, and Peter Stenzel, "An X-linked Syndrome of Diarrhea, Polyendocrinopathy, and Fatal Infection in Infancy," *Journal of Pediatrics* 100, no. 5 (1982).

14 *The gene was named FOXP3:* C. L. Bennett et al., "The Immune Dysregulation, Polyendocrinopathy, Enteropathy, X-linked Syndrome (IPEX) Is Caused by Mutations of FOXP3," *Nature Genetics* 27, no. 1 (2001).

14 *the finding upended:* I'm condensing lots of research here. For more see Leonie Taams, Arne N. Akbar, and Marca H. M. Wauben, *Regulatory T cells in Inflammation*, Progress in Inflammation Research (Basel and Boston: Birkhäuser, 2005), 3–17.

Chapter 2: *Homo Squalidus:* The Filthy Ape

21 *[O]ne can properly think of most human lives:* William McNeill, *Plagues and People,* 5.

21 *"Homo sapiens ranks":* R. W. Ashford and W. Crewe, *The Parasites of Homo Sapiens: An Annotated Checklist of the Protozoa, Helminths and Arthropods for Which We Are Home* (London: Taylor & Francis, 2003), ix.

22 *The first such transition occurred:* George J. Armelagos, "The Paleolithic Disease-scape, the Hygiene Hypothesis, and the Second Epidemiological Transition," in *The Hygiene Hypothesis and Darwinian Medicine*, ed. G. A. W. Rook (Basel and Boston: Birkhäuser, 2009).

23 *Modern-day hunter-gatherer groups:* All stats are from the discussion section of this review, which focuses on the Hadza of Tanzania, and also touches on the !Kung and others. F. J. Bennett et al., "Helminth and Protozoal Parasites of the Hadza of Tanzania," *Transactions of the Royal Society of Tropical Medicine and Hygiene* 64, no. 6 (1970).

23 *Our primate relatives:* This 1976 survey of wild chimps in Tanzania observed six different helminth species and two protozoa species. Of thirty-two individuals examined, all were infected by at least one, and most by several. S. K. File, W. C. McGrew, and C. E. Tutin, "The Intestinal Parasites of a Community of Feral Chimpanzees, *Pan Troglodytes Schweinfurthii,*" *Journal of Parasitology* 62, no. 2 (1976).

23 *the native human repertoire of parasites:* R. W. Ashford, "Parasites as Indicators of Human Biology and Evolution," *Journal of Medical Microbiology* 49, no. 9 (2000).

23 *Eric Hoberg, a scientist:* E. P. Hoberg, "Phylogeny of Taenia: Species Definitions and Origins of Human Parasites," *Parasitology International* 55 Suppl. (2006).

24 *David Reed at the Florida Museum:* David L. Reed et al., "Pair of Lice Lost or Parasites Regained: The Evolutionary History of Anthropoid Primate Lice," *BMC Biology* 5 (2007).

24 *"We'll never know if it was sex":* Nicholas Wade, "In Lice, Clues to Human Origin and Attire," *New York Times,* March 8, 2007.

24 *the head louse evolved into its new niche:* Robin A. Weiss, "Apes, Lice and Prehistory," *Journal of Biology* 8, no. 2 (2009).

24 *Melanesians and some Southeast Asians:* David Reich et al., "Genetic History of an Archaic Hominin Group from Denisova Cave in Siberia," *Nature* 468, no. 7327 (2010).

24 *A "race" of human head lice found:* David L. Reed et al., "Genetic Analysis of Lice Supports Direct Contact Between Modern and Archaic Humans," *PLoS Biology* 2, no. 11 (2004).

25 *Amerindians had ascaris worms:* Odile Loreille and Françoise Bouchet, "Evolution of Ascariasis in Humans and Pigs: A Multi-disciplinary Approach," *Memórias do Instituto Oswaldo Cruz* 98, Suppl. 1 (2003).

26 *Judging from the skeletons:* For a fascinating review of life in the late Paleolithic, see Brigitte M. Holt, "Hunters of the Ice Age: The Biology of Upper Paleolithic People," *American Journal of Physical Anthropology* Suppl. 47 (2008).

26 *Toe bones began to atrophy:* Erik Trinkaus and Hong Shang, "Anatomical Evidence for the Antiquity of Human Footwear: Tianyuan and Sunghir," *Journal of Archaeological Science* 35, no. 7 (2008).

26 *By the late Paleolithic:* Mary C. Stiner, "Carnivory, Coevolution, and the Geographic Spread of the Genus Homo," *Journal of Archaeological Research* 10, no. 1 (2002).

27 *Some 41,000 years ago:* Murray P. Cox et al., "Autosomal Resequence Data Reveal Late Stone Age Signals of Population Expansion in Sub-Saharan African Foraging and Farming Populations," *PLoS ONE* 4, no. 7 (2009).

27 *And 22,000 years ago:* Luisa Pereira et al., "Population Expansion in the North African Late Pleistocene Signalled by Mitochondrial DNA Haplogroup U6," *BMC Evolutionary Biology* 10, no. 1 (2010).

27 *However, between 100,000 and 60,000:* Yali Xue et al., "Spread of an Inactive Form of Caspase-12 in Humans Is Due to Recent Positive Selection," *American Journal of Human Genetics* 78, no. 4 (2006). M. Saleh et al., "Differential Modulation of Endotoxin Responsiveness by Human Caspase-12 Polymorphisms," *Nature* 429, no. 6987 (2004).

27 *The nonfunctional variant of a gene called CARD8:* Dennis C. Ko et al., "A Genome-wide In Vitro Bacterial-Infection Screen Reveals Human Variation in the Host Response Associated with Inflammatory Disease," *American Journal of Human Genetics* 85, no. 2 (2009).

29 *By 8,000 years ago:* For a good graphic of domestication, see M. A. Zeder, "Domestication and Early Agriculture in the Mediterranean Basin: Origins, Diffusion, and Impact," *Proceedings of the National Academy of Sciences* 105, no. 33 (2008). Brian L. Peasnall et al., "Hallan Çemi, Pig Husbandry, and Post-Pleistocene Adaptations Along the Taurus-Zagros Arc (Turkey)," *Paeléorient* 24, no. 1 (1998). G. Larson et al., "Patterns of East Asian Pig Domestication, Migration, and Turnover Revealed by Modern and Ancient DNA," *Proceedings of the National Academy of Sciences* 107, no. 17 (2010). Y. P. Liu et al., "Multiple Maternal Origins of Chickens: Out of the Asian Jungles," *Molecular Phylogenetic Evolution* 38, no. 1 (2006). A. K. Outram et al., "The Earliest Horse Harnessing and Milking," *Science* 323, no. 5919 (2009).

29 *Wolves, some think, began scavenging:* Carlos A. Driscoll, David W. Macdonald, and Stephen J. Obrien, "From Wild Animals to Domestic Pets, an Evolutionary View of Domestication," *Proceedings of the National Academy of Sciences* 106, Suppl. 1 (2009).

29 *Residents of Çatalhöyük:* Shahina Farid, "Çatalhöyük 2009 Archive Report" (catalhoyuk. com).

29 *Around the same time, the first verifiable:* Israel Hershkovitz et al., "Detection and Molecular Characterization of 9,000-Year-Old Mycobacterium Tuberculosis from a Neolithic Settlement in the Eastern Mediterranean," *PLoS ONE* 3, no. 10 (2008).

29 *early agriculturalists were generally:* Vered Eshed et al., "Paleopathology and the Origin of Agriculture in the Levant," *American Journal of Physical Anthropology* 143, no. 1 (2010).

30 *The broader diseasescape was shifting:* George J. Armelagos, "The Paleolithic Disease-scape, the Hygiene Hypothesis, and the Second Epidemiological Transition," in Rook, *The Hygiene Hypothesis.*

30 *By five millennia ago:* See Crawford, *Deadly Companions,* 65.

30 *Egyptian papyri from the same period:* The pictograms with poxlike disease date from 3730–1555 B.C. Mummies with pox date to between 1570 and 1085 B.C. See ibid., 107.

31 *By A.D. 100, malarial fevers:* J. N. Hays, *Epidemics and Pandemics: Their Impacts on Human History* (Santa Barbara, Calif.: ABC-CLIO, 2005).

31 *You can see it in their stature:* Nikola Koepke and Joerg Baten, "The Biological Standard of Living in Europe During the Last Two Millennia," *European Review of Economic History* 9 (2005).

32 *And in 1347, twelve Genoese sailing ships:* See Crawford, *Deadly Companions,* 84, 86.

32 *If you're a human-adapted parasite:* How narrow the human bottleneck really was, and whether it really was a single discrete event, is still hotly debated. For pro, see The Genographic Project and S. H. Ambrose, "Late Pleistocene Human Population Bottlenecks, Volcanic Winter, and Differentiation of Modern Humans," *Journal of Human Evolution* 34, no. 6 (1998). For con, see J. Hawks et al., "Population Bottlenecks and Pleistocene Human Evolution," *Molecular Biology and Evolution* 17, no. 1 (2000).

33 *the field of paleo-parasitology begins:* M. A. Ruffer, "Note on the Presence of 'Bilharzia Haematobia' in Egyptian Mummies of the Twentieth Dynasty [1250–1000 B.C.]," *British Medical Journal* 1, no. 2557 (1910).

33 *The salient symptom of schistosomiasis:* See Crawford, *Deadly Companions,* 69.

33 *lakeside communities in Switzerland and Germany:* Matthieu Le Bailly et al., "Diphyllobothrium: Neolithic Parasite?," *Journal of Parasitology* 91, no. 4 (2005).

33 *In both Gallo-Roman and medieval times:* Françoise Bouchet, Stéphanie Harter, and Matthieu Le Bailly, "The State of the Art of Paleoparasitological Research in the Old World," *Memórias do instituto Oswaldo Cruz* 98, Suppl. 1 (2003).

34 *The widespread prevalence of hookworm:* Hookworm eggs show up in 7,200-year-old coprolites in Brazil, and in Peruvian mummies from a thousand years ago. And frigid climates don't protect against worms. The remains of a seven-year-old child buried 840 years ago in Adak Island, Alaska, have ascaris and fish tapeworm. F. Bouchet et al., "Identification of Parasitoses in a Child Burial from Adak Island (Central Aleutian Islands, Alaska)," *Comptes Rendus de L'Académie des Sciences Série III, Sciences de la Vie* 324, no. 2 (2001). Adauto Araujo et al., "Parasites as Probes for Prehistoric Human Migrations?," *Trends in Parasitology* 24, no. 3 (2008).

34 *the frigid Bering Land Bridge:* T. Goebel, M. R. Waters, and D. H. O'Rourke, "The Late Pleistocene Dispersal of Modern Humans in the Americas," *Science* 319, no. 5869 (2008).

34 *And Chinese laborers:* Karl J. Reinhard et al., "Chinese Liver Flukes in Latrine Sediments from Wong Nim's Property, San Bernardino, California: Archaeoparasitology of the Caltrans District Headquarters," *Journal of Parasitology* 94, no. 1 (2008).

34 *Sediments laid down in colonial Albany:* Charles L. Fisher et al., "Privies and Parasites: The Archaeology of Health Conditions in Albany, New York," *Historical Archaeology* 41, no. 4 (2007).

35 *Erected in 1771, Cromford Mill:* I adopted this beginning to the Industrial Revolution, of the possible choices, from Gavin Weightman, *The Industrial Revolutionaries: The Making of the Modern World, 1776–1914* (New York: Grove Press, 2007).

35 *Towns and cities had been filthy:* Much is drawn from Martin V. Melosi, *Garbage in the Cities.*

35 *"Industrialism, the main creative force":* Lewis Mumford, *The City in History: Its Origins, Its Transformations, and Its Prospects* (New York: Harcourt, 1961), 447.

35 *In 1801, London had:* Richard H. Steckel and Roderick Floud, *Health and Welfare During Industrialization,* National Bureau of Economic Research project report (Chicago: University of Chicago Press, 1997), 93.

35 *In cities, however, it was considerably less:* Kinsgley Davis, "The Urbanization of the Human Population," *Scientific American* (September 1965), 41–53.

36 *the height of military recruits decreased:* Well-to-do boys also measured a full six inches taller than poorer counterparts of the same age. Roderick Floud, Kenneth W. Wachter, and Annabel Gregory, *Height, Health, and History: Nutritional Status in the United Kingdom, 1750–1980,* Cambridge Studies in Population, Economy, and Society in Past Time (Cambridge and New York: Cambridge University Press, 1990), 1.

36 *"The Thames is now made a great cesspool":* Thomas Cubitt made that observation. Much here is drawn from Halliday, *The Great Stink of London,* 35.

37 *just one of twenty Americans:* See Melosi, *Garbage in the Cities,* 10, and Suellen M. Hoy, *Chasing Dirt,* 4.

37 *"filthy, bordering on the beastly":* Hoy, *Chasing Dirt,* 4. Ibid., 8. The second quote is from William Cobbett, a British agricultural reformer.

37 *"dogs, undistinguishable in mire":* Charles Dickens, *Bleak House* (New York: Hurd and Houghton, 1870), 11.

37 *And superficial grime aside:* J. Komlos, *Stature, Living Standards, and Economic Development: Essays in Anthropometric History* (Chicago: University of Chicago Press, 1994), 157.

38 *Between 1820 and 1850:* See Melosi, *Garbage in the Cities,* 14.

38 *"It is a melancholy fact":* S. Smith, *The City That Was* (New York: Frank Allaben, 1911), 63.

38 *"The country is horrified":* Judith Walzer Leavitt and Ronald L. Numbers, *Sickness and Health in America: Readings in the History of Medicine and Public Health,* 3rd ed. (Madison: University of Wisconsin Press, 1997), 442.

38 *V. cholerae is really a hybrid organism:* Marie-Eve Val et al., "The Single-Stranded Genome of Phage CTX Is the Form Used for Integration into the Genome of *Vibrio Cholerae,*" *Molecular Cell* 19, no. 4 (2005).

39 *Perhaps cholera's foreignness:* See Christopher Hamlin, *Cholera: The Biography,* Biographies of Disease (Oxford and New York: Oxford University Press, 2009).

40 *Squads operating day and night:* See Hoy, *Chasing Dirt,* 64.

40 *the American president Zachary Taylor:* President Zachary Taylor declared August 3, 1849, a day of "fasting, humiliation and prayer" to check the wave of cholera washing across the nation. Zachary Taylor, "National Fast," *Journal of the American Temperance Union,* August 1, 1849.

40 *A 2007 survey:* Annabel Ferriman, "BMJ Readers Choose the 'Sanitary Revolution' as Greatest Medical Advance Since 1840," *British Medical Journal* 334, no. 7585 (2007).

41 *Someone born in western Europe:* Life expectancy figures for western Europe come from Sweden, one of the longest datasets available. They're available online at: John R. Wilmoth and Vladimir Shkolnikov, "The Human Mortality Database," UC Berkeley, http://demog.berkeley.edu/~bmd/Sweden/LifeTables/e0.1x1. http://demog.berkeley.edu/~bmd/Sweden/Rates/Mx.early.5x1.

41 *child mortality in the U.S. and U.K.:* UN, "World Population Prospects: The 2010 Revision Infant Mortality Rate (Both Sexes Combined) By Major Area, Region and Country, 1950–2100" (United Nations, Department of Economic and Social Affairs, 2011).

41 *Infant mortality among modern-day:* I'm extrapolating from Michael Gurven's comparative study on mortality among hunger-gatherers. Michael Gurven and Hillard Kaplan, "Longevity Among Hunter-Gatherers: A Cross-Cultural Examination," *Population and Development Review* 33, no. 2 (2007).

42 *The affliction was new in Britain:* M. B. Emanuel, "Hay Fever, a Post Industrial Revolution Epidemic: A History of Its Growth During the 19th Century," *Clinical & Experimental Allergy* 18, no. 3 (1988).

42 *the Manchester physician Charles Blackley:* Blackley's observations are written in an 1873 treatise entitled *Experimental Research on Hay Fever.* K. J. Waite, "Blackley and the Development of Hay Fever as a Disease of Civilization in the Nineteenth Century," *Medical History* 39, no. 2 (1995).

43 *By then, money and status:* Morrell Mackenzie also halfheartedly looked into the class issue around hay fever. He counted sixty-one hay fever cases in his private practice, but not a single case among his hospital patients, who came from less-affluent backgrounds. Gregg Mitman, "Hay Fever Holiday: Health, Leisure, and Place in Gilded-Age America," *Bulletin of the History of Medicine* 77, no. 3 (2003).

43 *"The fact of exemption":* The physician who noted the English excellence in hay fever was W. P. Dunbar. From Waite, "Blackley and the Development of Hay Fever as a Disease of Civilization in the Nineteenth Century."

44 *Scandinavia, Italy, Spain, Russia:* See A. Maddison, *The World Economy: A Millennial Perspective, Vol. 2: Historical Statistics* (Academic Foundation, 2007), 264. Available online at: theworldeconomy.org. (Accessed March 26, 2012.)

Chapter 3: Island of Autoimmunity

45 *Nothing makes sense in biology:* T. Dobzhansky, "Nothing in Biology Makes Sense Except in the Light of Evolution." *American Biology Teacher* 35 (1973)

46 *Sheep outnumber people:* Statistics on Italy and Sardinia come from ISTAT, *Istituto Nazionale Di Statistica.* http://www.istat.it/en/.

47 *Today, 1 in 430 Sardinians:* Maura Pugliatti et al., "Multiple Sclerosis in Sardinia, Insular Italy: The Highest Burden Worldwide?," *European Journal of Neurology* 17, Suppl. 3 (2010).

47 *One in every 270 Sardinians:* M. Songini and C. Lombardo, "The Sardinian Way to Type 1 Diabetes," *Journal of Diabetes Science and Technology* 4, no. 5 (2010).

48 *And by the time Geoffrey Chaucer:* Paul Reiter, "From Shakespeare to Defoe: Malaria in England in the Little Ice Age," *Emerging Infectious Diseases* 6, no. 1 (2000).

48 *And the death caused by:* P. falciparum: P. W. Hedrick, "Population Genetics of Malaria Resistance in Humans," *Heredity* 107, no. 4 (2011).

49 *In 1954, A. C. Allison:* A. C. Allison, "Protection Afforded by Sickle-cell Trait Against Subtertian Malarial Infection," *British Medical Journal* 1, no. 4857 (1954).

49 *"[T]he struggle against diseases":* See reprint of the original 1949 paper, J. B. S. Haldane, "Disease and Evolution," in Krishna R. Dronamraju and Paolo Arese, eds., *Emerging Infectious Diseases of the 21st Century* (New York: Springer, 2006).

49 *The higher one went in Sardinia:* Licinio Contu, Carlo Carcassi, and Sandro Orrù, "Malaria e genetica popolazionistica," in *Sardegna e Malaria: Un nuovo approccio a un antico malanno,* ed. Ugo Carcassi and Ida Mura (Sassari, Sardinia: Carlo Delfino Editore, 2009), 154.

49 *High altitude couldn't protect:* In Sardinia, the landscape is categorized by *malaricitá.* As a concept, *malaricitá* denotes how mosquito friendly a given area is. As recently as the 1930s, the coastal areas had more than 80 percent *malaricitá,* which means that nearly everyone living there was exposed to *P. falciparum.* At 600 meters (2,000 feet), some 45 percent of the landscape was malarial. And in the more mountainous areas, which reach 1,000 meters (3,300 feet) in altitude, the landscape was still somewhat malarial—about 10 percent *malaricitá.* Values are inferred from the charts in ibid., 151–52.

50 *During those periods:* Stephen L. Dyson and Robert J. Rowland, *Archaeology and History in Sardinia from the Stone Age to the Middle Ages: Shepherds, Sailors, and Conquerors* (Philadelphia: University of Pennsylvania Museum of Archaeology and Anthropology, 2007), 22.

50 *After the last Ice Age:* S. D. Matthews et al., "Evidence for Late Pleistocene Population Expansion of the Malarial Mosquitoes, *Anopheles Arabiensis* and *Anopheles Gambiae* in Nigeria," *Medical and Veterinary Entomology* 21, no. 4 (2007).

51 *A subsequent study:* Weimin Liu et al., "Origin of the Human Malaria Parasite *Plasmodium Falciparum* in Gorillas," *Nature* 467, no. 7314 (2010).

51 *Egyptian texts began referring:* Crawford, *Deadly Companions,* 68.

51 *The earliest direct evidence:* A. G. Nerlich et al., "*Plasmodium Falciparum* in Ancient Egypt," *Emerging Infectious Diseases* 14, no. 8 (2008).

51 *Deformed bones unearthed:* The submerged village is now called Atlit-Yam. I. Hershkovitz et al., "Possible Congenital Hemolytic Anemia in Prehistoric Coastal Inhabitants of Israel," *American Journal of Physical Anthropology* 85, no. 1 (1991).

52 *Comparative genetic studies find:* Also see S. Sotgiu et al., "Multiple Sclerosis Complexity in Selected Populations: The Challenge of Sardinia, Insular Italy," *European Journal of Neurology* 9, no. 4 (2002).

52 *The effort proved remarkably successful:* Eugenia Tognotti, "La lotta alla malaria in Sardegna. Dalle bonifiche al DDT," in *Sardegna e Malaria,* 69. E. Tognotti, "Program to Eradicate Malaria in Sardinia, 1946–1950," *Emerging Infectious Diseases* 15, no. 9 (2009). Stats from ISTAT.

52 *In Sardinia, the incidence:* M. Pugliatti et al., "Multiple Sclerosis Epidemiology in Sardinia: Evidence for a True Increasing Risk," *Acta Neurologica Scandinavica* 103, no. 1 (2001).

52 *it more than doubled:* G. Rosati et al., "Epidemiology of Multiple Sclerosis in Northwestern

Sardinia: Further Evidence for Higher Frequency in Sardinians Compared to Other Italians," *Neuroepidemiology* 15, no. 1 (1996).

52 *Then a paper in the mid-1990s:* W. McGuire et al., "Severe Malarial Anemia and Cerebral Malaria Are Associated with Different Tumor Necrosis Factor Promoter Alleles," *Journal of Infectious Diseases* 179, no. 1 (1999).

53 *In 2001, geneticists found:* P. P. Bitti et al., "Association Between the Ancestral Haplotype HLA A30B18DR3 and Multiple Sclerosis in Central Sardinia," *Genetic Epidemiology* 20, no. 2 (2001).

53 *the entire Sardinian population:* One variant, called TNF-376A, was carried by 6.5 percent of Sicilians. By contrast, 51 percent of Sardinians had it. Sebastian A. Wirz et al., "High Frequency of TNF Alleles -238A and -376A in Individuals from Northern Sardinia," *Cytokine* 26, no. 4 (2004).

53 *He mixed:* P. falciparum: Stefano Sotgiu et al., "Multiple Sclerosis and Anti-*Plasmodium Falciparum* Innate Immune Response," *Journal of Neuroimmunology* 185, no. 1 (2007).

54 *In Sardinia, they remained elevated:* Stefano Sotgiu et al., "Hygiene Hypothesis: Innate Immunity, Malaria and Multiple Sclerosis," *Medical Hypotheses* 70, no. 4 (2008).

54 *Of the 100,000 patients:* B. M. Greenwood, "Autoimmune Disease and Parasitic Infections in Nigerians," *Lancet* 2, no. 7564 (1968).

55 *Back in Britain, Greenwood:* B. M. Greenwood, E. M. Herrick, and A. Voller, "Can Parasitic Infection Suppress Autoimmune Disease?," *Proceedings of the Royal Society of Medicine* 63, no. 1 (1970). B. M. Greenwood and A. Voller, "Suppression of Autoimmune Disease in New Zealand Mice Associated with Infection with Malaria. I. (NZBxNZW) F1 Hybrid Mice," *Clinical and Experimental Immunology* 7, no. 6 (1970).

55 *the British researcher:* G. A. Butcher, "Does Malaria Select for Predisposition to Autoimmune Disease?," *Journal of the Royal Society of Medicine* 84, no. 8 (1991).

56 *Mice with a gene known to increase:* Borna Mehrad et al., "The Lupus-Susceptibility Locus, Sle3, Mediates Enhanced Resistance to Bacterial Infections," *Journal of Immunology* 176, no. 5 (2006).

56 *East Africans with two copies of this gene:* Lisa C. Willcocks et al., "A Defunctioning Polymorphism in FCGR2B Is Associated with Protection Against Malaria but Susceptibility to Systemic Lupus Erythematosus," *Proceedings of the National Academy of Sciences* 107, no. 17 (2010). Kenneth G. C. Smith and Menna R. Clatworthy, "FcgammaRIIB in Autoimmunity and Infection: Evolutionary and Therapeutic Implications," *Nature Reviews Immunology* 10, no. 5 (2010). Menna R. Clatworthy et al., "Systemic Lupus Erythematosus–associated Defects in the Inhibitory Receptor FcgammaRIIb Reduce Susceptibility to Malaria," *Proceedings of the National Academy of Sciences* 104, no. 17 (2007).

56 *Why genes that predispose:* Luis B. Barreiro and Lluís Quintana-Murci, "From Evolutionary Genetics to Human Immunology: How Selection Shapes Host Defence Genes," *Nature Reviews Genetics* 11, no. 1 (2010).

57 *In one analysis:* Erik Corona, Joel T. Dudley, and Atul J. Butte, "Extreme Evolutionary Disparities Seen in Positive Selection Across Seven Complex Diseases," *PLoS ONE* 5, no. 8 (2010).

57 *Finnish children with:* K. Sadeharju et al., "The HLA-DR Phenotype Modulates the Humoral Immune Response to Enterovirus Antigens," *Diabetologia* 46, no. 8 (2003).

58 *Recently, they examined a decade's worth:* Andrea L. Graham and Daniel H. Nussey, "Fitness Correlates of Heritable Variation in Antibody Responsiveness in a Wild Mammal," *Science* 330, no. 6004 (2010).

59 *You're giving them a free pass:* Michael Walther et al., "Upregulation of TGF-beta, FOXP3, and CD4+CD25+ Regulatory T Cells Correlates with More Rapid Parasite Growth in Human Malaria Infection," *Immunity* 23, no. 3 (2005).

59 *Consider the Fulani of Burkina Faso:* M. G. Torcia et al., "Functional Deficit of T Regulatory Cells in Fulani, an Ethnic Group with Low Susceptibility to *Plasmodium Falciparum* Malaria," *Proceedings of the National Academy of Sciences* 105, no. 2 (2008).

59 *Riley and her colleagues:* Michael Walther et al., "Distinct Roles for FOXP3 and FOXP3 CD4 T Cells in Regulating Cellular Immunity to Uncomplicated and Severe *Plasmodium falciparum* Malaria," *PLoS Pathogens* 5, no. 4 (2009).

59 *Sardinians offer further evidence of that balance:* The sickle-cell trait may work by promoting a semitolerant response. For decades, scientists have argued about how the trait protects those who possess it. Does it shore up the defenses of red blood cells, making them invulnerable to invasion? Does it eliminate the parasite more quickly? Does it help purge parasitized erythrocytes? Much evidence suggests none of the above: People with the sickle-cell trait are infected with malaria as often as those without, and they have plenty of parasites in their blood. What they avoid is cerebral malaria, the major complication of infection with *P. falciparum.* How? They survive by both controlling and tolerating multiple infections. Thomas N. Williams et al., "An Immune Basis for Malaria Protection by the Sickle Cell Trait," *PLoS Medicine* 2, no. 5 (2005). M. Cyrklaff et al., "Hemoglobins S and C Interfere with Actin Remodeling in *Plasmodium falciparum*–infected Erythrocytes," *Science* 334, no. 6060 (2011).

60 *Alessandro Mathieu and Rosa Sorrentino:* Alessandro Mathieu et al., "The Interplay Between the Geographic Distribution of HLA-B27 Alleles and Their Role in Infectious and Autoimmune Diseases: A Unifying Hypothesis," *Autoimmunity Reviews* 8, no. 5 (2009).

60 *Riley finds that both:* Olivia C. Finney et al., "Homeostatic Regulation of T Effector to Treg Ratios in an Area of Seasonal Malaria Transmission," *European Journal of Immunology* 39, no. 5 (2009). Olivia C. Finney, Eleanor M. Riley, and Michael Walther, "Regulatory T Cells in Malaria—Friend or Foe?," *Trends in Immunology* 31, no. 2 (2010).

60 *people with autoimmune and allergic diseases:* Marta Barreto et al., "Low Frequency of CD4+CD25+ Treg in SLE Patients: A Heritable Trait Associated with CTLA4 and TGFbeta Gene Variants," *BMC Immunology* 10 (2009).

61 *the term* allelic rheostats: Rick M. Maizels, "Parasite Immunomodulation and Polymorphisms of the Immune System," *Journal of Biology* 8, no. 7 (2009).

61 *A decade ago:* A. Paolillo et al., "The Effect of Bacille Calmette-Guérin on the Evolution of New Enhancing Lesions to Hypointense T1 Lesions in Relapsing Remitting MS," *Journal of Neurology* 250, no. 2 (2003).

Chapter 4: Parasites to Heal the Gut

62 *Discovery consists of:* Roger von Oech, *Whack on the Side of the Head* (New York: Fine Communications, 2001), 11.

63 *Geneticists were valiantly trying:* A gene that predisposed to Crohn's—a variant of the NOD2/CARD15 gene, which encodes for a microbial sensor—was identified in 2001. D. P. Mcgovern et al., "NOD2 (Card15), the First Susceptibility Gene for Crohn's Disease," *Gut* 49, no. 6 (2001). The gene's full name is nucleotide-binding oligomerization domain containing 2. It encodes for a receptor that helps sense microbes, including resident bacteria. Having one copy of the gene more than doubles the odds of developing IBD. Having two copies increases it by a factor of 17. Details on how this variant works, or doesn't, are fuzzy, but it presumably interferes with the immune system's ability to distinguish friend from foe. Shiva Yazdanyar et al., "Penetrance of NOD2/CARD15 Genetic Variants in the General Population," *Canadian Medical Association Journal* 182, no. 7 (2010).

63 *Sitting on the plane:* Much of this chapter is drawn from Elliott and Weinstock, "Inflammatory Bowel Disease and the Hygiene Hypothesis: An Argument for the Role of Helminths," in Rook, *The Hygiene Hypothesis and Darwinian Medicine.*

63 *The cleaner one's circumstances:* The risk of IBD increases fivefold with a cleaner upbringing. A. E. Gent et al., "Inflammatory Bowel Disease and Domestic Hygiene in Infancy," *Lancet* 343, no. 8900 (1994).

64 *Immunologically speaking, that worm infection:* Joel V. Weinstock et al., "The Possible Link Between De-worming and the Emergence of Immunological Disease," *Journal of Laboratory and Clinical Medicine* 139, no. 6 (2002). D. E. Elliott et al., "Does the Failure to Acquire Helminthic Parasites Predispose to Crohn's Disease?," *FASEB Journal* 14, no. 12 (2000).

65 *In 1859, a judge sentenced:* Much of this history of IBD is taken from Joseph B. Kirsner, "The Historical Basis of the Idiopathic Inflammatory Bowel Diseases," *Inflammatory Bowel Diseases,* no. 1 (1995); S. Wilks, "Morbid Appearances in the Intestine of Miss Bankes," *London Medical Gazette* 2 (1859).

66 *there was a marked acceleration:* J. F. Fielding, "Crohn's Disease and Dalziel's Syndrome: A History," *Journal of Clinical Gastroenterology* 10, no. 3 (1988).

66 *"well-to-do, well-nourished":* W. H. Allchin, "A Discussion on 'Ulcerative Colitis': Introductory Address," *Proceedings of the Royal Society of Medicine* 2, no. Med Sect (1909).

67 *In 1909, when the newly established:* J. A. Ferrell, *The Rural School and Hookworm Disease* (Washington, D.C.: General Printing Office, 1914), 42–43.

67 *"One cannot have experienced the war":* The combined weight of the worms themselves equaled that of 442,000 men, according to Stoll. "Speaking helminthologically, it may be referred to as the grave host rôle which the lives of men play in the lives of worms," he wrote. N. R. Stoll, "This Wormy World," *Journal of Parasitology* 33, no. 1 (1947).

68 *African Americans remained infested:* In the early 1970s, one-third of New Orleans's African American schoolchildren still had worms, although infections were light. Ditto for South Carolina. G. M. Jeffery et al., "Study of Intestinal Helminth Infections in a Coastal South Carolina Area," *Public Health Reports* 78 (1963). D. W. Hubbard et al., "Intestinal Parasite Survey of Kindergarten Children in New Orleans," *Pediatric Research* 8, no. 6 (1974).

68 *Cherokees living on a reservation:* G. R. Healy et al., "Prevalence of Ascariasis and Amebiasis in Cherokee Indian School Children," *Public Health Reports* 84, no. 10 (1969).

68 *In 1965, more than two-thirds:* H. S. Fulmer and H. R. Huempfner, "Intestinal Helminths in Eastern Kentucky: A Survey in Three Rural Counties," *American Journal of Tropical Medicine and Hygiene* 14 (1965).

68 *This pattern was roughly the inverse:* Veterans' medical records indicated that soldiers from the North tended to have more IBD than those from the South, and that whites had it more than African Americans. Younger veterans of all colors got IBD more often than older ones. But the risk was modifiable. Soldiers who served in Vietnam, and who were presumably exposed to its tropical parasites, got IBD less often than those who didn't. Former prisoners of war, generally held in filthy conditions, also got less IBD than those who never lived through the ordeal. See A. Sonnenberg and I. H. Wasserman, "Epidemiology of Inflammatory Bowel Disease Among U.S. Military Veterans," *Gastroenterology* 101, no. 1 (1991); F. Delcò and A. Sonnenberg, "Military History of Patients with Inflammatory Bowel Disease: An Epidemiological Study Among U.S. Veterans," *American Journal of Gastroenterology* 93, no. 9 (1998).

68 *Those who remained on the reservations:* Chris Green et al., "A Population-based Ecologic Study of Inflammatory Bowel Disease: Searching for Etiologic Clues," *American Journal of Epidemiology* 164, no. 7 (2006).

69 *the "first 46 patients":* I. Segal, "Ulcerative Colitis in a Developing Country of Africa: The Baragwanath Experience of the First 46 Patients," *International Journal of Colorectal Disease* 3, no. 4 (1988).

69 *Inject them first with:* David E. Elliott et al., "Exposure to Schistosome Eggs Protects Mice from TNBS-induced Colitis," *American Journal of Physiology: Gastrointestinal and Liver Physiology* 284, no. 3 (2003).

69 *They could prevent the mouse version:* Diane Sewell et al., "Immunomodulation of Experimental Autoimmune Encephalomyelitis by Helminth Ova Immunization," *International Immunology* 15, no. 1 (2003).

70 *Beginning in 2004:* R. W. Summers et al., "*Trichuris Suis* Therapy in Crohn's Disease," *Gut* 54, no. 1 (2005). R. W. Summers et al., "*Trichuris Suis* Therapy for Active Ulcerative Colitis: A Randomized Controlled Trial," *Gastroenterology* 128, no. 4 (2005).

72 *There was some evidence:* G. L. Radford-Smith, "Will Worms Really Cure Crohn's Disease?," *Gut* 54, no. 1 (2005).

72 *"There is no predicting":* Herbert J. Van Kruiningen and A. Brian West, "Potential Danger in the Medical Use of *Trichuris suis* for the Treatment of Inflammatory Bowel Disease," *Inflammatory Bowel Diseases* 11, no. 5 (2005).

73 *But in 2006 came a case report:* Richard L. Kradin et al., "Iatrogenic *Trichuris suis* Infection in a Patient with Crohn Disease," *Archives of Pathology & Laboratory Medicine* 130, no. 5 (2006).

73 *Weinstock and Elliott again thought:* Robert W. Summers, David E. Elliott, and Joel V.

Weinstock, "Why *Trichuris Suis* Should Prove Safe for Use in Inflammatory Bowel Diseases," *Inflammatory Bowel Diseases* 11, no. 8 (2005).

78 *"It's ridiculous and incredibly inappropriate"*: Elizabeth Cohen, "Man Finds Extreme Healing Eating Parasitic Worms," CNN, http://www.cnn.Com/2010/HEALTH/12/09/worms.health/index.html. (Accessed January 31, 2012.)

78 *East Asians, formerly considered:* Suk-Kyun Yang et al., "Epidemiology of Inflammatory Bowel Disease in the Songpa-Kangdong District, Seoul, Korea, 1986–2005: A KASID Study," *Inflammatory Bowel Diseases* 14, no. 4 (2008).

78 *South Asian immigrants to the U.K.:* I. Carr and J. F. Mayberry, "The Effects of Migration on Ulcerative Colitis: A Three-Year Prospective Study Among Europeans and First- and Second-Generation South Asians in Leicester (1991–1994)," *American Journal of Gastroenterol* 94, no. 10 (1999).

78 *And India, which has developed:* H. G. Desai and P. A. Gupte, "Increasing Incidence of Crohn's Disease in India: Is It Related to Improved Sanitation?," *Indian Journal of Gastroenterology* 24, no. 1 (2005).

79 *People who develop Crohn's disease:* J. Kabeerdoss et al., "Exposure to Hookworms in Patients with Crohn's Disease: A Case-Control Study," *Alimentary Pharmacology & Therapeutics* 34, no. 10 (2011).

79 *Southern Indian patients:* Vivekanandhan Aravindhan et al., "Decreased Prevalence of Lymphatic Filariasis Among Subjects with Type-1 Diabetes," *American Journal of Tropical Medicine and Hygiene* 83, no. 6 (2010).

Chapter 5: What Is Asthma For?

80 *We should think of each host:* Joshua Lederberg, "Infectious History," *Science* 288, no. 5464 (2000).

81 *He couldn't locate a single:* R. C. Godfrey, "Asthma and IgE Levels in Rural and Urban Communities of the Gambia," *Clinical Allergy* 5, no. 2 (1975).

82 *He alternately exposed lung fragments:* R. C. Godfrey and C. F. Gradidge, "Allergic Sensitisation of Human Lung Fragments Prevented by Saturation of IgE Binding Sites," *Nature* 259, no. 5543 (1976).

82 *"One theoretical approach to prevention":* "Editorial: IgE, Parasites, and Allergy," *Lancet* 1, no. 7965 (1976).

83 *He'd infected himself:* J. A. Turton, "Letter: IgE, Parasites, and Allergy," *Lancet* 2, no. 7987 (1976).

83 *Pritchard and his team collected:* Elizabeth Svoboda, "The Worms Crawl In," *New York Times,* July 1, 2008.

84 *"Allergy could be regarded":* D. I. Pritchard, C. Hewitt, and R. Moqbel, "The Relationship Between Immunological Responsiveness Controlled by T-Helper 2 Lymphocytes and Infections with Parasitic Helminths," *Parasitology* 115 Suppl. (1997).

84 *And if we required contact:* D. I. Pritchard and A. Brown, "Is *Necator americanus* Approaching a Mutualistic Symbiotic Relationship with Humans?," *Trends in Parasitology* 17, no. 4 (2001).

84 *Researchers in Thailand observed:* For more on how worms may temper the response to malaria, see Mathieu Nacher, "Interactions Between Worm Infections and Malaria," *Clinical Reviews in Allergy & Immunology* 26, no. 2 (2004).

86 *Or maybe not. Observations in rural areas:* H. Yemaneberhan et al., "Prevalence of Wheeze and Asthma and Relation to Atopy in Urban and Rural Ethiopia," *Lancet* 350, no. 9071 (1997).

87 *Necator americanus halved the odds of wheeze:* S. Scrivener et al., "Independent Effects of Intestinal Parasite Infection and Domestic Allergen Exposure on Risk of Wheeze in Ethiopia: A Nested Case-Control Study," *Lancet* 358, no. 9292 (2001).

88 *The researchers arranged a randomized:* Anita H. J. van den Biggelaar et al., "Long-term Treatment of Intestinal Helminths Increases Mite Skin-Test Reactivity in Gabonese Schoolchildren," *Journal of Infectious Diseases* 189, no. 5 (2004).

89 *The posttreatment allergic response:* Maria Yazdanbakhsh wasn't the first to conduct

experiments of this sort. Beginning in the 1980s, Neil Lynch at the Institute of Biomedicine at Central Caracas University had conducted what was the first deliberate intervention, with the aim of studying allergy, on a worm-infested human population. He'd already noted that the prevalence of allergy varied considerably between socioeconomic groups in Caracas, the Venezuelan capital. The poorer one was, the fewer allergies one had. Poor *caraqueños* had a relatively high worm burden. They also had IgE levels ten times higher than their well-to-do, allergy-prone counterparts. Like David Pritchard, Lynch had concluded that worm infection protected against allergy, and that those who best fought off worms were also the most likely to become allergic in their absence.

Working in a Caracas slum called Barrio Los Erasos, where parasite infestation was common, Lynch treated 107 children with a two-year regimen of deworming drugs. At the study's close, he measured allergic skin reactivity, compared it with reactivity at the study's outset, and with an untreated group. With worms absent, allergic reactivity had increased substantially. By removing worms, in other words, he'd made the children more allergic.

The study had its flaws. The researchers knew who was treated, and the control group, children who'd declined deworming medication, was self-selected. What's more, political and economic turmoil had gripped Venezuela during the late 1980s and early '90s. Conditions had deteriorated in the slum. As a result, the control group had acquired more parasites over the course of the study, making comparisons difficult.

Nonetheless, this first intervention study indicated that worms manipulated the human immune system in a way that was immediate and ongoing. When they disappeared, the allergic response briskly rebounded. Lynch ascribed helminths' power over the allergic response to a version of the mast cell saturation hypothesis: a flurry of pointless IgE inundated mast cells, reducing sensitivity to any specific allergen. N. R. Lynch et al., "Effect of Anthelmintic Treatment on the Allergic Reactivity of Children in a Tropical Slum," *Journal of Allergy and Clinical Immunology* 92, no. 3 (1993). N. R. Lynch et al., "Allergic Reactivity and Socio-economic Level in a Tropical Environment," *Clinical Allergy* 17, no. 3 (1987).

90 *Maizels transferred regulatory T cells:* Mark S. Wilson et al., "Suppression of Allergic Airway Inflammation by Helminth-Induced Regulatory T Cells," *Journal of Experimental Medicine* 202, no. 9 (2005). Mark S. Wilson and Rick M. Maizels, "Regulatory T Cells Induced by Parasites and the Modulation of Allergic Responses," *Chemical Immunology and Allergy* 90 (2006).

90 *the evolutionary biologist Leigh Van Valen:* L. van Valen, "A New Evolutionary Law," *Evolutionary Theory* 1 (1973).

92 *The answer:* A description of Curt Lively's work on snails can be found on his Indiana University website at: http://www.indiana.edu/~curtweb/Research/cost%20of%20males. html. (Accessed November 30, 2011.) Much of this section is informed by Zuk, *Riddled with Life.*

92 *Dominant males in chimpanzee troops:* Michael P. Muehlenbein and David P. Watts, "The Costs of Dominance: Testosterone, Cortisol and Intestinal Parasites in Wild Male Chimpanzees," *Biopsychosocial Medicine* 4 (2010).

93 *For Julian Hopkin at the University of Wales:* J. Hopkin, "Immune and Genetic Aspects of Asthma, Allergy and Parasitic Worm Infections: Evolutionary Links," *Parasite Immunology* 31, no. 5 (2009).

94 *The geneticists Matteo Fumagalli and Manuela Sironi:* Matteo Fumagalli et al., "Parasites Represent a Major Selective Force for Interleukin Genes and Shape the Genetic Predisposition to Autoimmune Conditions," *Journal of Experimental Medicine* 206, no. 6 (2009). Matteo Fumagalli et al., "The Landscape of Human Genes Involved in the Immune Response to Parasitic Worms," *BMC Evolutionary Biology* 10 (2010).

95 *pathogens, and especially worms:* Matteo Fumagalli et al., "Signatures of Environmental Genetic Adaptation Pinpoint Pathogens as the Main Selective Pressure Through Human Evolution," *PLoS Genetics* 7, no. 11 (2011). Gopalakrishnan Netuveli et al., "Ethnic Variations in UK Asthma Frequency, Morbidity, and Health-Service Use: A Systematic Review and Meta-analysis," *Lancet* 365, no. 9456 (2005).

95 *The enrichment of problematic genes:* P. N. Le Souëf, P. Candelaria, and J. Goldblatt, "Evolution and Respiratory Genetics," *European Respiratory Journal* 28, no. 6 (2006).

95 *British food researcher John Jenkins:* John A. Jenkins, Heimo Breiteneder, and E. N. Clare Mills, "Evolutionary Distance from Human Homologs Reflects Allergenicity of Animal Food Proteins," *Journal of Allergy and Clinical Immunology* 120, no. 6 (2007).

95 *Swedish biochemist Michael Spangfort:* Cecilia Emanuelsson and Michael D. Spangfort, "Allergens as Eukaryotic Proteins Lacking Bacterial Homologues," *Molecular Immunology* 44, no. 12 (2007).

96 *British chemist Colin Fitzsimmons:* Colin M. Fitzsimmons and David W. Dunne, "Survival of the Fittest: Allergology or Parasitology?," *Trends in Parasitology* 25, no. 10 (2009).

96 *The single volunteer who:* Kevin Mortimer et al., "Dose-Ranging Study for Trials of Therapeutic Infection with *Necator americanus* in Humans," *American Journal of Tropical Medicine and Hygiene* 75, no. 5 (2006).

97 *They recruited thirty hay fever:* J. Feary et al., "Safety of Hookworm Infection in Individuals with Measurable Airway Responsiveness: A Randomized Placebo-Controlled Feasibility Study," *Clinical and Experimental Allergy* 39, no. 7 (2009).

97 *placebo-controlled study:* J. R. Feary et al., "Experimental Hookworm Infection: A Randomized Placebo-Controlled Trial in Asthma," *Clinical and Experimental Allergy* 40, no. 2 (2010).

Chapter 6: Missing "Old Friends"

99 *It is now widely appreciated:* William Parker, "Reconstituting the Depleted Biome to Prevent Immune Disorders," *Evolution & Medicine Review* (2010): http://evmedreview. com/?p=457. (Accessed February 12, 2012.)

99 *By the late 1980s, the German:* Erika von Mutius's story is based on author interview. Also see Rob Hyde, "Erika von Mutius: Reshaping the Landscape of Asthma Research," *Lancet* 372, no. 9643 (2008).

100 *Working with East German colleagues:* Like Strachan, von Mutius found that older siblings protected younger ones from allergy. This is a fairly consistent finding. T. Nicolai and E. von Mutius, "Respiratory Hypersensitivity and Environmental Factors: East and West Germany," *Toxicology Letters* 86, nos. 2–3 (1996). E. von Mutius et al., "Prevalence of Asthma and Allergic Disorders Among Children in United Germany: A Descriptive Comparison," *British Medical Journal (Clinical Research Ed.)* 305, no. 6866 (1992).

101 *Strachan thought that early-life:* D. P. Strachan, "Hay Fever, Hygiene, and Household Size," *British Medical Journal (Clinical Research Ed.)* 299, no. 6710 (1989).

102 *In 2000, Fernando Martinez's group:* T. M. Ball et al., "Siblings, Day-Care Attendance, and the Risk of Asthma and Wheezing During Childhood," *New England Journal of Medicine* 343, no. 8 (2000).

103 *Early on, measles seemed:* S. O. Shaheen et al., "Measles and Atopy in Guinea-Bissau," *Lancet* 347, no. 9018 (1996).

103 *Among six thousand British children:* S. A. Lewis and J. R. Britton, "Measles Infection, Measles Vaccination and the Effect of Birth Order in the Aetiology of Hay Fever," *Clinical and Experimental Allergy* 28, no. 12 (1998).

103 *A look at half a million Finns:* M. Paunio et al., "Measles History and Atopic Diseases: A Population-based Cross-sectional Study," *JAMA* 283, no. 3 (2000).

103 *a study of nearly two thousand children:* C. Bodner, D. Godden, and A. Seaton, "Family Size, Childhood Infections and Atopic Diseases. The Aberdeen WHEASE Group," *Thorax* 53, no. 1 (1998).

103 *Paolo Matricardi looked at allergic disease:* P. M. Matricardi et al., "Cross Sectional Retrospective Study of Prevalence of Atopy Among Italian Military Students with Antibodies Against Hepatitis A Virus," *British Medical Journal (Clinical Research Ed.)* 314, no. 7086 (1997).

103 *Infection with chicken pox:* P. M. Matricardi et al., "Sibship Size, Birth Order, and Atopy in 11,371 Italian Young Men," *Journal of Allergy and Clinical Immunology* 101, no. 4, pt. 1 (1998). P. M. Matricardi et al., "Exposure to Foodborne and Orofecal Microbes Versus

Airborne Viruses in Relation to Atopy and Allergic Asthma: Epidemiological Study," *British Medical Journal (Clinical Research Ed.)* 320, no. 7232 (2000).

104 *Matricardi parsed the records:* Paolo Maria Matricardi et al., "Hay Fever and Asthma in Relation to Markers of Infection in the United States," *Journal of Allergy and Clinical Immunology* 110, no. 3 (2002).

105 *Not only were farming children:* C. Braun-Fahrländer et al., "Prevalence of Hay Fever and Allergic Sensitization in Farmers' Children and Their Peers Living in the Same Rural Community," *Clinical and Experimental Allergy* 29, no. 1 (1999).

105 *Von Mutius corroborated the relationship:* O. S. von Ehrenstein et al., "Reduced Risk of Hay Fever and Asthma Among Children of Farmers," *Clinical and Experimental Allergy* 30, no. 2 (2000).

105 *numerous studies:* Charlotte Braun-Fahrländer, "Environmental Exposure to Endotoxin and Other Microbial Products and the Decreased Risk of Childhood Atopy: Evaluating Developments Since April 2002," *Current Opinion in Allergy and Clinical Immunology* 3, no. 5 (2003). F. Horak et al., "Parental Farming Protects Children Against Atopy: Longitudinal Evidence Involving Skin Prick Tests," *Clinical and Experimental Allergy* 32, no. 8 (2002).

105 *They were generally small:* Even where farming failed to protect, the reasons were instructive. In Crete, animal farmers had the same allergy as others living in the immediate vicinity. Both groups had about half as much asthma as Crete's city-dwellers. In the Mediterranean clime, however, animals remained outside year-round, limiting the close-quarters contact characteristic of barns and cowsheds in temperate Switzerland and Germany. In New Zealand, animal husbandry not only didn't protect, it slightly increased the risk of allergy. But Kiwi farmers tended to raise sheep. For whatever reason, sheep weren't protective, and they also stayed outside year round. Close contact was limited. Even in New Zealand, however, consumption of yogurt and unpasteurized milk—and the close contact with milk-giving animals that implied—reduced one's chances of allergy by two-thirds. K. Wickens et al., "Farm Residence and Exposures and the Risk of Allergic Diseases in New Zealand Children," *Allergy* 57, no. 12 (2002). M. Barnes et al., "Crete: Does Farming Explain Urban and Rural Differences in Atopy?," *Clinical and Experimental Allergy* 31, no. 12 (2001).

106 *Children who accompanied their parents:* U. Gehring et al., "Exposure to Endotoxin Decreases the Risk of Atopic Eczema in Infancy: A Cohort Study," *Journal of Allergy and Clinical Immunology* 108, no. 5 (2001).

106 *Scientists observed a major difference:* Roger P. Lauener et al., "Expression of CD14 and Toll-like Receptor 2 in Farmers' and Non-Farmers' Children," *Lancet* 360, no. 9331 (2002).

106 *Paradoxically, this enhanced microbe-sensing machinery:* Charlotte Braun-Fahrländer et al., "Environmental Exposure to Endotoxin and Its Relation to Asthma in School-Age Children," *New England Journal of Medicine* 347, no. 12 (2002).

107 *But even here, if you compared:* Matthew S. Perzanowski et al., "Endotoxin in Inner-City Homes: Associations with Wheeze and Eczema in Early Childhood," *Journal of Allergy and Clinical Immunology* 117, no. 5 (2006).

107 *A study in Colorado:* Benjamin J. Song and Andrew H. Liu, "Metropolitan Endotoxin Exposure, Allergy and Asthma," *Current Opinion in Allergy and Clinical Immunology* 3, no. 5 (2003). J. E. Gereda et al., "Metropolitan Home Living Conditions Associated with Indoor Endotoxin Levels," *Journal of Allergy and Clinical Immunology* 107, no. 5 (2001).

108 *Timing was similarly important:* M. K. Tulic, P. G. Holt, and P. D. Sly, "Modification of Acute and Late-phase Allergic Responses to Ovalbumin with Lipopolysaccharide," *International Archives of Allergy Immunology* 129, no. 2 (2002).

108 *a potpourri of microbes and other detritus:* Steve M. Lee et al., "Regulatory T Cells Contribute to Allergen Tolerance Induced by Daily Airway Immunostimulant Exposures," *American Journal of Respiratory Cell and Molecular Biology* 44, no. 3 (2011).

109 *A scientist named Meri Tulic:* Meri K. Tulic et al., "Role of Toll-like Receptor 4 in Protection by Bacterial Lipopolysaccharide in the Nasal Mucosa of Atopic Children but Not Adults," *Lancet* 363, no. 9422 (2004).

109 *Dale Umetsu, a researcher at Stanford University:* O. L. Frick et al., "Allergen Immunotherapy

with Heat-Killed *Listeria monocytogenes* Alleviates Peanut and Food-induced Anaphylaxis in Dogs," *Allergy* 60, no. 2 (2005).

109 *Participants received injections:* F. Estelle Simons et al., "Selective Immune Redirection in Humans with Ragweed Allergy by Injecting Amb a 1 Linked to Immunostimulatory DNA," *Journal of Allergy and Clinical Immunology* 113, no. 6 (2004).

109 *placebo-controlled trial at Johns Hopkins:* Peter S. Creticos et al., "Immunotherapy with a Ragweed-toll-like Receptor 9 Agonist Vaccine for Allergic Rhinitis," *New England Journal of Medicine* 355, no. 14 (2006).

109 *But a third study, on forty asthmatics:* Gail M. Gauvreau et al., "Immunostimulatory Sequences Regulate Interferon-Inducible Genes but Not Allergic Airway Responses," *American Journal of Respiratory and Critical Care Medicine* 174, no. 1 (2006).

111 *And studies in Japan:* T. Shirakawa et al., "The Inverse Association Between Tuberculin Responses and Atopic Disorder," *Science* 275, no. 5296 (1997). T. Annus et al., "Atopic Disorders Among Estonian Schoolchildren in Relation to Tuberculin Reactivity and the Age at BCG Vaccination," *Allergy* 59, no. 10 (2004). C. C. Obihara et al., "Inverse Association Between *Mycobacterium tuberculosis* Infection and Atopic Rhinitis in Children," *Allergy* 60, no. 9 (2005). Graham A. W. Rook, Eckard Hamelmann, and L. Rosa Brunet, "Mycobacteria and Allergies," *Immunobiology* 212, no. 6 (2007).

111 *scientists' attention eventually fell on:* S. L. Young et al., "Environmental Strains of *Mycobacterium avium* Interfere with Immune Responses Associated with *Mycobacterium bovis* BCG Vaccination," *Infection and Immunity* 75, no. 6 (2007).

112 *And from the shores of Lake Kyoga:* Sachs, *Good Germs, Bad Germs,* 94–95.

112 *Both improved by 50 percent:* P. D. Arkwright and T. J. David, "Effect of *Mycobacterium vaccae* on Atopic Dermatitis in Children of Different Ages," *British Journal of Dermatology* 149, no. 5 (2003).

112 *Mice treated with:* Claudia Zuany-Amorim et al., "Suppression of Airway Eosinophilia by Killed *Mycobacterium vaccae*–induced Allergen-Specific Regulatory T-Cells," *Nature Medicine* 8, no. 6 (2002). J. R. F. Hunt et al., "Intragastric Administration of *Mycobacterium vaccae* Inhibits Severe Pulmonary Allergic Inflammation in a Mouse Model," *Clinical and Experimental Allergy* 35, no. 5 (2005).

113 *around the world—in Vietnam:* For the studies that have found untreated water protective, see Carsten Flohr et al., "Poor Sanitation and Helminth Infection Protect Against Skin Sensitization in Vietnamese Children: A Cross-sectional Study," *Journal of Allergy and Clinical Immunology* 118, no. 6 (2006).

114 *Russian Karelians score dramatically lower:* T. Laatikainen et al., "Allergy Gap Between Finnish and Russian Karelia on Increase," *Allergy* 66, no. 7 (2011). Anita Kondrashova et al., "A Six-fold Gradient in the Incidence of Type 1 Diabetes at the Eastern Border of Finland," *Annals of Medicine* 37, no. 1 (2005). Anita Kondrashova et al., "Lower Economic Status and Inferior Hygienic Environment May Protect Against Celiac Disease," *Annals of Medicine* 40, no. 3 (2008).

115 *both Finns and Russians have:* Hanna Viskari et al., "Circulating Vitamin D Concentrations in Two Neighboring Populations with Markedly Different Incidence of Type 1 Diabetes," *Diabetes Care* 29, no. 6 (2006).

115 *That's when they noted:* Erkki Vartiainen et al., "Allergic Diseases, Skin Prick Test Responses, and IgE Levels in North Karelia, Finland, and the Republic of Karelia, Russia," *Journal of Allergy and Clinical Immunology* 109, no. 4 (2002).

115 *Russian Karelians were rife:* Leena C. von Hertzen et al., "Infectious Burden as a Determinant of Atopy: A Comparison Between Adults in Finnish and Russian Karelia," *International Archives of Allergy and Immunology* 140, no. 2 (2006).

116 *As a result, Russian water:* L. von Hertzen et al., "Microbial Content of Drinking Water in Finnish and Russian Karelia: Implications for Atopy Prevalence," *Allergy* 62, no. 3 (2007).

117 *In Finnish homes:* L. von Hertzen et al., "Risk of Atopy Associated with Microbial Components in House Dust," *Annals of Allergy, Asthma & Immunology* 104, no. 3 (2010).

117 *scientists sprayed mice with both dusts:* Harri Alenius et al., "Contrasting Immunological Effects of Two Disparate Dusts: Preliminary Observations," *International Archives of Allergy and Immunology* 149, no. 1 (2009).

117 *Frequently hanging out in an animal shed:* Markus Johannes Ege et al., "Not All Farming Environments Protect Against the Development of Asthma and Wheeze in Children," *Journal of Allergy and Clinical Immunology* 119, no. 5 (2007).

117 *Microbiologists, meanwhile, were dissecting:* Leonid Gorelik et al., "Modulation of Dendritic Cell Function by Cowshed Dust Extract," *Innate Immunity* 14, no. 6 (2008). M. Peters et al., "Inhalation of Stable Dust Extract Prevents Allergen Induced Airway Inflammation and Hyperresponsiveness," *Thorax* 61, no. 2 (2006).

117 *About 13 percent of barn dust:* Marcus Peters et al., "Arabinogalactan Isolated from Cowshed Dust Extract Protects Mice from Allergic Airway Inflammation and Sensitization," *Journal of Allergy and Clinical Immunology* 126, no. 3 (2010).

118 *Diversity took center stage:* Markus J. Ege et al., "Exposure to Environmental Microorganisms and Childhood Asthma," *New England Journal of Medicine* 364, no. 8 (2011). Robert Theodoor van Strien et al., "Microbial Exposure of Rural School Children, as Assessed by Levels of N-acetyl-muramic Acid in Mattress Dust, and Its Association with Respiratory Health," *Journal of Allergy and Clinical Immunology* 113, no. 5 (2004).

118 *The least allergic farming children:* Caroline Roduit et al., "Prenatal Animal Contact and Gene Expression of Innate Immunity Receptors at Birth Are Associated with Atopic Dermatitis," *Journal of Allergy and Clinical Immunology* 127, no. 1 (2011).

119 *But his first study, a comparative survey:* This study also compared Poland with Sweden and Estonia. Only one in six Poles was sensitized. L. Bråbäck et al., "Risk Factors for Respiratory Symptoms and Atopic Sensitisation in the Baltic Area," *Archives of Disease in Childhood* 72, no. 6 (1995).

120 *Estonian children harbored:* B. Björkstén et al., "The Intestinal Microflora in Allergic Estonian and Swedish 2-Year-Old Children," *Clinical and Experimental Allergy* 29, no. 3 (1999). E. Sepp et al., "Intestinal Microflora of Estonian and Swedish Infants," *Acta Paediatrica* 86, no. 9 (1997).

120 *Like Estonians, Ethiopian newborns:* R. Bennet et al., "Intestinal Bacteria of Newborn Ethiopian Infants in Relation to Antibiotic Treatment and Colonisation by Potentially Pathogenic Gram-Negative Bacteria," *Scandinavian Journal of Infectious Diseases* 23, no. 1 (1991). I. Adlerberth et al., "Intestinal Colonization with Enterobacteriaceae in Pakistani and Swedish Hospital-delivered Infants," *Acta Paediatrica Scandinavica* 80, nos. 6–7 (1991).

121 *He followed children:* B. Björkstén et al., "Allergy Development and the Intestinal Microflora During the First Year of Life," *Journal of Allergy and Clinical Immunology* 108, no. 4 (2001).

121 *Finnish newborns who went on to develop:* M. Kalliomaki et al., "Distinct Patterns of Neonatal Gut Microflora in Infants in Whom Atopy Was and Was Not Developing," *Journal of Allergy and Clinical Immunology* 107, no. 1 (2001).

121 *Early colonization by a bacterium:* M. M. Grönlund et al., "Importance of Intestinal Colonisation in the Maturation of Humoral Immunity in Early Infancy: A Prospective Follow-up Study of Healthy Infants Aged 0–6 Months," *Archives of Disease in Childhood: Fetal and Neonatal Edition* 83, no. 3 (2000). A. E. Wold, "The Hygiene Hypothesis Revised: Is the Rising Frequency of Allergy Due to Changes in the Intestinal Flora?," *Allergy* 53, no. 46 Suppl. (1998).

121 *Estonian house dust had about:* M. F. Böttcher et al., "Endotoxin Levels in Estonian and Swedish House Dust and Atopy in Infancy," *Clinical and Experimental Allergy* 33, no. 3 (2003).

122 *"abnormally stable":* For more on this idea, see I. Adlerberth and A. E. Wold, "Establishment of the Gut Microbiota in Western Infants," *Acta Paediatrica* 98, no. 2 (2009).

122 *From a very early age:* K. Julge, M. Vasar, and B. Björkstén, "Development of Allergy and IgE Antibodies During the First Five Years of Life in Estonian Children," *Clinical and Experimental Allergy* 31, no. 12 (2001).

122 *In the first two years of life:* M. F. Böttcher et al., "Cytokine Responses to Allergens During the First 2 Years of Life in Estonian and Swedish Children," *Clinical and Experimental Allergy* 36, no. 5 (2006).

122 *In Estonia, the very first milk:* Sara Tomicić et al., "Breast Milk Cytokine and IgA Composition Differ in Estonian and Swedish Mothers: Relationship to Microbial Pressure and Infant Allergy," *Pediatric Research* 68, no. 4 (2010).

123 *The immune profile of the treated Swedish:* Malin Fagerås Böttcher et al., "Low Breast Milk TGF-beta2 Is Induced by *Lactobacillus reuteri* Supplementation and Associates with Reduced Risk of Sensitization During Infancy," *Pediatric Allergy and Immunology* 19, no. 6 (2008).

123 *several studies found probiotics:* Vibeke Rosenfeldt et al., "Effect of Probiotic Lactobacillus Strains in Children with Atopic Dermatitis," *Journal of Allergy and Clinical Immunology* 111, no. 2 (2003).

123 *A follow-up at seven years of age:* Marko Kalliomäki et al., "Probiotics During the First 7 Years of Life: A Cumulative Risk Reduction of Eczema in a Randomized, Placebo-controlled Trial," *Journal of Allergy and Clinical Immunology* 119, no. 4 (2007).

123 *They treated both pregnant mothers:* Matthias Volkmar Kopp et al., "Randomized, Double-blind, Placebo-controlled Trial of Probiotics for Primary Prevention: No Clinical Effects of Lactobacillus GG Supplementation," *Pediatrics* 121, no. 4 (2008).

123 *And Australian scientists observed:* Angie L. Taylor, Janet A. Dunstan, and Susan L. Prescott, "Probiotic Supplementation for the First 6 Months of Life Fails to Reduce the Risk of Atopic Dermatitis and Increases the Risk of Allergen Sensitization in High-Risk Children: A Randomized Controlled Trial," *Journal of Allergy and Clinical Immunology* 119, no. 1 (2007).

124 *Analysis showed that the least allergic:* Mei Wang et al., "Reduced Diversity in the Early Fecal Microbiota of Infants with Atopic Eczema," *Journal of Allergy and Clinical Immunology* 121, no. 1 (2008). Y. M. Sjögren et al., "Altered Early Infant Gut Microbiota in Children Developing Allergy Up to 5 Years of Age," *Clinical and Experimental Allergy* 39, no. 4 (2009).

124 *They also excreted the most:* Anna Sandin et al., "High Salivary Secretory IgA Antibody Levels Are Associated with Less Late-onset Wheezing in IgE-sensitized Infants," *Pediatric Allergy and Immunology* (2011).

125 *Consider the case of children:* L. A. M. Smit et al., "Atopy and New-onset Asthma in Young Danish Farmers and CD14, TLR2, and TLR4 Genetic Polymorphisms: A Nested Case-control Study," *Clinical and Experimental Allergy* 37, no. 11 (2007).

125 *The same principle:* Adnan Custovic et al., "Effect of Day Care Attendance on Sensitization and Atopic Wheezing Differs by Toll-like Receptor 2 Genotype in 2 Population-Based Birth Cohort Studies," *Journal of Allergy and Clinical Immunology* 127, no. 2 (2011).

125 *And then there's Barbados:* April Zambelli-Weiner et al., "Evaluation of the CD14/-260 Polymorphism and House Dust Endotoxin Exposure in the Barbados Asthma Genetics Study," *Journal of Allergy and Clinical Immunology* 115, no. 6 (2005).

125 *compared with 28 percent in Manchester:* Angela Simpson et al., "Endotoxin Exposure, CD14, and Allergic Disease: An Interaction Between Genes and the Environment," *American Journal of Respiratory and Critical Care Medicine* 174, no. 4 (2006).

128 *Rodents raised in germ-free conditions:* Torsten Olszak et al., "Microbial Exposure During Early Life Has Persistent Effects on Natural Killer T Cell Function," *Science* (2012).

Chapter 7: Mom Matters Most

129 *Genes are not Stalinist dictators:* The David Barker quote is borrowed from Stephen S. Hall, "Small and Thin," *New Yorker,* November 19, 2007. Much about Barker comes from Hall.

129 *Erika von Mutius and her colleagues:* J. Riedler et al., "Exposure to Farming in Early Life and Development of Asthma and Allergy: A Cross-sectional Survey," *Lancet* 358, no. 9288 (2001).

130 *they threatened to reframe:* J. Douwes et al., "Farm Exposure In Utero May Protect Against Asthma, Hay Fever and Eczema," *European Respiratory Journal* 32, no. 3 (2008).

130 *Expectant mothers who regularly spent:* Ege et al., "Prenatal Farm Exposure Is Related to the Expression of Receptors of the Innate Immunity and to Atopic Sensitization in School-Age Children," *Journal of Allergy and Clinical Immunology* 117, no. 4 (2006).

130 *"[M]aternal farm exposure":* Bianca Schaub et al., "Maternal Farm Exposure Modulates Neonatal Immune Mechanisms Through Regulatory T Cells," *Journal of Allergy and Clinical Immunology* 123, no. 4 (2009).

131 *Many initially scoffed:* For a good review, see Alexander Vaiserman, "Early-life Origin of

NOTES

Adult Disease: Evidence from Natural Experiments," *Experimental Gerontology* 46, nos. 2–3 (2011).

132 *For Swedes born in the late nineteenth:* G. Kaati, L. O. Bygren, and S. Edvinsson, "Cardiovascular and Diabetes Mortality Determined by Nutrition During Parents' and Grandparents' Slow Growth Period," *European Journal of Human Genetics* 10, no. 11 (2002).

132 *Feeding pregnant rats:* Graham C. Burdge et al., "Dietary Protein Restriction of Pregnant Rats in the F0 Generation Induces Altered Methylation of Hepatic Gene Promoters in the Adult Male Offspring in the F1 and F2 Generations," *British Journal of Nutrition* 97, no. 3 (2007).

132 *She gauged the divergence:* Bianca Schaub et al., "Neonatal Immune Responses to TLR2 Stimulation: Influence of Maternal Atopy on FOXP3 and IL-10 Expression," *Respiratory Research* 7 (2006). Bianca Schaub et al., "Impairment of T-regulatory Cells in Cord Blood of Atopic Mothers," *Journal of Allergy and Clinical Immunology* 121, no. 6 (2008).

133 *The FOXP3 gene so important:* Jing Liu et al., "T Regulatory Cells in Cord Blood: FOXP3 Demethylation as Reliable Quantitative Marker," *PLoS ONE* 5, no. 10 (2010).

133 *the researcher Harald Renz:* Melanie L. Conrad et al., "Maternal TLR Signaling Is Required for Prenatal Asthma Protection by the Nonpathogenic Microbe *Acinetobacter lwoffii* F78," *Journal of Experimental Medicine* 206, no. 13 (2009).

133 *"soothing signals":* Patrick G. Holt and Deborah H. Strickland, "Soothing Signals: Transplacental Transmission of Resistance to Asthma and Allergy," *Journal of Experimental Medicine* 206, no. 13 (2009).

133 *"[W]e seem to be at the dawn":* Catherine A. Thornton, Trisha V, Macfarlane, and Patrick G. Holt, "The Hygiene Hypothesis Revisited: Role of Materno-Fetal Interactions," *Current Allergy and Asthma Reports* (2010).

134 *a scientist named Alison Elliott:* Alison M. Elliott et al., "Helminth Infection During Pregnancy and Development of Infantile Eczema," *JAMA* 294, no. 16 (2005). Alison M. Elliott et al., "A Randomised Controlled Trial of the Effects of Albendazole in Pregnancy on Maternal Responses to Mycobacterial Antigens and Infant Responses to Bacille Calmette-Guérin (BCG) Immunisation [ISRCTN32849447]," *BMC Infectious Diseases* 5 (2005).

134 *Elliott found that deworming:* Michael Brown et al., "Helminth Infection Is Not Associated with Faster Progression of HIV Disease in Coinfected Adults in Uganda," *Journal of Infectious Diseases* 190, no. 10 (2004).

134 *Elliott in fact observed:* Alison M. Elliott et al., "Effects of Maternal and Infant Co-infections, and of Maternal Immunisation, on the Infant Response to BCG and Tetanus Immunisation," *Vaccine* 29, no. 2 (2010).

134 *two-fifths of some 2,500 pregnant women:* Lawrence Muhangi et al., "Associations Between Mild-to-Moderate Anaemia in Pregnancy and Helminth, Malaria and HIV Infection in Entebbe, Uganda," *Transactions of the Royal Society of Tropical Medicine and Hygiene* 101, no. 9 (2007).

134 *Elliott and her colleagues were finding:* Emily L. Webb et al., "Effect of Single-Dose Anthelmintic Treatment During Pregnancy on an Infant's Response to Immunisation and on Susceptibility to Infectious Diseases in Infancy: A Randomised, Double-blind, Placebo-controlled Trial," *Lancet* 377, no. 9759 (2011).

134 *Two-thirds of women carried:* Harriet Mpairwe et al., "Skin Prick Test Reactivity to Common Allergens Among Women in Entebbe, Uganda," *Transactions of the Royal Society of Tropical Medicine and Hygiene* 102, no. 4 (2008).

134 *came the results of the double-blind:* Harriet Mpairwe et al., "Anthelminthic Treatment During Pregnancy Is Associated with Increased Risk of Infantile Eczema: Randomised-controlled Trial Results," *Pediatric Allergy and Immunology* 22, no. 3 (2011).

135 *"The risks and benefits of routine":* Alison M. Elliott et al., "Treatment with Anthelminthics During Pregnancy: What Gains and What Risks for the Mother and Child?," *Parasitology* 138, no. 12 (October 2011): 1499–1507.

135 *At least since the late 1990s:* For children under age five, an allergic father increases your risk of wheezing by 1.6, while an allergic mother increases it fivefold. A. A. Litonjua et al., "Parental History and the Risk for Childhood Asthma: Does Mother Confer More

329

Risk than Father?," *American Journal of Respiratory and Critical Care Medicine* 158, no. 1 (1998).

135 *A fever episode while pregnant:* B. Xu et al., "Maternal Infections in Pregnancy and the Development of Asthma Among Offspring," *International Journal of Epidemiology* 28, no. 4 (1999).

135 *Contracting the flu during pregnancy:* M. Calvani et al., "Infectious and Uterus Related Complications During Pregnancy and Development of Atopic and Nonatopic Asthma in Children," *Allergy* 59, no. 1 (2004).

135 *An extremely preterm birth:* Casey Crump et al., "Risk of Asthma in Young Adults Who Were Born Preterm: A Swedish National Cohort Study," *Pediatrics* 127, no. 4 (2011). Bradford D. Gessner and Marc-Andre R. Chimonas, "Asthma Is Associated with Preterm Birth but Not with Small for Gestational Age Status Among a Population-Based Cohort of Medicaid-enrolled Children 10 Years of Age," *Thorax* 62, no. 3 (2007).

135 *Rajesh Kumar, a scientist:* Rajesh Kumar et al., "Prematurity, Chorioamnionitis, and the Development of Recurrent Wheezing: A Prospective Birth Cohort Study," *Journal of Allergy and Clinical Immunology* 121, no. 4 (2008).

136 *the Maastricht University scientist Boris Kramer:* Boris W. Kramer, "Chorioamnionitis: New Ideas from Experimental Models," *Neonatology* 99, no. 4 (2011).

136 *Danish researchers following:* Bisgaard, Hans, et al., "Interaction Between Asthma and Lung Function Growth in Early Life," *American Journal of Respiratory and Critical Care Medicine* (March 29, 2012).

136 *After five years:* Miranda Smith et al., "Children with Egg Allergy Have Evidence of Reduced Neonatal CD4(+)CD25(+)CD127(lo/-) Regulatory T Cell Function," *Journal of Allergy and Clinical Immunology* 121, no. 6 (2008).

137 *their white blood cells:* Pia Reece et al., "Maternal Allergy Modulates Cord Blood Hematopoietic Progenitor Toll-like Receptor Expression and Function," *Journal of Allergy and Clinical Immunology* 127, no. 2 (2011).

138 *By contrast, these children:* Meri K. Tulic et al., "Differences in Innate Immune Function Between Allergic and Nonallergic Children: New Insights into Immune Ontogeny," *Journal of Allergy and Clinical Immunology* 127, no. 2 (2011).

138 *The scientists had preserved:* Susan L. Prescott et al., "Reduced Placental FOXP3 Associated with Subsequent Infant Allergic Disease," *Journal of Allergy and Clinical Immunology* 128, no. 4 (October 2011): 886–87.

138 *Communication between mother and fetus:* J. M. Harris et al., "New Pregnancies and Loss of Allergy," *Clinical and Experimental Allergy* 34, no. 3 (2004).

138 *By their calculations:* S. Upchurch, J. M. Harris, and P. Cullinan, "Temporal Changes in UK Birth Order and the Prevalence of Atopy," *Allergy* 65, no. 8 (2010).

138 *the average household had nearly:* Hobbs and Stoops, "Demographic Trends in the 20th Century." Washington, D.C.: U.S. Census Bureau, 2002, 141–42.

139 *Expectant mothers who stuck:* L. Chatzi et al., "Mediterranean Diet in Pregnancy Is Protective for Wheeze and Atopy in Childhood," *Thorax* 63, no. 6 (2008).

139 *Von Mutius and her colleagues suspected:* C. Braun-Fahrländer and E. von Mutius, "Can Farm Milk Consumption Prevent Allergic Diseases?," *Clinical and Experimental Allergy* 41, no. 1 (2011).

139 *Indeed, there's some evidence:* Janet A. Dunstan et al., "Fish Oil Supplementation in Pregnancy Modifies Neonatal Allergen-specific Immune Responses and Clinical Outcomes in Infants at High Risk of Atopy: A Randomized, Controlled Trial," *Journal of Allergy and Clinical Immunology* 112, no. 6 (2003).

140 *Between 1997 and 2008:* Scott H. Sicherer et al., "US Prevalence of Self-reported Peanut, Tree Nut, and Sesame Allergy: 11-Year Follow-up," *Journal of Allergy and Clinical Immunology* 125, no. 6 (2010). Ruchi S. Gupta et al., "The Prevalence, Severity, and Distribution of Childhood Food Allergy in the United States," *Pediatrics* 128, no. 1 (2011).

140 *Prescott calls food allergy:* Susan Prescott and Katrina J. Allen, "Food Allergy: Riding the Second Wave of the Allergy Epidemic," *Pediatric Allergy and Immunology* 22, no. 2 (2011).

141 *An Ohio State University study:* Christa Lynn Whitney-Miller, David Katzka, and Emma Elizabeth Furth, "Eosinophilic Esophagitis: A Retrospective Review of Esophageal Biopsy

Specimens from 1992 to 2004 at an Adult Academic Medical Center," *American Journal of Clinical Pathology* 131, no. 6 (2009). Not all the studies found the increase real. See Charles W. Debrosse et al., "Identification, Epidemiology, and Chronicity of Pediatric Esophageal Eosinophilia, 1982–1999," *Journal of Allergy and Clinical Immunology* 126, no. 1 (2010).

141 *After a decade, some 60 percent:* R. Leung, "Asthma and Migration," *Respirology* 1, no. 2 (1996).

141 *For Asian teens, having lived:* C. V. Powell et al., "Respiratory Symptoms and Duration of Residence in Immigrant Teenagers Living in Melbourne, Australia," *Archives of Disease in Childhood* 81, no. 2 (1999).

141 *Studies on immigrants to Sweden:* A. F. Kalyoncu and G. Stålenheim, "Serum IgE Levels and Allergic Spectra in Immigrants to Sweden," *Allergy* 47, no. 4, pt. 1 (1992), Netuveli, Hurwitz, and Sheikh, "Ethnic Variations in Incidence of Asthma Episodes in England & Wales: National Study of 502,482 Patients in Primary Care," *Respiratory Research* 6, no. 1 (2005).

142 *"In contrast to the challenge":* Duane Alexander, "FY 2002 Hearing on Life Span. Witness Appearing Before the House Subcommittee on Labor-HHS-Education Appropriations. April 4, 2001," ed. National Institutes of Health (National Institute of Child Health & Human Development, 2001).

142 *"[I]nvestments targeting fetal health":* Douglas Almond, "Is the 1918 Influenza Pandemic Over? Long-Term Effects of In Utero Influenza Exposure in the Post-1940 U.S. Population," *Journal of Political Economy* 114, no. 4 (2006).

143 *Annie Murphy Paul writes:* Annie Murphy Paul, *Origins: How the Nine Months Before Birth Shape the Rest of Our Lives* (New York: Free Press, 2010), 218.

Chapter 8: The Disappearing Microbiota

144 *I will be an enemy: The Thompson Chain-Reference Bible,* ed. Frank Charles Thompson (Indianapolis: B. B. Kirkbride Bible Co., 1984). Sentiment borrowed from Martin J. Blaser and Stanley Falkow.

144 *The extinction spasm:* E. O. Wilson, *Consilience: The Unity of Knowledge* (New York: Vintage Books, 1999), 321.

145 *So in 1983, when Warren and Marshall:* See their story here: B. Marshall, "*Helicobacter pylori*—a Nobel Pursuit?," *Canadian Journal of Gastroenterology* 22, no. 11 (2008).

146 *the poor everywhere disproportionately bore:* Mohammed Mahdy Khalifa, Radwa Raed Sharaf, and Ramy Karam Aziz, "*Helicobacter pylori*: A Poor Man's Gut Pathogen?," *Gut Pathogens* 2, no. 1 (2010).

146 *Countries in transition usually fell somewhere in between:* In the 1990s, three-quarters of South Koreans had *H. pylori,* for example, but the bacterium didn't infect all Koreans equally. More than 40 percent of poor Korean children had *H. pylori,* while just 12 percent of their middle- and upper-class counterparts did. In the U.S., similar patterns were evident, but they occurred along racial and ethnic lines. In Houston, Texas, African Americans were twice as likely to harbor *H. pylori* as whites (70 percent versus 34 percent). Hispanic immigrants were also more likely to carry the bug than white Americans. H. M. Malaty et al., "Prevalence of *Helicobacter pylori* Infection in Korean Children: Inverse Relation to Socioeconomic Status Despite a Uniformly High Prevalence in Adults," *American Journal of Epidemiology* 143, no. 3 (1996). D. Y. Graham et al., "Epidemiology of *Helicobacter pylori* in an Asymptomatic Population in the United States. Effect of Age, Race, and Socioeconomic Status," *Gastroenterology* 100, no. 6 (1991).

146 *Childred who acquired it:* J. E. Thomas et al., "Early *Helicobacter pylori* Colonisation: The Association with Growth Faltering in the Gambia," *Archives of Disease in Childhood* 89, no. 12 (2004).

146 *"I believe H.* pylori *are very likely":* M. J. Blaser, "The Bacteria Behind Ulcers," *Scientific American* 274, no. 2 (1996).

147 *H.* pylori *is blamed:* Sam M. Mbulaiteye, Michie Hisada, and Emad M. El-Omar, "*Helicobacter pylori* Associated Global Gastric Cancer Burden," *Frontiers in Bioscience* 14 (2009).

148 *In Australia, Barry Marshall:* See Ondek Biologic Delivery Systems at http://www.ondek.com/.

149 *His work on a cohort:* A. Nomura et al., "*Helicobacter pylori* Infection and Gastric Carcinoma Among Japanese Americans in Hawaii," *New England Journal of Medicine* 325, no. 16 (1991).

149 *And in 1993, along with an Italian group:* M. J. Blaser et al., "Infection with *Helicobacter pylori* Strains Possessing CagA Is Associated with an Increased Risk of Developing Adenocarcinoma of the Stomach," *Cancer Research* 55, no. 10 (1995).

150 *the prevalence of Barrett's esophagus:* J. E. Richter, G. W. Falk, and M. F. Vaezi, "*Helicobacter pylori* and Gastroesophageal Reflux Disease: The Bug May Not Be All Bad," *American Journal of Gastroenterology* 93, no. 10 (1998).

150 *GERD and esophageal cancer:* J. Powell and C. C. McConkey, "The Rising Trend in Oesophageal Adenocarcinoma and Gastric Cardia," *European Journal of Cancer Prevention* 1, no. 3 (1992).

150 *those with* H. pylori *infections:* J. J. Vicari et al., "The Seroprevalence of CagA-positive *Helicobacter pylori* Strains in the Spectrum of Gastroesophageal Reflux Disease," *Gastroenterology* 115, no. 1 (1998).

151 *scientist Maria Gloria Domínguez-Bello isolated:* Chandrabali Ghose et al., "East Asian Genotypes of *Helicobacter pylori* Strains in Amerindians Provide Evidence for Its Ancient Human Carriage," *Proceedings of the National Academy of Sciences* 99, no. 23 (2002).

151 *Seven founding strains of* H. pylori *existed:* Daniel Falush et al., "Traces of Human Migrations in *Helicobacter pylori* Populations," *Science* 299, no. 5612 (2003).

151 *a white Tennessee man harbored:* Maria-Gloria Dominguez-Bello and Martin J. Blaser, "The Human Microbiota as a Marker for Migrations of Individuals and Populations," *Annual Review of Anthropology* 40 (2011).

151 *In 2007, scientists:* Bodo Linz et al., "An African Origin for the Intimate Association Between Humans and *Helicobacter pylori*," *Nature* 445, no. 7130 (2007).

152 *He conducted three studies:* Yu Chen and Martin J. Blaser, "Inverse Associations of *Helicobacter pylori* with Asthma and Allergy," *Archives of Internal Medicine* 167, no. 8 (2007); J. Reibman et al., "Asthma Is Inversely Associated with *Helicobacter pylori* Status in an Urban Population," *PLoS ONE* 3, no. 12 (2008); Yu Chen and Martin J. Blaser, "*Helicobacter pylori* Colonization Is Inversely Associated with Childhood Asthma," *Journal of Infectious Diseases* 198, no. 4 (2008).

153 *In Finland, scientists examined:* T. U. Kosunen et al., "Increase of Allergen-specific Immunoglobulin E Antibodies from 1973 to 1994 in a Finnish Population and a Possible Relationship to *Helicobacter pylori* Infections," *Clinical and Experimental Allergy* 32, no. 3 (2002).

153 *Cross-sectional "snapshot" studies:* Anne McCune et al., "Reduced Risk of Atopic Disorders in Adults with *Helicobacter pylori* Infection," *European Journal of Gastroenterology & Hepatology* 15, no. 6 (2003). Olf Herbarth et al., "*Helicobacter pylori* Colonisation and Eczema," *Journal of Epidemiology & Community Health* 61, no. 7 (2007). Shigeyoshi Imamura et al., "Inverse Association Between *Helicobacter pylori* Infection and Allergic Rhinitis in Young Japanese," *Journal of Gastroenterology and Hepatology* 25, no. 7 (2010).

153 *"erroneous attempts to blame":* David Y. Graham, Yoshio Yamaoka, and Hoda M. Malaty, "Contemplating the Future without *Helicobacter pylori* and the Dire Consequences Hypothesis," *Helicobacter* 12, Suppl. 2 (2007).

153 *the people of northern peninsular Malaysia:* S. M. Raj et al., "Evidence Against *Helicobacter pylori* Being Related to Childhood Asthma," *Journal of Infectious Diseases* 199, no. 6 (2009).

155 *Successful carriers greeted:* K. Robinson et al., "*Helicobacter pylori*–induced Peptic Ulcer Disease Is Associated with Inadequate Regulatory T Cell Responses," *Gut* 57, no. 10 (2008).

155 H. pylori *infection lowered the risk:* This paper remains unpublished, but has been presented. D. P. Letley et al., "*Helicobacter pylori*–induced Protection from Allergy Is Associated with Peripheral Blood Regulatory Cells," in Regulatory T Cells in Inflammatory and Infectious Diseases: New Horizons for Old Friends (Kennedy Lecture Theatre, UCL Institute of Child Health, London, 2010).

155 *Her Ethiopian colleagues followed:* A. Amberbir et al., "Effects of *Helicobacter pylori*, Geohelminth Infection and Selected Commensal Bacteria on the Risk of Allergic Disease

and Sensitization in 3-Year-Old Ethiopian Children," *Clinical and Experimental Allergy* 41, no. 10 (2011).

155 *the earlier* H. pylori *colonized:* Isabelle C. Arnold et al., "Tolerance Rather Than Immunity Protects from *Helicobacter pylori*–induced Gastric Preneoplasia," *Gastroenterology* 140, no. 1 (2011). Isabelle C. Arnold et al., "*Helicobacter pylori* Infection Prevents Allergic Asthma in Mouse Models Through the Induction of Regulatory T Cells," *Journal of Clinical Investigation* (2011).

156 *In 1992, the British scientist:* C. Holcombe, "*Helicobacter pylori*: The African Enigma," *Gut* 33, no. 4 (1992).

157 *Africans responded to* H. pylori *infections:* For more on the different immune response to *H. pylori* in the developing world, see Sam M. Mbulaiteye et al., "*H. pylori*–infection and Antibody Immune Response in a Rural Tanzanian Population," *Infectious Agents and Cancer* 1 (2006); H. M. Mitchell et al., "Major Differences in the IgG Subclass Response to *Helicobacter pylori* in the First and Third Worlds," *Scandinavian Journal of Gastroenterology* 37, no. 5 (2002).

157 *"large, free-living feline enigma"*: K. A. Terio et al., "Comparison of *Helicobacter* spp. in Cheetahs (*Acinonyx jubatus*) with and without Gastritis," *Journal of Clinical Microbiology* 43, no. 1 (2005).

157 *the epidemiologist Amnon Sonnenberg:* Amnon Sonnenberg, "Differences in the Birth-Cohort Patterns of Gastric Cancer and Peptic Ulcer," *Gut* 59, no. 6 (2010).

158 *Sonnenberg traced this same upsurge:* "Everything which is thrown out from the human body is unclean and becomes at once dangerous to human life," wrote the sanitarian Mary Armstrong in 1878. These sentiments, indicative of a new cultural obsession with cleanliness and purity, first gained currency in the late eighteenth century. It's possible that new mores began altering the transmission of *H. pylori* before sewers, clean water, and public hygiene did. See G. H. Pelto, Y. Zhang, and J. P. Habicht, "Premastication: The Second Arm of Infant and Young Child Feeding for Health and Survival?," *Maternal & Child Nutrition* 6, no. 1 (2010).

159 *James Fox, a scientist:* J. G. Fox et al., "Concurrent Enteric Helminth Infection Modulates Inflammation and Gastric Immune Responses and Reduces Helicobacter-Induced Gastric Atrophy," *Nature Medicine* 6, no. 5 (2000). Mark T. Whary et al., "Intestinal Helminthiasis in Colombian Children Promotes a Th2 Response to *Helicobacter pylori*: Possible Implications for Gastric Carcinogenesis," *Cancer Epidemiology, Biomarkers & Prevention* 14, no. 6 (2005).

160 *The scientists confirmed that harboring:* Sharon Perry et al., "Infection with *Helicobacter pylori* Is Associated with Protection Against Tuberculosis," *PLoS ONE* 5, no. 1 (2010).

161 *And Israeli soldiers hosting:* D. Cohen et al., "An Inverse and Independent Association Between *Helicobacter pylori* Infection and the Incidence of Shigellosis and Other Diarrheal Diseases," *Clinical Infectious Diseases* (2011).

161 *It will probably come as no surprise:* Mieczysław Wender, "[Prevalence of *Helicobacter pylori* Infection Among Patients with Multiple Sclerosis]," *Neurologia i neurochirurgia polska* 37, no. 1 (2003); Wei Li et al., "*Helicobacter pylori* Infection Is a Potential Protective Factor Against Conventional Multiple Sclerosis in the Japanese Population," *Journal of Neuroimmunology* 184, nos. 1–2 (2007). Amr H. Sawalha et al., "Association Between Systemic Lupus Erythematosus and *Helicobacter pylori* Seronegativity," *Journal of Rheumatology* 31, no. 8 (2004). Yoshihiro Matsukawa et al., "Exacerbation of Rheumatoid Arthritis Following *Helicobacter pylori* Eradication: Disruption of Established Oral Tolerance Against Heat Shock Protein?," *Medical Hypotheses* 64, no. 1 (2005). Ai Tada et al., "Eradication of *Helicobacter pylori* May Trigger Onset of Type 1 Diabetes: A Case Report," *Annals of the New York Academy of Sciences* 1150 (2008). Antonio Tursi, "Onset of Crohn's Disease After *Helicobacter pylori* Eradication," *Inflammatory Bowel Diseases* 12, no. 10 (2006). W. Li et al., "Association of Anti–*Helicobacter pylori* Neutrophil-activating Protein Antibody Response with Anti-aquaporin-4 Autoimmunity in Japanese Patients with Multiple Sclerosis and Neuromyelitis Optica," *Multiple Sclerosis* 15, no. 12 (2009).

162 *Blaser frets most about:* M. J. Blaser and S. Falkow, "What Are the Consequences of the Disappearing Human Microbiota?," *Nature Reviews Microbiology* 7, no. 12 (2009).

162 *"Each generation could be beginning":* M. J. Blaser, "Antibiotic Overuse: Stop the Killing of Beneficial Bacteria," *Nature* 476, no. 7361 (2011).

162 *By one count, children:* Anders Hviid, Henrik Svanström, and Morten Frisch, "Antibiotic Use and Inflammatory Bowel Diseases in Childhood," *Gut* 60, no. 1 (2011).

162 *by strengthening those regulatory:* Anna Lundgren et al., "Mucosal FOXP3-Expressing CD4+ CD25 High Regulatory T Cells in *Helicobacter pylori*–infected Patients," *Infection and Immunity* 73, no. 1 (2005).

163 *And among the Venezuelan Warao:* E. Marini et al., "*Helicobacter pylori* and Intestinal Parasites Are Not Detrimental to the Nutritional Status of Amerindians," *American Journal of Tropical Medicine and Hygiene* 76, no. 3 (2007).

163 *the imported African and European strains:* M. G. Dominguez-Bello et al., "Amerindian *Helicobacter pylori* Strains Go Extinct, as European Strains Expand Their Host Range," *PLoS ONE* 3, no. 10 (2008).

Chapter 9: Community-Wide Derangement

164 *There is a dimension to human evolution:* Jeffrey I. Gordon and Todd R. Klaenhammer, "A Rendezvous with Our Microbes," *Proceedings of the National Academy of Sciences* 108, Suppl. 1 (2011).

164 *Scientists conducted this experiment:* I. Mulder et al., "Environmentally-acquired Bacteria Influence Microbial Diversity and Natural Innate Immune Responses at Gut Surfaces," *BMC Biology* 7, no. 79 (2009).

167 *Indeed, Margaret McFall-Ngai's research:* Margaret McFall-Ngai, "Adaptive Immunity: Care for the Community," *Nature* 445, no. 7124 (2007).

167 *Humans "can be regarded":* Justin Sonnenburg was paraphrasing/quoting an earlier pioneer in the field, microbiologist Dwayne Savage.

167 *Let's review the basic facts:* David A. Hill and David Artis, "Intestinal Bacteria and the Regulation of Immune Cell Homeostasis," *Annual Review of Immunology* 28 (2010).

168 *If you unrolled and hammered flat:* A singles court measures twenty-seven feet by seventy-eight feet. Measurement from Kevin J. Maloy and Fiona Powrie, "Intestinal Homeostasis and Its Breakdown in Inflammatory Bowel Disease," *Nature* 474, no. 7351 (2011).

169 *They delivered mice by C-section:* Karen Smith, Kathy D. McCoy, and Andrew J. Macpherson, "Use of Axenic Animals in Studying the Adaptation of Mammals to Their Commensal Intestinal Microbiota," *Seminars in Immunology* 19, no. 2 (2007).

169 *But they looked really weird:* Pasteur's quote and much more were borrowed from B. S. Wostmann, "The Germfree Animal in Nutritional Studies," *Annual Review of Nutrition* 1 (1981).

170 *Introducing B. fragilis into this human-created:* June L. Round and Sarkis K. Mazmanian, "Inducible FOXP3+ Regulatory T-cell Development by a Commensal Bacterium of the Intestinal Microbiota," *Proceedings of the National Academy of Sciences* 107, no. 27 (2010).

170 *"This raises the possibility":* June L. Round and Sarkis K. Mazmanian, "The Gut Microbiota Shapes Intestinal Immune Responses During Health and Disease," *Nature Reviews Immunology* 9, no. 5 (2009).

170 *Dan Littman and Ivaylo Ivanov:* I. I. Ivanov and D. R. Littman, "Segmented Filamentous Bacteria Take the Stage," *Mucosal Immunology* 3, no. 3 (2010).

171 *these same segmented bacteria:* Ivaylo I. Ivanov et al., "Gut-residing Segmented Filamentous Bacteria Drive Autoimmune Arthritis via T Helper 17 Cells," *Immunity* 32, no. 6 (2010).

171 *Sarkis Mazmanian showed:* Yun Kyung Lee et al., "Microbes and Health Sackler Colloquium: Proinflammatory T-cell Responses to Gut Microbiota Promote Experimental Autoimmune Encephalomyelitis," *Proceedings of the National Academy of Sciences* 108, Suppl. 1 (2011).

171 *But when Daniel Frank and Norman Pace:* Daniel N. Frank et al., "Molecular-phylogenetic Characterization of Microbial Community Imbalances in Human Inflammatory Bowel Diseases," *Proceedings of the National Academy of Sciences* 104, no. 34 (2007).

172 *Alexander Swidsinski at the Charité Hospital:* Alexander Swidsinski et al., "Active Crohn's

Disease and Ulcerative Colitis Can Be Specifically Diagnosed and Monitored Based on the Biostructure of the Fecal Flora," *Inflammatory Bowel Diseases* 14, no. 2 (2008).

172 *a single species with the immense name:* Harry Sokol et al., "*Faecalibacterium prausnitzii* Is an Anti-inflammatory Commensal Bacterium Identified by Gut Microbiota Analysis of Crohn's Disease Patients," *Proceedings of the National Academy of Sciences* 105, no. 43 (2008).

172 *could predict which twin:* Ben Willing et al., "Twin Studies Reveal Specific Imbalances in the Mucosa-associated Microbiota of Patients with Ileal Crohn's Disease," *Inflammatory Bowel Diseases* 15, no. 5 (2009).

172 *They took a top-down approach:* Koji Atarashi et al., "Induction of Colonic Regulatory T Cells by Indigenous Clostridium Species," *Science* 331, no. 6015 (2011).

173 *before rheumatoid arthritis set in:* Jussi Vaahtovuo et al., "Fecal Microbiota in Early Rheumatoid Arthritis," *Journal of Rheumatology* 35, no. 8 (2008).

173 *children who later developed diabetes:* Adriana Giongo et al., "Toward Defining the Autoimmune Microbiome for Type 1 Diabetes," *ISME Journal* 5, no. 1 (2011).

173 *children with celiac disease:* Giada de Palma et al., "Intestinal Dysbiosis and Reduced Immunoglobulin-coated Bacteria Associated with Coeliac Disease in Children," *BMC Microbiology* 10 (2010).

174 *With bifidobacteria present:* Jana Cinova et al., "Role of Intestinal Bacteria in Gliadin-induced Changes in Intestinal Mucosa: Study in Germ-free Rats," *PLoS ONE* 6, no. 1 (2011).

175 *During late pregnancy:* Michael Wilson, *Bacteriology of Humans: An Ecological Perspective* (Malden, Mass.: Blackwell Publishers, 2008), 202.

175 *Colonies of bifidobacteria sprout:* Rocío Martín et al., "Isolation of Bifidobacteria from Breast Milk and Assessment of the Bifidobacterial Population by PCR-denaturing Gradient Gel Electrophoresis and Quantitative Real-time PCR," *Applied and Environmental Microbiology* 75, no. 4 (2009).

175 *at least in mice, white blood cells:* Pablo F. Perez et al., "Bacterial Imprinting of the Neonatal Immune System: Lessons from Maternal Cells?," *Pediatrics* 119, no. 3 (2007).

175 *healthy placentas, long thought sterile:* R. Satokari et al., "Bifidobacterium and Lactobacillus DNA in the Human Placenta," *Letters in Applied Microbiology* 48, no. 1 (2009).

175 *It contains a dizzying number:* J. Bruce German et al., "Human Milk Oligosaccharides: Evolution, Structures and Bioselectivity as Substrates for Intestinal Bacteria," *Nestlé Nutrition Workshop Series Paediatric Programme* 62 (2008).

176 *In a study of 2,800 Norwegian children:* M. Eggesbø et al., "Cesarean Delivery and Cow Milk Allergy/Intolerance," *Allergy* 60, no. 9 (2005); Merete Eggesbø et al., "Is Delivery by Cesarean Section a Risk Factor for Food Allergy?," *Journal of Allergy and Clinical Immunology* 112, no. 2 (2003). This study finds no association between food allergy and C-section delivery. Bente Kvenshagen, Ragnhild Halvorsen, and Morten Jacobsen, "Is There an Increased Frequency of Food Allergy in Children Delivered by Caesarean Section Compared to Those Delivered Vaginally?," *Acta Paediatrica* 98, no. 2 (2009).

176 *Two allergic parents plus a C-section:* C. Roduit et al., "Asthma at 8 Years of Age in Children Born by Caesarean Section," *Thorax* 64, no. 2 (2009).

176 *A metanalysis of twenty studies:* C. R. Cardwell et al., "Caesarean Section Is Associated with an Increased Risk of Childhood-onset Type 1 Diabetes Mellitus: A Meta-analysis of Observational Studies," *Diabetologia* 51, no. 5 (2008).

176 *After one month, Finnish children:* M. M. Grönlund et al., "Mode of Delivery Directs the Phagocyte Functions of Infants for the First 6 Months of Life," *Clinical and Experimental Immunology* 116, no. 3 (1999); M. M. Grönlund et al., "Fecal Microflora in Healthy Infants Born by Different Methods of Delivery: Permanent Changes in Intestinal Flora After Cesarean Delivery," *Journal of Pediatric Gastroenterology and Nutrition* 28, no. 1 (1999).

176 *alterations to the microbiota:* S. Salminen et al., "Influence of Mode of Delivery on Gut Microbiota Composition in Seven Year Old Children," *Gut* 53, no. 9 (2004).

176 *children born at home had half:* Frederika A. van Nimwegen et al., "Mode and Place of Delivery, Gastrointestinal Microbiota, and Their Influence on Asthma and Atopy," *Journal of Allergy and Clinical Immunology* 128, no. 5 (2011).

176 *In Venezuela, whereas vaginally born:* C-section babies were colonized by bacteria native to the skin. Maria G. Dominguez-Bello et al., "Delivery Mode Shapes the Acquisition and Structure of the Initial Microbiota Across Multiple Body Habitats in Newborns," *Proceedings of the National Academy of Sciences* (2010).

177 *The broader point is not:* It's possible that the differences observed in people born by C-section have nothing to do with the microbiota. Instead, they may stem from the relatively stress-free birth by surgery, a mode of delivery that apparently leaves some genes "off" that should be "on." T. Schlinzig et al., "Epigenetic Modulation at Birth—Altered DNA-methylation in White Blood Cells after Caesarean Section," *Acta Paediatrica* 98, no. 7 (2009).

177 *Gary Huffnagle and the postdoc Mairi Noverr:* Andrew Shreiner, Gary B. Huffnagle, and Mairi C. Noverr, "The 'Microflora Hypothesis' of Allergic Disease," *Advances in Experimental Medicine and Biology* 635 (2008).

178 *For every four allergic Swedish children:* J. S. Alm et al., "Atopy in Children of Families with an Anthroposophic Lifestyle," *Lancet* 353, no. 9163 (1999). T. Alfvén et al., "Allergic Diseases and Atopic Sensitization in Children Related to Farming and Anthroposophic Lifestyle—the Parsifal Study," *Allergy* 61, no. 4 (2006).

178 *The Steiner children had measurably:* Johan S. Alm et al., "An Anthroposophic Lifestyle and Intestinal Microflora in Infancy," *Pediatric Allergy and Immunology* 13, no. 6 (2002).

178 *A survey of twenty published studies:* William Murk, Kari R. Risnes, and Michael B. Bracken, "Prenatal or Early-life Exposure to Antibiotics and Risk of Childhood Asthma: A Systematic Review," *Pediatrics* 127, no. 6 (2011).

178 *The diversity of bacteroides species:* Cecilia Jernberg et al., "Long-term Ecological Impacts of Antibiotic Administration on the Human Intestinal Microbiota," *ISME Journal* 1, no. 1 (2007).

179 *a broad-spectrum antibiotic:* Les Dethlefsen et al., "The Pervasive Effects of an Antibiotic on the Human Gut Microbiota, as Revealed by Deep 16S rRNA Sequencing," *PLoS Biology* 6, no. 11 (2008).

179 *If scientists housed an untreated:* Dionysios A. Antonopoulos et al., "Reproducible Community Dynamics of the Gastrointestinal Microbiota Following Antibiotic Perturbation," *Infection and Immunity* 77, no. 6 (2009).

179 *children and teens were averaging:* L. F. McCaig and J. M. Hughes, "Trends in Antimicrobial Drug Prescribing Among Office-based Physicians in the United States," *JAMA* 273, no. 3 (1995).

179 *had lessened the prescription rate:* Mike Sharland and SACAR Paediatric Subgroup, "The Use of Antibacterials in Children: A Report of the Specialist Advisory Committee on Antimicrobial Resistance (SACAR) Paediatric Subgroup," *Journal of Antimicrobial Chemotherapy* 60, Suppl. 1 (2007).

180 *Roughly 70 percent in the U.S.:* Margaret Mellon, Charles Benbrook, and Karen Lutz Benbrook, "Hogging It: Estimates of Antimicrobial Abuse in Livestock" (Cambridge, Mass.: Union of Concerned Scientists, January 2001).

180 *A scathing 2010 report:* Gill H. Harden, "FSIS National Residue Program for Cattle Audit Report 24601-08-KC March 2010" (Washington, D.C.: USDA Office of Inspector General, 2010).

180 *Between three-quarters and half:* Joanne C. Chee-Sanford et al., "Fate and Transport of Antibiotic Residues and Antibiotic Resistance Genes Following Land Application of Manure Waste," *Journal of Environmental Quality* 38, no. 3 (2009).

180 *Carrots absorbed:* Alistair B. A. Boxall et al., "Uptake of Veterinary Medicines from Soils into Plants," *Journal of Agricultural and Food Chemistry* 54, no. 6 (2006).

180 *Staple crops, such as wheat:* Tammy L. Jones-Lepp et al., "Method Development and Application to Determine Potential Plant Uptake of Antibiotics and Other Drugs in Irrigated Crop Production Systems," *Journal of Agricultural and Food Chemistry* 58, no. 22 (2010).

180 *Just over half of the sludge:* Chad A. Kinney et al., "Survey of Organic Wastewater Contaminants in Biosolids Destined for Land Application," *Environmental Science & Technology* 40, no. 23 (2006).

180 *Soybeans fertilized with human biosolids:* Chenxi Wu et al., "Uptake of Pharmaceutical and

Personal Care Products by Soybean Plants from Soils Applied with Biosolids and Irrigated with Contaminated Water," *Environmental Science & Technology* 44, no. 16 (2010).

181 *These drugs diffused:* Gretchen M. Bruce, Richard C. Pleus, and Shane A. Snyder, "Toxicological Relevance of Pharmaceuticals in Drinking Water," *Environmental Science & Technology* 44, no. 14 (2010); Mark J. Benotti et al., "Pharmaceuticals and Endocrine Disrupting Compounds in U.S. Drinking Water," *Environmental Science & Technology* 43, no. 3 (2009).

181 *sediments laid down in the 1970s:* For more on triclosan, see H. Singer et al., "Triclosan: Occurrence and Fate of a Widely Used Biocide in the Aquatic Environment: Field Measurements in Wastewater Treatment Plants, Surface Waters, and Lake Sediments," *Environmental Science & Technology* 36, no. 23 (2002). K. McClellan and R. U. Halden, "Pharmaceuticals and Personal Care Products in Archived U.S. Biosolids from the 2001 EPA National Sewage Sludge Survey," *Water Research* 44, no. 2 (2010).

181 *Antibiotic-resistant bacteria:* Jason K. Blackburn et al., "Evidence of Antibiotic Resistance in Free-Swimming, Top-Level Marine Predatory Fishes," *Journal of Zoo and Wildlife Medicine* 41, no. 1 (2010).

181 *Giving broad-spectrum antibiotics:* Shannon L. Russell et al., "Early Life Antibiotic-Driven Changes in Microbiota Enhance Susceptibility to Allergic Asthma," *EMBO Reports* (2012).

182 *They contrasted the microbiota:* Carlotta De Filippo et al., "Impact of Diet in Shaping Gut Microbiota Revealed by a Comparative Study in Children from Europe and Rural Africa," *Proceedings of the National Academy of Sciences* 107, no. 33 (Aug. 17, 2010).

182 *Although Amerindians and Africans:* Yatsunenko, Tanya, et al., "Human Gut Microbiome Viewed Across Age and Geography," *Nature* (May 3, 2012).

182 *"Cultural differences in diet":* Justin L. Sonnenburg, "Microbiology: Genetic Pot Luck," *Nature* 464, no. 7290 (2010).

183 *"the globalization of the microbiota":* D. Raoult, "The Globalization of Intestinal Microbiota," *European Journal of Clinical Microbiology & Infectious Diseases* 29, no. 9 (2010).

183 *During just the twentieth century:* For more on the Western diet, see Loren Cordain et al., "Origins and Evolution of the Western Diet: Health Implications for the 21st Century," *American Journal of Clinical Nutrition* 81, no. 2 (2005).

184 *The altered community improved:* Reiner Jumpertz et al., "Energy-balance Studies Reveal Associations Between Gut Microbes, Caloric Load, and Nutrient Absorption in Humans," *American Journal of Clinical Nutrition* 94, no. 1 (2011).

184 *inflammation drove the syndrome:* Patrice D. Cani and Nathalie M. Delzenne, "The Gut Microbiome as Therapeutic Target," *Pharmacology & Therapeutics* 130, no. 2 (2011).

185 *Paresh Dandona at the State University of New York:* Husam Ghanim et al., "Orange Juice Neutralizes the Proinflammatory Effect of a High-Fat, High-Carbohydrate Meal and Prevents Endotoxin Increase and Toll-like Receptor Expression," *American Journal of Clinical Nutrition* 91, no. 4 (2010).

185 *Wendy Garrett at the Harvard School of Public Health:* Wendy S. Garrett, Jeffrey I. Gordon, and Laurie H. Glimcher, "Homeostasis and Inflammation in the Intestine," *Cell* 140, no. 6 (2010). Wendy S. Garrett et al., "Communicable Ulcerative Colitis Induced by T-bet Deficiency in the Innate Immune System," *Cell* 131, no. 1 (2007).

186 *In one, scientists incapacitated:* Matam Vijay-Kumar et al., "Metabolic Syndrome and Altered Gut Microbiota in Mice Lacking Toll-like Receptor 5," *Science* 328, no. 6975 (2010).

186 *Daniel Frank, the scientist who noted the peculiar absence:* D. N. Frank et al., "Disease Phenotype and Genotype Are Associated with Shifts in Intestinal-associated Microbiota in Inflammatory Bowel Diseases," *Inflammatory Bowel Disease* 17, no. 1 (2011).

186 *Yolanda Sanz observed something similar:* Ester Sánchez et al., "Influence of Environmental and Genetic Factors Linked to Celiac Disease Risk on Infant Gut Colonization by Bacteroides Species," *Applied and Environmental Microbiology* 77, no. 15 (2011).

187 *Vincent Young at the University of Michigan:* Seth T. Walk et al., "Alteration of the Murine Gut Microbiota During Infection with the Parasitic Helminth *Heligmosomoides polygyrus*," *Inflammatory Bowel Diseases* 16, no. 11 (2010).

188 *In 2008, Alexander Khoruts:* Carl Zimmer, "How Microbes Defend and Define Us," *New York Times*, July 12, 2010.

188 *He sampled her microflora:* Alexander Khoruts, Johan Dicksved, and Janet K. Jansson, "Changes in the Composition of the Human Fecal Microbiome After Bacteriotherapy for Recurrent *Clostridium difficile*–associated Diarrhea," *Journal of Clinical Gastroenterology* 44, no. 5 (2010).

188 *In the U.S., deaths from* C. difficile: J. A. Karas, D. A. Enoch, and S. H. Aliyu, "A Review of Mortality Due to *Clostridium difficile* Infection," *Journal of Infection* 61, no. 1 (2010).

188 *The bacterium is naturally present:* B. Aronsson, R. Mollby, and C. E. Nord, "Antimicrobial Agents and *Clostridium difficile* in Acute Enteric Disease: Epidemiological Data from Sweden, 1980–1982," *Journal of Infectious Diseases* 151, no. 3 (1985).

189 *One prospective study:* Amee R. Manges et al., "Comparative Metagenomic Study of Alterations to the Intestinal Microbiota and Risk of Nosocomial *Clostridum difficile*–associated Disease," *Journal of Infectious Diseases* 202, no. 12 (2010).

189 *has successfully treated people:* Thomas J. Borody et al., "Treatment of Ulcerative Colitis Using Fecal Bacteriotherapy," *Journal of Clinical Gastroenterology* 37, no. 1 (2003).

189 *Dutch scientists tried:* A. Vrieze et al., "Metabolic Effects of Transplanting Gut Microbiota from Lean Donors to Subjects with Metabolic Syndrome," *European Association for the Study of Diabetes* 90 (2010).

189 *we have fecotypes:* M. Arumugam et al., "Enterotypes of the Human Gut Microbiome," *Nature* 473, no. 7346 (2011).

190 *The microbes may have to match:* We should take encouragement that we're on the right track with fecal transplants from our own anatomy. Consider the appendix, long viewed as a useless evolutionary vestige in humans. William Parker at Duke University thinks that the fingerlike organ reseeded the gut with friendly microbes after catastrophes. If a spell of diarrhea depleted the microbiota, the appendix repopulated gut microbes from its own backup supply. R. Randal Bollinger et al., "Biofilms in the Large Bowel Suggest an Apparent Function of the Human Vermiform Appendix," *Journal of Theoretical Biology* 249, no. 4 (2007).

190 *And as it becomes more apparent:* Steven A. Frese et al., "The Evolution of Host Specialization in the Vertebrate Gut Symbiont *Lactobacillus reuteri*," *PLoS Genetics* 7, no. 2 (2011).

190 *Pets, which protect against:* R. M. Maier et al., "Environmental Determinants of and Impact on Childhood Asthma by the Bacterial Community in Household Dust," *Applied and Environmental Microbiology* 76, no. 8 (2010).

190 *gut microbial communities grow:* Ruth E. Ley et al., "Evolution of Mammals and Their Gut Microbes," *Science* 320, no. 5883 (2008).

191 *Generally, more microbial diversity:* Another study pointing to the importance of microbial diversity: Hans Bisgaard et al., "Reduced Diversity of the Intestinal Microbiota During Infancy Is Associated with Increased Risk of Allergic Disease at School Age," *Journal of Allergy and Clinical Immunology* 128, no. 3 (2011).

191 *farmers, the least-allergic:* A. Sonnenberg and J. T. Walker, "Occupational Mortality Associated with Inflammatory Bowel Disease in the United States 1984–1998," *Inflammatory Bowel Diseases* (2011).

191 *having the right "keystone" species:* Johan Dicksved et al., "Molecular Fingerprinting of the Fecal Microbiota of Children Raised According to Different Lifestyles," *Applied and Environmental Microbiology* 73, no. 7 (2007).

191 *The Bedouin, a nomadic people:* R. A. Lewin, "More on Merde," *Perspectives in Biology and Medicine* 44, no. 4 (2001).

191 *And the Inuit of the Arctic often consumed:* Otto Schaefer, "Eskimos (Inuit)," in *Western Diseases: Their Emergence and Prevention,* eds. H. Trowell and D. Burkitt, Harvard University Press, 1981.

191 *The evolutionary biologist Michael Lombardo:* Michael P. Lombardo, "Access to Mutualistic Endosymbiotic Microbes: An Underappreciated Benefit of Group Living," *Behavioral Ecology and Sociobiology* 62, no. 4 (2008).

192 *Follow-up studies on those farm pigs:* Schmidt, Bettina, et al., "Establishment of Normal Gut Microbiota Is Compromised Under Excessive Hygiene Conditions," *PLoS ONE* 6, no. 12 (2011).

Chapter 10: Multiple Sclerosis: Worms That Never Come,
a Tardy Virus, and a Brain That Degenerates

193   *If the tree of life metaphor:* Luis P. Villarreal and Guenther Witzany, "Viruses Are Essential Agents Within the Roots and Stem of the Tree of Life," *Journal of Theoretical Biology* 262, no. 4 (2010).

194   *Her name was Lidwina:* Lidwina fell in 1396. R. Medaer, "Does the History of Multiple Sclerosis Go Back as Far as the 14th Century?," *Acta Neurologica Scandinavica* 60, no. 3 (1979).

194   *Eventually, she became:* T. Jock Murray, "The History of Multiple Sclerosis: The Changing Frame of the Disease over the Centuries," *Journal of the Neurological Sciences* 277, Suppl. 1 (2009).

195   *In 1822, Augustus d'Esté:* Anne-Marie Landtblom et al., "The First Case History of Multiple Sclerosis: Augustus D'Esté (1794–1848)," *Neurological Sciences* 31, no. 1 (2010).

195   *The first description of what, biologically speaking:* T. Jock Murray, "Robert Carswell: The First Illustrator of MS," *International MS Journal* 16 (2009).

196   *South Asian immigrants to the U.K.:* M. Elian et al., "Multiple Sclerosis Among United Kingdom–born Children of Immigrants from the Indian Subcontinent, Africa and the West Indies," *Journal of Neurology, Neurosurgery, and Psychiatry* 53, no. 10 (1990).

196   *identical-twin studies showed:* T. Hansen et al., "Concordance for Multiple Sclerosis in Danish Twins: An Update of a Nationwide Study," *Multiple Sclerosis* 11, no. 5 (2005).

196   *Beginning in the mid-twentieth century:* U. Leibowitz et al., "Epidemiological Study of Multiple Sclerosis in Israel. II. Multiple Sclerosis and Level of Sanitation," *Journal of Neurology, Neurosurgery, and Psychiatry* 29, no. 1 (1966).

196   *In South Africa, native-born Afrikaners:* G. Dean, "Annual Incidence, Prevalence, and Mortality of Multiple Sclerosis in White South African–born and in White Immigrants to South Africa," *British Medical Journal* 2, no. 5554 (1967). G. Dean and J. F. Kurtzke, "On the Risk of Multiple Sclerosis According to Age at Immigration to South Africa," *British Medical Journal* 3, no. 5777 (1971).

197   *the nucleus was Olmsted County, Minnesota:* W. T. Mayr et al., "Incidence and Prevalence of Multiple Sclerosis in Olmsted County, Minnesota, 1985–2000," *Neurology* 61, no. 10 (2003).

197   *In a district-by-district:* C. Green et al., "A Population-based Ecologic Study of Inflammatory Bowel Disease: Searching for Etiologic Clues," *American Journal of Epidemiology* 164, no. 7 (2006); K. Kimura et al., "Concurrence of Inflammatory Bowel Disease and Multiple Sclerosis," *Mayo Clinic Proceedings* 75, no. 8 (2000).

198   *John Fleming, a neurologist:* John O. Fleming and Thomas D. Cook, "Multiple Sclerosis and the Hygiene Hypothesis," *Neurology* 67, no. 11 (2006).

199   *In 2007, Correale published:* Jorge Correale and Mauricio Farez, "Association Between Parasite Infection and Immune Responses in Multiple Sclerosis," *Annals of Neurology* 61, no. 2 (2007).

202   *that was the conclusion of scientists:* M. T. Wallin, A. Heltberg, and J. F. Kurtzke, "Multiple Sclerosis in the Faroe Islands. 8. Notifiable Diseases," *Acta Neurologica Scandinavica* 122, no. 2 (2010).

202   *How and why would a virus:* Michael P. Pender, "The Essential Role of Epstein-Barr Virus in the Pathogenesis of Multiple Sclerosis," *Neuroscientist* 17, no. 4 (2011). Adam E. Handel et al., "An Updated Meta-analysis of Risk of Multiple Sclerosis Following Infectious Mononucleosis," *PLoS ONE* 5, no. 9 (2010).

202   *Danish scientists sought 6,800 people:* S. Haahr et al., "Increased Risk of Multiple Sclerosis After Late Epstein-Barr Virus Infection: A Historical Prospective Study," *Multiple Sclerosis* 1, no. 2 (1995).

202   *In the U.K., having had mononucleosis:* Michael J. Goldacre et al., "Multiple Sclerosis After Infectious Mononucleosis: Record Linkage Study," *Journal of Epidemiology & Community Health* 58, no. 12 (2004).

202   *In the U.S., among a cohort:* M. A. Hernán et al., "Multiple Sclerosis and Age at Infection with Common Viruses," *Epidemiology* 12, no. 3 (2001).

203 *those few lucky individuals:* Sven Haahr and Per Höllsberg, "Multiple Sclerosis Is Linked to Epstein-Barr Virus Infection," *Reviews in Medical Virology* 16, no. 5 (2006).

203 *In the U.K., for example:* Sreeram V. Ramagopalan et al., "Geography of Hospital Admissions for Multiple Sclerosis in England and Comparison with the Geography of Hospital Admissions for Infectious Mononucleosis: A Descriptive Study," *Journal of Neurology, Neurosurgery, and Psychiatry* 82, no. 6 (2011).

203 *but when scientists tried to isolate:* Hans Lassmann et al., "Epstein-Barr Virus in the Multiple Sclerosis Brain: A Controversial Issue—Report on a Focused Workshop Held in the Centre for Brain Research of the Medical University of Vienna, Austria," *Brain: A Journal of Neurology* 134, no. 9 (2011); Hans Helmut Niller et al., "Epigenetic Dysregulation of Epstein-Barr Virus Latency and Development of Autoimmune Disease," *Advances in Experimental Medicine and Biology* 711 (2011).

205 *She and her colleagues followed:* Shanie Saghafian-Hedengren et al., "Early-life EBV Infection Protects Against Persistent IgE Sensitization," *Journal of Allergy and Clinical Immunology* 125, no. 2 (2010).

206 *"silent commensals":* Michel Brahic, "Multiple Sclerosis and Viruses," *Annals of Neurology* 68, no. 1 (2010).

206 Homo sapiens, *freshly arrived from Africa:* Bent Sørensen, "Demography and the Extinction of European Neanderthals," *Journal of Anthropological Archaeology* 30 (2011).

206 *African and Asian elephants:* L. K. Richman et al., "Novel Endotheliotropic Herpes Viruses Fatal for Asian and African Elephants," *Science* 283, no. 5405 (1999).

207 *He infected mice with two herpes viruses:* Erik S. Barton et al., "Herpes Virus Latency Confers Symbiotic Protection from Bacterial Infection," *Nature* 447, no. 7142 (2007).

207 *Another group replicated Barton's results:* Eric J. Yager et al., "Gamma-Herpes Virus-Induced Protection Against Bacterial Infection Is Transient," *Viral Immunology* 22, no. 1 (2009).

208 *Indeed, some look to these associations:* For more on beneficial viruses, see Marilyn J. Roossinck, "The Good Viruses: Viral Mutualistic Symbioses," *Nature Reviews Microbiology* 9, no. 2 (2011).

208 *Take our adaptive immune system:* Luis P. Villarreal, "The Source of Self: Genetic Parasites and the Origin of Adaptive Immunity," *Annals of the New York Academy of Sciences* 1178 (2009).

208 *Or consider the placenta:* L. P. Villarreal, "On Viruses, Sex, and Motherhood," *Journal of Virology* 71, no. 2 (1997).

209 *Malaria plus Epstein-Barr:* Eriko Maeda et al., "Spectrum of Epstein-Barr Virus-related Diseases: A Pictorial Review," *Japanese Journal of Radiology* 27, no. 1 (2009).

209 *Italian scientists achieved:* A. Paolillo et al., "The Effect of Bacille Calmette-Guérin on the Evolution of New Enhancing Lesions to Hypointense T1 Lesions in Relapsing Remitting MS," *Journal of Neurology* 250, no.2 (2002). G. Ristori et al., "Use of Bacille Calmette-Guérin (BCG) in Multiple Sclerosis," *Neurology* 53, no. 7 (1999).

210 *Prior infection with* Mycobacterium bovis: JangEun Lee et al., "Mycobacteria-induced Suppression of Autoimmunity in the Central Nervous System," *Journal of Neuroimmune Pharmacology* 5, no. 2 (2010).

210 *One 1981 study found that:* E. Andersen, H. Isager, and K. Hyllested, "Risk Factors in Multiple Sclerosis: Tuberculin Reactivity, Age at Measles Infection, Tonsillectomy and Appendectomy," *Acta Neurologica Scandinavica* 63, no. 2 (1981).

210 *people who moved from the north:* C. R. Gale and C. N. Martyn, "Migrant Studies in Multiple Sclerosis," *Progress in Neurobiology* 47, nos. 4–5 (1995).

211 *Consider the dramatic increase of MS:* P. Cabre et al., "Role of Return Migration in the Emergence of Multiple Sclerosis in the French West Indies," *Brain* 128, no. 12 (2005).

212 *Four patients had begun to suffer:* Jorge Correale and Mauricio F. Farez, "The Impact of Parasite Infections on the Course of Multiple Sclerosis," *Journal of Neuroimmunology* 233 (2011).

212 *John Fleming's early results:* J. Fleming et al., "Probiotic Helminth Administration in Relapsing-remitting Multiple Sclerosis: A Phase 1 Study," *Multiple Sclerosis* 17, no. 6 (2011).

212 *an EBV vaccine could quite literally eradicate:* Some point out that if and when an EBV

vaccine arrives, it better have lifelong efficacy. If not, and it wears off in adulthood, then it will have the paradoxical effect of increasing both the risk of infectious mononucleosis and MS. This chapter is greatly informed by Alberto Ascherio and Kassandra L Munger, "Epstein-Barr Virus Infection and Multiple Sclerosis: A Review," *Journal of Neuroimmune Pharmacology* 5, no. 3 (2010).

212 *Erik Barton's concerns:* For an excellent review of his ideas, see D. W. White, R. Suzanne Beard, and E. S. Barton, "Immune Modulation During Latent Herpesvirus Infection," *Immunological Reviews* 245, no. 1 (2012).

212 *As if to underscore:* Silverberg, Jonathan et al. "Chickenpox in Childhood Is Associated with Decreased Atopic Disorders, IgE, Allergic Sensitization, and Leukocyte Subsets," *Pediatric Allergy and Immunology* 23, no. 1 (February 2012).

Chapter 11: Modernity's Developmental Disorder:
Autism and the Superorganism

214 *The immune system:* Betty Diamond et al., "It Takes Guts to Grow a Brain," *BioEssays: News and Reviews in Molecular, Cellular and Developmental Biology* 33, no. 8 (2011).

216 *In 1943, the Austrian-born psychiatrist:* Leo Kanner, "Autistic Disturbances of Affective Contact," *Nervous Child* 2 (1943). For more on Hans Asperger, see J. H. Baskin, M. Sperber, and B. H. Price, "Asperger Syndrome Revisited," *Reviews in Neurological Diseases* 3, no. 1 (2006).

217 *Others—maybe 40 percent:* S. Ozonoff et al., "A Prospective Study of the Emergence of Early Behavioral Signs of Autism," *Journal of the American Academy of Childhood Adolescent Psychiatry* 49, no. 3 (2010).

217 *And in 2009, the CDC revised:* Catherine Rice, "Prevalence of Autism Spectrum Disorders—Autism and Developmental Disabilities Monitoring Network, United States, 2006" (Atlanta: Center for Disease Control MMWR, 2009).

217 *about $3.2 million for each autistic person:* Estimates for costs come from Michael Ganz, "The Costs of Autism," in *Understanding Autism: From Basic Neuroscience to Treatment,* ed. Steven O. Moldin and John L. R. Rubenstein (Boca Raton, Fla.: CRC/Taylor & Frances, 2006).

218 *But one careful study in California:* Irva Hertz-Picciotto and Lora Delwiche, "The Rise in Autism and the Role of Age at Diagnosis," *Epidemiology* 20, no. 1 (2009).

218 *In the prestigious journal* Lancet: A. J. Wakefield et al., "Ileal-lymphoid-nodular Hyperplasia, Non-specific Colitis, and Pervasive Developmental Disorder in Children," *Lancet* 351, no. 9103 (1998).

218 *nor could epidemiologists show:* Frank DeStefano et al., "Age at First Measles-mumps-rubella Vaccination in Children with Autism and School-matched Control Subjects: A Population-based Study in Metropolitan Atlanta," *Pediatrics* 113, no. 2 (2004).

218 *In California, where autism:* L. Dales et al., "Time Trends in Autism and in MMR Immunization Coverage in California," *JAMA* 285, no. 9 (2001).

219 *In Japan, where authorities ended the MMR vaccine:* In Japan, autism more than quadrupled from 19.6 per 10,000 in those born in 1988, to 81.4 per 10,000 among those born in 1996. Note that in absolute terms, the rate is still far below the West—such as asthma, IBD, and the other immune-mediated problems. But also like these disorders, it's rising steadily. Hideo Honda, Yasuo Shimizu, and Michael Rutter, "No Effect of MMR Withdrawal on the Incidence of Autism: A Total Population Study," *Journal of Child Psychology and Psychiatry, and Allied Disciplines* 46, no. 6 (2005).

219 *And finally, in Poland:* Dorota Mrozek-Budzyn, Agnieszka Kiełtyka, and Renata Majewska, "Lack of Association Between Measles-mumps-rubella Vaccination and Autism in Children: A Case-control Study," *Pediatric Infectious Disease Journal* 29, no. 5 (2010).

219 *A later investigation:* Fiona Godlee, Jane Smith, and Harvey Marcovitch, "Wakefield's Article Linking MMR Vaccine and Autism Was Fraudulent," *British Medical Journal* 342 (2011).

219 *the Johns Hopkins psychiatrist John Money:* J. Money, N. A. Bobrow, and F. C. Clarke, "Autism and Autoimmune Disease: A Family Study," *Journal of Autism and Childhood Schizophrenia* 1, no. 2 (1971).

341

220 *Thirty-odd years later:* Anne Comi et al., "Familial Clustering of Autoimmune Disorders and Evaluation of Medical Risk Factors in Autism," *Journal of Child Neurology* 14, no. 6 (1999).

220 *The most authoritative study to date:* Hjördís O. Atladóttir et al., "Association of Family History of Autoimmune Diseases and Autism Spectrum Disorders," *Pediatrics* 124, no. 2 (2009).

220 *Since the late 1980s:* R. P. Warren et al., "Immune Abnormalities in Patients with Autism," *Journal of Autism and Developmental Disorders* 16, no. 2 (1986). W. W. Daniels et al., "Increased Frequency of the Extended or Ancestral Haplotype B44-SC30-DR4 in Autism," *Neuropsychobiology* 32, no. 3 (1995). Anthony R. Torres et al., "The Transmission Disequilibrium Test Suggests That HLA-DR4 and DR13 Are Linked to Autism Spectrum Disorder," *Human Immunology* 63, no. 4 (2002). William G. Johnson et al., "HLA-DR4 as a Risk Allele for Autism Acting in Mothers of Probands Possibly During Pregnancy," *Archives of Pediatrics & Adolescent Medicine* 163, no. 6 (2009).

220 *Again, a mother's having the variants:* Li-Ching Lee et al., "HLA-DR4 in Families with Autism," *Pediatric Neurology* 35, no. 5 (2006).

221 *a mother's having these antibodies:* Daniel Braunschweig et al., "Autism: Maternally Derived Antibodies Specific for Fetal Brain Proteins," *Neurotoxicology* 29, no. 2 (2008).

221 *the scientists injected those fetus-directed antibodies:* Loren A. Martin et al., "Stereotypes and Hyperactivity in Rhesus Monkeys Exposed to IgG from Mothers of Children with Autism," *Brain, Behavior, and Immunity* 22, no. 6 (2008).

221 *they entered a period of frenzied cerebral expansion:* E. Courchesne et al., "Unusual Brain Growth Patterns in Early Life in Patients with Autistic Disorder: An MRI Study," *Neurology* 57, no. 2 (2001).

221 *Soon after that peak:* Eric Courchesne, Kathleen Campbell, and Stephanie Solso, "Brain Growth Across the Life Span in Autism: Age-specific Changes in Anatomical Pathology," *Brain Research* 1380 (2011).

222 *The amygdala, which, in addition to initiating:* Cynthia Mills Schumann et al., "Amygdala Enlargement in Toddlers with Autism Related to Severity of Social and Communication Impairments," *Biological Psychiatry* 66, no. 10 (2009).

222 *"a brain disorder, or a disorder that affects the brain?":* Martha Herbert, "Autism: A Brain Disorder, or a Disorder That Affects the Brain?" *Clinical Neuropsychiatry* 2, no. 6 (2005).

222 *The scientists located eleven samples:* Diana L. Vargas et al., "Neuroglial Activation and Neuroinflammation in the Brain of Patients with Autism," *Annals of Neurology* 57, no. 1 (2005).

223 *He injected pregnant mice:* Harvey S. Singer et al., "Prenatal Exposure to Antibodies from Mothers of Children with Autism Produces Neurobehavioral Alterations: A Pregnant Dam Mouse Model," *Journal of Neuroimmunology* 211, nos. 1–2 (2009).

224 *In a 2007 paper:* Kevin G. Becker, "Autism, Asthma, Inflammation, and the Hygiene Hypothesis," *Medical Hypotheses* 69, no. 4 (2007).

224 *There was, it turned out, an intriguing:* For an enlightening review, see Nicolette A. Hodyl et al., "Prenatal Programming of the Innate Immune Response Following in Utero Exposure to Inflammation: A Sexually Dimorphic Process?," *Expert Review of Clinical Immunology* 7, no. 5 (2011).

227 *Between 20,000 and 30,000 children:* Alan S. Brown and Ezra S. Susser, "In Utero Infection and Adult Schizophrenia," *Mental Retardation and Developmental Disabilities Research Reviews* 8, no. 1 (2002).

227 *Later analysis found that:* Some material is drawn from P. H. Patterson, *Infectious Behavior: Brain-Immune Connections in Autism, Schizophrenia, and Depression,* 70. Original citation is: S. Chess, "Autism in Children with Congenital Rubella," *Journal of Autism and Childhood Schizophrenia* 1, no. 1 (1971).

228 *Patterson injected just interleukin-6:* Elaine Y. Hsiao and Paul H. Patterson, "Activation of the Maternal Immune System Induces Endocrine Changes in the Placenta via IL-6," *Brain, Behavior, and Immunity* 25 (2010). Other studies note that IL-6 is elevated in autistic people, and that high IL-6 alters neural migration. H. Wei et al., "IL-6 Is Increased in the Cerebellum of Autistic Brain and Alters Neural Cell Adhesion, Migration and Synaptic Formation," *Journal of Neuroinflammation* 8 (2011).

228   *Alan Brown at Columbia University found:* Alan S. Brown et al., "Serologic Evidence of Prenatal Influenza in the Etiology of Schizophrenia," *Archives of General Psychiatry* 61, no. 8 (2004). Alan S. Brown and Paul H. Patterson, "Maternal Infection and Schizophrenia: Implications for Prevention," *Schizophrenia Bulletin* 37, no. 2 (2011).

229   *a pregnant mother's being diagnosed:* Lisa A. Croen et al., "Maternal Autoimmune Diseases, Asthma and Allergies, and Childhood Autism Spectrum Disorders: A Case-control Study," *Archives of Pediatrics & Adolescent Medicine* 159, no. 2 (2005). The connection between autism and maternal infection has been made more recently than the rubella epidemics of the 1960s. The largest study to date, an analysis of 1.6 million expectant mothers and their children born in Denmark between 1980 and 2005, showed that a mother's admission to the hospital for viral infection during the first trimester tripled the chance for an autistic child. Bacterial infection during the second trimester increased that chance by nearly half. No particular virus or bacterium predicted the later diagnosis of autism; just having sought hospitalization for an infection increased the risk. Hjördis O. Atladóttir et al., "Maternal Infection Requiring Hospitalization During Pregnancy and Autism Spectrum Disorders," *Journal of Autism and Developmental Disorders* 40, no. 12 (2010).

229   *They looked at 365 mothers:* L. Heuer et al., "Association of an MET Genetic Variant with Autism-associated Maternal Autoantibodies to Fetal Brain Proteins and Cytokine Expression," *Translational Psychiatry* 1(2011).

229   *The scientists genetically programmed:* U. Meyer et al., "Adult Behavioral and Pharmacological Dysfunctions Following Disruption of the Fetal Brain Balance Between Pro-inflammatory and IL-10-Mediated Anti-inflammatory Signaling," *Molecular Psychiatry* 13, no. 2 (2008).

230   *Chris Coe and colleagues were replicating:* Auriel A. Willette et al., "Brain Enlargement and Increased Behavioral and Cytokine Reactivity in Infant Monkeys Following Acute Prenatal Endotoxemia," *Behavioural Brain Research* 219, no. 1 (2011).

230   *This is Meyer and Feldon's argument:* Here's Meyer and Feldon's argument in full: slightly different immune failures predispose to either autism or schizophrenia in the developing fetus. If a mother-to-be fails to turn off inflammation, then autism results. If the mother experiences an acute insult—like a viral infection—eradicates the threat, then returns to immune homeostasis, that leads not to autism, but to a vulnerability to schizophrenia. In adulthood, these individuals may then encounter a trigger—a "second hit"—that sends them spiraling into psychosis. Urs Meyer, Joram Feldon, and Olaf Dammann, "Schizophrenia and Autism: Both Shared and Disorder-specific Pathogenesis via Perinatal Inflammation?," *Pediatric Research* 69 (2011).

231   *some think that schizophrenia:* P. Munk-Jørgensen and P. B. Mortensen, "Is Schizophrenia Really on the Decrease?," *European Archives of Psychiatry and Clinical Neuroscience* 242, no. 4 (1993).

231   *As polio cases declined:* J. M. Suvisaari et al., "Decline in the Incidence of Schizophrenia in Finnish Cohorts Born from 1954 to 1965," *Archives of General Psychiatry* 56, no. 8 (1999); J. Suvisaari et al., "Association Between Prenatal Exposure to Poliovirus Infection and Adult Schizophrenia," *American Journal of Psychiatry* 156, no. 7 (1999).

231   *In 1980, after a viral infection:* R. M. Cotterill, "Fever in Autistics," *Nature* 313, no. 6002 (1985).

232   *"Of course, all children quiet down":* Gary Brown, "The Sometimes Son: Fever May Temporarily Cure Autism," *The Humanist* (1999).

232   *Zimmerman and Curran distributed:* Laura K. Curran et al., "Behaviors Associated with Fever in Children with Autism Spectrum Disorders," *Pediatrics* 120, no. 6 (2007).

233   *incapacitating mice's T cells messes:* Jonathan Kipnis et al., "T Cell Deficiency Leads to Cognitive Dysfunction: Implications for Therapeutic Vaccination for Schizophrenia and Other Psychiatric Conditions," *Proceedings of the National Academy of Sciences* 101, no. 21 (2004).

233   *When he knocks out just one:* Noël C. Derecki et al., "Regulation of Learning and Memory by Meningeal Immunity: A Key Role for IL-4," *Journal of Experimental Medicine* 207, no. 5 (2010).

233   *They have more TNF alpha:* Michael G. Chez et al., "Elevation of Tumor Necrosis Factor-alpha in Cerebrospinal Fluid of Autistic Children," *Pediatric Neurology* 36, no. 6 (2007).

233 *They have elevated levels of a hormone:* Paul Ashwood et al., "Brief Report: Plasma Leptin Levels Are Elevated in Autism: Association with Early Onset Phenotype?," *Journal of Autism and Developmental Disorders* 38, no. 1 (2008).

233 *When stimulated with endotoxin:* Amanda M. Enstrom et al., "Differential Monocyte Responses to TLR Ligands in Children with Autism Spectrum Disorders," *Brain, Behavior, and Immunity* 24, no. 1 (2010).

233 *They have less anti-inflammatory TGF beta:* Paul Ashwood et al., "Decreased Transforming Growth Factor Beta1 in Autism: A Potential Link Between Immune Dysregulation and Impairment in Clinical Behavioral Outcomes," *Journal of Neuroimmunology* 204, nos. 1–2 (2008).

233 *And they also have fewer:* Gehan A. Mostafa, Abeer Al Shehab, and Nermeen R. Fouad, "Frequency of CD4+CD25high Regulatory T Cells in Peripheral Blood of Egyptian Children with Autism," *Journal of Child Neurology* 25, no. 3 (2009).

233 *scientists have reported good results:* Michael G. Chez and Natalie Guido-Estrada, "Immune Therapy in Autism: Historical Experience and Future Directions with Immunomodulatory Therapy," *Neurotherapeutics* 7, no. 3 (2010).

233 *Intravenous immunoglobulin treatment:* Sudhir Gupta, Daljeet Samra, and Sudhanshu Agrawal, "Adaptive and Innate Immune Responses in Autism: Rationale for Therapeutic Use of Intravenous Immunoglobulin," *Journal of Clinical Immunology* 30, Suppl. 1 (2010).

233 *but the antibodies both tamp down:* M. S. Maddur et al., "Immunomodulation by Intravenous Immunoglobulin: Role of Regulatory T Cells," *Journal of Clinical Immunology* 30, Suppl. 1 (2010).

236 *40 percent of regressive autistic children:* For more on gut dysfunction in autism, see Valicenti-McDermott et al., "Gastrointestinal Symptoms in Children with an Autism Spectrum Disorder and Language Regression," *Pediatric Neurology* 39, no. 6 (December 2008): 392–98.

236 *For those with a family history:* For an argument that GI dysfunction is real in an autistic subset, see Amy C. Brown and Lewis Mehl-Madrona, "Autoimmune and Gastrointestinal Dysfunctions: Does a Subset of Children with Autism Reveal a Broader Connection?," *Expert Review of Gastroenterology & Hepatology* 5, no. 4 (2011).

236 *Still others have noted:* Harumi Jyonouchi et al., "Evaluation of an Association Between Gastrointestinal Symptoms and Cytokine Production Against Common Dietary Proteins in Children with Autism Spectrum Disorders," *Journal of Pediatrics* 146, no. 5 (2005).

236 *a 2010 consensus report on gut issues:* Timothy Buie et al., "Evaluation, Diagnosis, and Treatment of Gastrointestinal Disorders in Individuals with ASDs: A Consensus Report," *Pediatrics* 125, Suppl. 1 (2010).

237 *In one, a fourteen-year-old boy presented:* R. H. Sandler et al., "Relief of Psychiatric Symptoms in a Patient with Crohn's Disease After Metronidazole Therapy," *Clinical Infectious Diseases* 30, no. 1 (2000).

237 *a five-year-old boy with a diagnosis of autism:* Stephen J. Genuis and Thomas P. Bouchard, "Celiac Disease Presenting as Autism," *Journal of Child Neurology* 25, no. 1 (2010). S. J. Genuis, "Is Autism Reversible?," *Acta Paediatrica* 98, no. 10 (2009).

237 *There, celiac disease in particular:* Giulia Barcia et al., "Autism and Coeliac Disease," *Journal of Autism and Developmental Disorders* 38, no. 2 (2008).

237 *Autistic children had much more endotoxin:* Enzo Emanuele et al., "Low-grade Endotoxemia in Patients with Severe Autism," *Neuroscience Letters* 471, no. 3 (2010).

238 *he treated ten autistic children:* R. H. Sandler et al., "Short-term Benefit from Oral Vancomycin Treatment of Regressive-onset Autism," *Journal of Child Neurology* 15, no. 7 (2000).

238 *Most notable, perhaps, autistic children:* Sydney M. Finegold, "Desulfovibrio Species Are Potentially Important in Regressive Autism," *Medical Hypotheses* 77, no. 2 (2011); Sydney M. Finegold, "State of the Art; Microbiology in Health and Disease. Intestinal Bacterial Flora in Autism," *Anaerobe* (2011); Sydney M. Finegold et al., "Pyrosequencing Study of Fecal Microflora of Autistic and Control Children," *Anaerobe* 16, no. 4 (2010).

239 *Injecting rats with the stuff:* K. A. Foley et al., "Propionic Acid Alters Startle Response Magnitude in Adolescent Rats" (2011).

240 *In 2011, Swedish scientists:* Rochellys Diaz Heijtz et al., "Normal Gut Microbiota Modulates Brain Development and Behavior," *Proceedings of the National Academy of Sciences* 108, no. 7 (2011).
241 *One scientist has proposed giving:* M. R. Buehler, "A Proposed Mechanism for Autism: An Aberrant Neuroimmune Response Manifested as a Psychiatric Disorder," *Medical Hypotheses* 76, no. 6 (2011).

Chapter 12: Beyond Allergy and Autoimmunity:
Inflammation and the Diseases of Civilization

242 *In the wake of immunologically new humankind:* T. Haahtela, "Allergy Is Rare Where Butterflies Flourish in a Biodiverse Environment," *Allergy* 64, no. 12 (2009).
244 *Worm infection, it turns out:* For a good review, see E. Magen et al., "Can Worms Defend Our Hearts? Chronic Helminthic Infections May Attenuate the Development of Cardiovascular Diseases," *Medical Hypotheses* 64, no. 5 (2005).
244 *And when British scientists infected:* M. J. Doenhoff et al., "An Anti-atherogenic Effect of *Schistosoma mansoni* Infections in Mice Associated with a Parasite-induced Lowering of Blood Total Cholesterol," *Parasitology* 125, no. 5 (2002).
245 *In India, China, and Africa:* V. S. Zhdanov et al., "Development of Atherosclerosis over a 25-Year Period: An Epidemiological Autopsy Study in Males of 11 Towns," *International Journal of Cardiology* 68, no. 1 (1999). K. S. Woo et al., "Westernization of Chinese Adults and Increased Subclinical Atherosclerosis," *Arteriosclerosis, Thrombosis, and Vascular Biology* 19, no. 10 (1999).
245 *Take South Africa in the early 1970s:* A. R. Walker, "Survival Rate at Middle Age in Developing and Western Populations," *Postgraduate Medical Journal* 50, no. 579 (1974).
245 *how much of that first-rate immune:* In Finland, scientists found that early-life infection could protect against later heart disease. While Finns with antibodies to adulthood viruses did have more heart trouble—a support of the infection-equals-heart disease paradigm—those who'd suffered through more childhood illnesses had a decreased risk. And the greater the number of illnesses they contracted in childhood, the less their risk of heart disease later in life. Erkki Pesonen et al., "Dual Role of Infections as Risk Factors for Coronary Heart Disease," *Atherosclerosis* 192, no. 2 (2007).
246 *Analyzing data gathered from this cohort:* Thomas W. McDade et al., "Early Origins of Inflammation: Microbial Exposures in Infancy Predict Lower Levels of C-reactive Protein in Adulthood," *Proceedings of the Royal Society B: Biological Sciences* 277, no. 1684 (2010).
246 *Looking at two immune-system:* Thomas W. McDade, Stacy Tessler Lindau, and Kristen Wroblewski, "Predictors of C-reactive Protein in the National Social Life, Health, and Aging Project," *Journals of Gerontology: Series B, Psychological Sciences and Social Sciences* 66, no. 1 (2011).
247 *gene variants that boost production:* Yael T. Joffe et al., "Tumor Necrosis Factor-alpha Gene -308 G/A Polymorphism Modulates the Relationship Between Dietary Fat Intake, Serum Lipids, and Obesity Risk in Black South African Women," *Journal of Nutrition* 140, no. 5 (2010).
248 *The UCSF scientists found:* Davina Wu et al., "Eosinophils Sustain Adipose Alternatively Activated Macrophages Associated with Glucose Homeostasis," *Science* 332, no. 6026 (2011). For more on helminths and tissue repair, see F. Chen et al., "An Essential Role for T_H2-type Responses in Limiting Acute Tissue Damage During Experimental Helminth Infection," *Nature Medicine* 18, no. 2 (2012).
248 *"Can the rising incidence of metabolic syndrome":* Rick M. Maizels and Judith E. Allen, "Immunology: Eosinophils Forestall Obesity," *Science* 332, no. 6026 (2011).
248 *pinworms largely disappeared:* S. H. Vermund and S. MacLeod, "Is Pinworm a Vanishing Infection? Laboratory Surveillance in a New York City Medical Center from 1971 to 1986," *American Journal of Diseases of Children* 142, no. 5 (1988).
249 *Studies from Taiwan associate:* S-L. Huang, P-F. Tsai, and Y-F. Yeh, "Negative Association of *Enterobius* Infestation with Asthma and Rhinitis in Primary School Children in Taipei," *Clinical and Experimental Allergy* 32, no. 7 (2002).

249 *scientists find they can cure:* Kathrin Eller et al., "Potential Role of Regulatory T Cells in Reversing Obesity-Linked Insulin Resistance and Diabetic Nephropathy," *Diabetes* 60, no. 11 (2011).

249 *Mothers who are overweight during pregnancy:* Interestingly, high maternal BMI only predicted wheeze for children of parents who didn't already have allergies. Swatee P. Patel et al., "Associations Between Pre-pregnancy Obesity and Asthma Symptoms in Adolescents," *Journal of Epidemiology & Community Health* (2011).

249 *Communites in the central and eastern Arctic:* O. Schaefer, "When the Eskimo Comes to Town," *Nature Today* 6 (1971); H. Herxheimer and O. Schaefer, "Letter: Asthma in Canadian Eskimos," *New England Journal of Medicine* 291, no. 26 (1974).

249 *First came a wave of appendicitis:* O. Schaefer, "Aetiology of Appendicitis," *British Medical Journal* 1, no. 6172 (1979).

249 *And the types of cancers shifted:* J. A. Hildes and O. Schaefer, "The Changing Picture of Neoplastic Disease in the Western and Central Canadian Arctic (1950–1980)," *Canadian Medical Association Journal* 130, no. 1 (1984).

250 *And if you look at mortality statistics:* "Occupation, Health, and Mortality in the Light of the Registrar-General's Decennial Supplement," *Lancet* 211, no. 5449 (1928).

250 *between 15 and 20 percent of all cancers:* Francesco Colotta et al., "Cancer-related Inflammation, the Seventh Hallmark of Cancer: Links to Genetic Instability," *Carcinogenesis* 30, no. 7 (2009).

250 *But some scientists now argue:* Graham A. W. Rook and Angus Dalgleish, "Infection, Immunoregulation, and Cancer," *Immunological Reviews* 240, no. 1 (2011); Leena C. von Hertzen, H. Joensuu, and T. Haahtela, "Microbial Deprivation, Inflammation and Cancer," *Cancer Metastasis Reviews* 30, no. 2 (2011).

250 *breast cancer risk follows the incidence:* Anamaria Savu et al., "Breast Cancer and Microbial Cancer Incidence in Female Populations Around the World: A Surprising Hyperbolic Association," *International Journal of Cancer* 123, no. 5 (2008).

251 *when immigrants move from the developing world:* Paulo S. Pinheiro et al., "Cancer Incidence in First Generation U.S. Hispanics: Cubans, Mexicans, Puerto Ricans, and New Latinos," *Cancer Epidemiology, Biomarkers & Prevention* 18, no. 8 (2009).

251 *In a series of studies on female textile workers:* W. Li et al., "Occupational Risk Factors for Pancreatic Cancer Among Female Textile Workers in Shanghai, China," *Occupational and Environmental Medicine* 63, no. 12 (2006). George Astrakianakis et al., "Lung Cancer Risk Among Female Textile Workers Exposed to Endotoxin," *Journal of the National Cancer Institute* 99, no. 5 (2007). Roberta M. Ray et al., "Occupational Exposures and Breast Cancer Among Women Textile Workers in Shanghai," *Epidemiology* 18, no. 3 (2007). Karen J. Wernli et al., "Occupational Exposures and Ovarian Cancer in Textile Workers," *Epidemiology* 19, no. 2 (2008).

251 *not only did working in a cotton mill:* G. Mastrangelo et al., "Lung and Other Cancer Site Mortality in a Cohort of Italian Cotton Mill Workers," *Occupational and Environmental Medicine* 65, no. 10 (2008). G. Mastrangelo, V. Marzia, and G. Marcer, "Reduced Lung Cancer Mortality in Dairy Farmers: Is Endotoxin Exposure the Key Factor?," *American Journal of Industrial Medicine* 30, no. 5 (1996).

251 *Work by Susan Erdman:* Susan E. Erdman and Theofilos Poutahidis, "Cancer Inflammation and Regulatory T Cells," *International Journal of Cancer* 127, no. 4 (2010); Susan E. Erdman et al., "Unifying Roles for Regulatory T Cells and Inflammation in Cancer," *International Journal of Cancer* 126, no. 7 (2010).

252 *people who take nonsteroidal anti-inflammatory drugs:* P. M. Rothwell et al., "Effect of Daily Aspirin on Long-term Risk of Death Due to Cancer: Analysis of Individual Patient Data from Randomised Trials," *Lancet* 377, no. 9759 (2011).

252 *"Coley's Toxin":* Edward F. McCarthy, "The Toxins of William B. Coley and the Treatment of Bone and Soft-tissue Sarcomas," *Iowa Orthopaedic Journal* 26(2006).

252 *To this day, however, oncologists:* U. Hobohm, "Fever Therapy Revisited," *British Journal of Cancer* 92, no. 3 (2005).

253 *In general, people who've had more infections:* K. F. Kölmel et al., "Infections and Melanoma

Risk: Results of a Multicentre EORTC Case-control Study," *Melanoma Research* 9, no. 5 (1999).

253 *Melanoma is among the fastest-increasing cancers:* C. Garbe and U. Leiter, "Melanoma Epidemiology and Trends," *Clinical Dermatology* 27, no. 1 (2009).

253 *The scientists Bernd Krone and John Grange:* B. Krone and J. M. Grange, "Melanoma, Darwinian Medicine and the Inner World," *Journal of Cancer Research and Clinical Oncology* 136, no. 12 (2010).

254 *Younger siblings get it less often:* Ellen T. Chang et al., "Childhood Social Environment and Hodgkin's Lymphoma: New Findings from a Population-based Case-control Study," *Cancer Epidemiology, Biomarkers & Prevention* 13, no. 8 (2004).

254 *Firstborn children have it more than later-born:* Mel Greaves, "Infection, Immune Responses and the Aetiology of Childhood Leukaemia," *Nature Reviews Cancer* 6, no. 3 (2006).

254 *Early day-care attendance protects:* C. Gilham et al., "Day Care in Infancy and Risk of Childhood Acute Lymphoblastic Leukaemia: Findings from UK Case-control Study," *British Medical Journal (Clinical Research Ed.)* 330, no. 7503 (2005).

254 *The difference in risk was stark:* S. J. O'Keefe et al., "Rarity of Colon Cancer in Africans Is Associated with Low Animal Product Consumption, Not Fiber," *American Journal of Gastroenterology* 94, no. 5 (1999).

254 *the rate at which these cells proliferate:* S. J. O'Keefe et al., "Products of the Colonic Microbiota Mediate the Effects of Diet on Colon Cancer Risk," *Journal of Nutrition* 139, no. 11 (2009).

254 *The black South African microbiota:* J. B. Greer and S. J. O'Keefe, "Microbial Induction of Immunity, Inflammation, and Cancer," *Frontiers in Physiology* 1 (2011).

254 *the allergic response is really:* M. El-Zein et al., "History of Asthma or Eczema and Cancer Risk Among Men: A Population-based Case-control Study in Montreal, Quebec, Canada," *Annals of Allergy, Asthma & Immunology* 104, no. 5 (2010).

255 *eosinophils, the cells that help expel worms:* Fanny Legrand et al., "Human Eosinophils Exert TNF-A and Granzyme A-Mediated Tumoricidal Activity toward Colon Carcinoma Cells," *Journal of Immunology* 185, no. 12 (2010).

255 *The more one has, the less the chance:* A. E. Prizment et al., "Inverse Association of Eosinophil Count with Colorectal Cancer Incidence: Atherosclerosis Risk in Communities Study," *Cancer Epidemiology Biomarkers and Prevention* 20, no. 9 (2011).

255 *Elevated levels of all these hormones:* People who have gene variants that increase growth-hormone levels have a greater risk of cancer. Elizabeth C. Leroy et al., "Genes in the Insulin and Insulin-like Growth Factor Pathway and Odds of Metachronous Colorectal Neoplasia," *Human Genetics* 129, no. 5 (2011). On the other hand, people whose growth-hormone signaling is disrupted are remarkably free of cancer. A study of people living in remote Ecuadorian villages who have a mutation in the growth-hormone receptor that disables it—and who are short as a result—finds them not to suffer from cancer. Jaime Guevara-Aguirre et al., "Growth Hormone Receptor Deficiency Is Associated with a Major Reduction in Pro-aging Signaling, Cancer, and Diabetes in Humans," *Science: Translational Medicine* 3, no. 70 (2011).

255 *Westernized populations have skyrocketing levels:* Many of the ideas on cancer and hormone levels presented here come from Tessa M. Pollard, *Western Diseases: An Evolutionary Perspective* (Cambridge and New York: Cambridge University Press, 2008), 75–98. For hormone levels in men in different environments, see P. T. Ellison et al., "Population Variation in Age-related Decline in Male Salivary Testosterone," *Human Reproduction* 17, no. 12 (2002).

256 *experimentally infecting rodents with parasites:* For an excellent review on testosterone and immune function, see Michael P. Muehlenbein and Richard G. Bribiescas, "Testosterone-mediated Immune Functions and Male Life Histories," *American Journal of Human Biology* 17, no. 5 (2005).

256 *children grow less hurriedly in parasite-ridden:* T. W. McDade et al., "Maintenance Versus Growth: Investigating the Costs of Immune Activation Among Children in Lowland

Bolivia," *American Journal of Physical Anthropology* 136, no. 4 (2008). Aaron D. Blackwell et al., "Life History, Immune Function, and Intestinal Helminths: Trade-offs Among Immunoglobulin E, C-reactive Protein, and Growth in an Amazonian Population," *American Journal of Human Biology* 22, no. 6 (2010).

256 *Both are risk factors for metabolic syndrome:* Lise Aksglaede et al., "Age at Puberty and the Emerging Obesity Epidemic," *PLoS ONE* 4, no. 12 (2009).

256 *Bangladeshi women who immigrate to the U.K.:* A. Nuñez-de la Mora et al., "Childhood Conditions Influence Adult Progesterone Levels," *PLoS Medicine* 4, no. 5 (2007).

256 *Your average sixty-five-year-old American woman:* D. M. Parkin et al., "Global Cancer Statistics, 2002," *CA: A Cancer Journal for Clinicians* 55, no. 2 (2005).

256 *The children of East Asians:* J. E. Dunn, "Cancer Epidemiology in Populations of the United States—with Emphasis on Hawaii and California—and Japan," *Cancer Research* 35, no. 11, pt. 2 (1975); J. Lee et al., "Cancer Incidence Among Korean-American Immigrants in the United States and Native Koreans in South Korea," *Cancer Control* 14, no. 1 (2007).

257 *American-born women whose grandparents:* R. G. Ziegler et al., "Migration Patterns and Breast Cancer Risk in Asian-American Women," *Journal of the National Cancer Institute* 85, no. 22 (1993).

257 *Lowry wanted to understand:* C. A. Lowry et al., "Identification of an Immune-responsive Mesolimbocortical Serotonergic System: Potential Role in Regulation of Emotional Behavior," *Neuroscience* 146, no. 2 (2007).

257 *M. vaccae–treated patients with adenocarcinoma:* J. Stanford, "Successful Immunotherapy with *Mycobacterium vaccae* in the Treatment of Adenocarcinoma of the Lung," *European Journal of Cancer* 44, no. 2 (2008).

258 *He was exploring why up to half:* Charles L. Raison, Lucile Capuron, and Andrew H. Miller, "Cytokines Sing the Blues: Inflammation and the Pathogenesis of Depression," *Trends in Immunology* 27, no. 1 (2006). Charles L. Raison et al., "Depressive Symptoms and Viral Clearance in Patients Receiving Interferon-alpha and Ribavirin for Hepatitis C," *Brain, Behavior and Immunity* 19, no. 1 (2005).

258 *Chronic fatigue, he found, was also associated:* Andrew H. Miller, Vladimir Maletic, and Charles L. Raison, "Inflammation and Its Discontents: The Role of Cytokines in the Pathophysiology of Major Depression," *Biological Psychiatry* 65, no. 9 (2009).

258 *Patients undergoing this therapy:* B. E. Leonard, "The Immune System, Depression and the Action of Antidepressants," *Progress in Neuro-Psychopharmacology & Biological Psychiatry* 25, no. 4 (2001).

258 *Depression was a common "comorbidity":* Graham A. W. Rook, Christopher A. Lowry, and Charles L. Raison, "Lymphocytes in Neuroprotection, Cognition and Emotion: Is Intolerance Really the Answer?," *Brain, Behavior and Immunity* 25, no. 4 (2011).

258 *they found that inflammatory markers:* Julie A. Pasco et al., "Association of High-sensitivity C-reactive Protein with de Novo Major Depression," *British Journal of Psychiatry* 197 (2010).

258 *Asthma in childhood increased the risk:* M. R. Hayatbakhsh et al., "Association of Psychiatric Disorders, Asthma and Lung Function in Early Adulthood," *Journal of Asthma* 47, no. 7 (2010).

258 *Even serotonin-reuptake inhibitors:* Michael Diamond, John P. Kelly, and Thomas J. Connor, "Antidepressants Suppress Production of the Th1 Cytokine Interferon-gamma, Independent of Monoamine Transporter Blockade," *European Neuropsychopharmacology* 16, no. 7 (2006).

260 *the elite group of people who survive to very old age:* E. Naumova et al., "'Immunogenetics of Aging': Report on the Activities of the 15th International HLA and Immunogenetics Working Group and 15th International HLA and Immunogenetics Workshop," *Tissue Antigens* 77, no. 3 (2011).

260 *The same is true in Sicily:* Carmela Rita Balistreri et al., "LPS-Mediated Production of Pro/anti-inflammatory Cytokines and Eicosanoids in Whole Blood Samples: Biological Effects of +896A/G TLR4 Polymorphism in a Sicilian Population of Healthy Subjects," *Mechanisms of Ageing and Development* 132, no. 3 (2011).

260 *Among Ghanaians living in a highly infectious:* Maris Kuningas et al., "Selection for

Genetic Variation Inducing Pro-inflammatory Responses Under Adverse Environmental Conditions in a Ghanaian Population," *PLoS ONE* 4, no. 11 (2009).

261 *Among Italian, Greek, and Tunisian:* Lucia Malaguarnera et al., "Human Chitotriosidase Polymorphism Is Associated with Human Longevity in Mediterranean Nonagenarians and Centenarians," *Journal of Human Genetics* 55, no. 1 (2010).

261 *probiotics can correct the natural shift of the microbiota:* Yu-Rong Fu et al., "Effects of *Bifidobacterium bifidum* on Adaptive Immune Senescence in Aging Mice," *Microbiology and Immunology* 54, no. 10 (2010).

262 *and even acne:* Rural Africans don't get acne either. See A. A. Hogewoning et al., "Prevalence and Risk Factors of Inflammatory Acne Vulgaris in Rural and Urban Ghanaian Schoolchildren," *British Journal of Dermatology* 161, no. 2 (2009).

Chapter 13: The Hookworm Underground: Crowdsourcing "The Cure" in an Age of Immune Dysfunction

263 *The caution is understandable:* "Let them Eat Worms," *New Scientist,* August 3, 2011.

263 *For Jasper Lawrence, trouble began:* Based on author interviews. Jasper Lawrence has also told his story in various places on the Web. Here's one: http://wormtherapy.wordpress.com/2011/01/19/the-story-of-how-i-decided-to-infect-myself-with-hookworms-and-the-founding-of-autoimmune-therapies-2-2-2/. (Accessed January 1, 2012.)

264 *he posted online a version of the story:* Jasper Lawrence posted his Cameroon story at: http://www.kuro5hin.org/story/2006/4/30/91945/8971. (Accessed January 1, 2012.)

266 *ABC wrote about him on its website:* Lauren Cox, "Allergy Desperation: I'll Take a Parasite, Please," ABC News, http://abcnews.go.com/Health/AllergiesNews/story?id=8114307&page=1&singlePage=true#.TwMqUUoyc7g. (Accessed January 31, 2012.)

266 *Even the* New York Times *mentioned:* Elizabeth Svoboda, "David Pritchard: The Worms Crawl In," *New York Times,* July 1, 2008.

266 *And then near midday on November 3, 2009:* Date comes from FDA documents attained by FOI request.

267 *And tomorrow another favorable:* Tim Adams, "Gut Instinct: The Miracle of the Parasitic Hookworm," *Guardian/Observor*, May 22, 2010.

268 *First, I discovered that Lawrence had posed:* You can see Jasper Lawrence's ruse at: http://curezone.com/forums/am.asp?i=1025654&s=2#i21. (Accessed January 1, 2012.) He has used the username "FQ1513" here and elsewhere on the Web.

269 *The trial results, published in 2009:* J. R. Feary et al., "Experimental Hookworm Infection: A Randomized Placebo-controlled Trial in Asthma," *Clinical and Experimental Allergy* 40, no. 2 (2010).

272 *He heard Jasper Lawrence's story on Radiolab:* There are two Radiolab pieces on Jasper Lawrence. The first is as http://www.radiolab.org/2009/sep/07/sculptors-of-monumental-narrative/. (Accessed January 1, 2012.) The follow-up story is at http://www.radiolab.org/blogs/radiolab-blog/2010/apr/02/an-update-on-hookworms/. (Accessed January 1, 2012.)

278 *one could only wish for greater harmony:* FDA agents interviewed some of Lawrence's clients after he fled. One mentioned that he would have preferred buying an FDA-approved helminth, but there weren't any.

Chapter 14: Life with the American Murderer, and the Body's Largest Organ

280 *Life is naturally tattered:* M. Zuk, *Riddled with Life* (Orlando, Fla.: Harcourt, 2007)7.

283 *By directly examining live infection:* John Croese et al., "Allergy Controls the Population Density of *Necator americanus* in the Small Intestine," *Gastroenterology* 131, no. 2 (2006); John Croese and Richard Speare, "Intestinal Allergy Expels Hookworms: Seeing Is Believing," *Trends in Parasitology* 22, no. 12 (2006).

284 *Light, transitory infections may prime:* Scientists less frequently find *Ascaris lumbricoides* to be protective, and often find that it exacerbates allergy. See J. Calvert and P. Burney, "Ascaris, Atopy, and Exercise-induced Bronchoconstriction in Rural and Urban South African Children," *Journal of Allergy and Clinical Immunology* 125, no. 1 (2010). Dog-

adapted species that colonize humans reliably worsen allergy, perhaps because the infection is light and transient. E. Pinelli et al., "Infection with the Roundworm *Toxocara canis* Leads to Exacerbation of Experimental Allergic Airway Inflammation," *Clinical and Experimental Allergy* 38, no. 4 (2008).

285 *and a study wherein scientists kept children away:* R. S. Zeiger et al., "Effect of Combined Maternal and Infant Food-Allergen Avoidance on Development of Atopy in Early Infancy: A Randomized Study," *Journal of Allergy and Clinical Immunology* 84, no. 1 (1989).

285 *But the recommendations never sat well:* Much of Gideon Lack's story is told in Jerome Groopman, "The Peanut Puzzle," *The New Yorker,* February 2, 2011.

286 *Jews in Israel and London:* George du Toit et al., "Early Consumption of Peanuts in Infancy Is Associated with a Low Prevalence of Peanut Allergy," *Journal of Allergy and Clinical Immunology* 122, no. 5 (2008).

286 *Mothers who used these ointments:* Gideon Lack et al., "Factors Associated with the Development of Peanut Allergy in Childhood," *New England Journal of Medicine* 348, no. 11 (2003).

287 *food avoidance failed to curb the increase:* Bianca E. P. Snijders et al., "Age at First Introduction of Cow Milk Products and Other Food Products in Relation to Infant Atopic Manifestations in the First 2 Years of Life: The KOALA Birth Cohort Study," *Pediatrics* 122, no. 1 (2008).

287 *the American Academy of Pediatricians backtracked:* Ananth Thygarajan and Arvil W. Burks, "American Academy of Pediatrics Recommendations on the Effects of Early Nutritional Interventions on the Development of Atopic Disease," *Current Opinion in Pediatrics* 20, no. 6 (2008).

287 *The gene they pinpointed encoded a protein:* Irwin McLean tells much of the story in Irwin Mclean, "The Allergy Gene," *Scientist,* December 1, 2010.

288 *They were twice as likely to have allergies:* John Henderson et al., "The Burden of Disease Associated with Filaggrin Mutations: A Population-based, Longitudinal Birth Cohort Study," *Journal of Allergy and Clinical Immunology* 121, no. 4 (2008).

288 *In Denmark, the gene predicted:* K. Bonnelykke et al., "Filaggrin Gene Variants and Atopic Diseases in Early Childhood Assessed Longitudinally from Birth," *Pediatric Allergy and Immunology* 21, no. 6 (2010).

288 *Japanese populations had a different mutation:* Toshifumi Nomura et al., "Specific Filaggrin Mutations Cause Ichthyosis Vulgaris and Are Significantly Associated with Atopic Dermatitis in Japan," *Journal of Investigative Dermatology* 128, no. 6 (2008).

288 *carriers of these genes:* Alan Irvine, Irwin McLean, and Donald Leung, "Filaggrin Mutations Associated with Skin and Allergic Diseases," *New England Journal of Medicine* 365 (2011).

288 *the risk of peanut allergies:* S. J. Brown et al., "Loss-of-function Variants in the Filaggrin Gene Are a Significant Risk Factor for Peanut Allergy," *Journal of Allergy and Clinical Immunology* 127, no. 3 (2011).

288 *experiments firmly established:* A. de Benedetto et al., "Tight Junction Defects in Patients with Atopic Dermatitis," *Journal of Allergy and Clinical Immunology* 127, no. 3 (2011).

288 *Just applying egg protein:* Padraic G. Fallon et al., "A Homozygous Frameshift Mutation in the Mouse Flg Gene Facilitates Enhanced Percutaneous Allergen Priming," *Nature Genetics* 41, no. 5 (2009).

288 *"stripping" rodent skin with adhesive tape:* J. Strid et al., "Epicutaneous Exposure to Peanut Protein Prevents Oral Tolerance and Enhances Allergic Sensitization," *Clinical and Experimental Allergy* 35, no. 6 (2005). Thierry Olivry et al., "Stratum Corneum Removal Facilitates Experimental Sensitization to Mite Allergens in Atopic Dogs," *Veterinary Dermatology* 22, no. 2 (2011).

288 *McLean and Irvine hypothesize::* Alan D. Irvine and W. H. Irwin McLean, "Breaking the (Un)sound Barrier: Filaggrin Is a Major Gene for Atopic Dermatitis," *Journal of Investigative Dermatology* 126, no. 6 (2006).

289 *But is the deciding factor:* Sarita Sehra et al., "IL-4 Regulates Skin Homeostasis and the Predisposition Toward Allergic Skin Inflammation," *Journal of Immunology* 184, no. 6 (2010); Michael D. Howell et al., "Cytokine Modulation of Atopic Dermatitis Filaggrin Skin Expression," *Journal of Allergy and Clinical Immunology* 124, no. 3 (2009).

290 *Farming protects against eczema:* Carsten Flohr and Lindsey Yeo, "Atopic Dermatitis and the Hygiene Hypothesis Revisited," *Current Problems in Dermatology* 41 (2011).

290 *German schoolchildren who were positive:* O. Herbarth et al., "*Helicobacter pylori* Colonisation and Eczema," *Journal of Epidemiology and Community Health* 89

290 *less-diverse microbiota at a younger age:* Sunia Foliaki et al., "Antibiotic Use in Infancy and Symptoms of Asthma, Rhinoconjunctivitis, and Eczema in Children 6 and 7 Years Old: International Study of Asthma and Allergies in Childhood Phase III," *Journal of Allergy and Clinical Immunology* 124, no. 5 (2009).

290 *Mothers who regularly tend:* C. Roduit et al., "Prenatal Animal Contact and Gene Expression of Innate Immunity Receptors at Birth Are Associated with Atopic Dermatitis," *Journal of Allergy and Clinical Immunology* 127, no. 1 (2011).

290 *results of the largest placebo-controlled:* C. Flohr et al., "Reduced Helminth Burden Increases Allergen Skin Sensitization but Not Clinical Allergy: A Randomized, Double-blind, Placebo-controlled Trial in Vietnam," *Clinical and Experimental Allergy* 40, no. 1 (2010).

Chapter 15: The Collapse of the Superorganism, and What to Do About It

292 *When we try to pick out:* J. Muir, *My First Summer in the Sierra* (Houghton Mifflin, 1911), 211. Thanks to the filmmaker Sharon Shattuck for bringing it to my attention.

293 *your immune system grew better:* For a discussion on developing immunity to gut parasites, see Maria Yazdanbakhsh and David L. Sacks, "Why Does Immunity to Parasites Take So Long to Develop?," *Nature Reviews Immunology* 10, no. 2 (2010).

293 *Roughly 25 percent of you died:* I gave a slightly optimistic number for premodern infant mortality, based on numbers from eighteenth-century Sweden, the longest dataset available. http://www.mortality.org/hmd/SWE/STATS/cMx_5x10.txt. (Accessed January 1, 2012.)

293 *You reached puberty later:* Some peoples who live in infectious environments, such as the pygmy of central African rain forest, reach puberty sooner than other peoples. This is one strategy for coping with a highly infectious environment: Have children as soon as you can. But that's an exception to the rule. For more on the trade-offs between growth and immune-system development, see Robert Walker et al., "Growth Rates and Life Histories in Twenty-two Small-scale Societies," *American Journal of Human Biology* 18, no. 3 (2006).

293 *You were probably shorter:* Today, rural Nigerians exposed to parasites don't necessarily end up shorter, but they grow more gradually. Omolola Ayoola et al., "Relative Height and Weight Among Children and Adolescents of Rural Southwestern Nigeria," *Annals of Human Biology* 36, no. 4 (2009).

294 *The tuberculosis epidemics:* J. M. Grange et al., "Historical Declines in Tuberculosis: Nature, Nurture and the Biosocial Model," *International Journal of Tuberculosis and Lung Disease* 5, no. 3 (2001).

295 *The polio epidemics that began:* For more on polio, see N. Nathanson and O. M. Kew, "From Emergence to Eradication: The Epidemiology of Poliomyelitis Deconstructed," *American Journal of Epidemiology* 172, no. 11 (20110). For a mid-twentieth-century study on polio incidence according to class in Morocco, see J. R. Paul and D. M. Horstmann, "A Survey of Poliomyelitis Virus Antibodies in French Morocco," *American Journal of Tropical Medicine and Hygiene* 4, no. 3 (1955).

296 *The more people use it:* J. H. Savage et al., "Triclosan, a Common Ingredient in Household Products, Is Associated with Allergic Sensitization," *Journal of Allergy and Clinical Immunology* 129, no. 2 (2012).

296 *In rats, for example, triclosan:* K. B. Paul et al., "Developmental Triclosan Exposure Decreases Maternal and Neonatal Thyroxine in Rats," *Environmental Toxicology and Chemistry* 29, no. 12 (2010).

296 *It has been linked to asthma:* E. M. Clayton et al., "The Impact of Bisphenol A and Triclosan on Immune Parameters in the U.S. Population, NHANES 2003–2006," *Environmental Health Perspectives* 119, no. 3 (2011); J. M. Braun et al., "Impact of Early-life Bisphenol A Exposure on Behavior and Executive Function in Children," *Pediatrics* 128, no. 5 (2011).

296 *Children exposed in utero to excess BPA:* Adam J. Spanier et al., "Prenatal Bisphenol A Is a Risk Factor for Early Transient Wheeze" (presented at the Pediatric Academic Societies Annual Meeting, 2011).

296 *Mice prenatally exposed develop:* T. Midoro-Horiuti et al., "Maternal Bisphenol A Exposure Promotes the Development of Experimental Asthma in Mouse Pups," *Environmental Health Perspectives* 118, no. 2 (2010).

296 *Female rats develop inflamed guts:* V. Braniste et al., "Impact of Oral Bisphenol A at Reference Doses on Intestinal Barrier Function and Sex Differences After Perinatal Exposure in Rats," *Proceedings of the National Academy of Sciences* 107, no. 1 (2010).

296 *at this point, more than twenty studies:* Christie Aschwanden, "Studies Suggest an Acetaminophen-Asthma Link," *New York Times*, December 19, 2011.

296 *Prenatal exposure seems most important:* Richard W. Beasley et al., "Acetaminophen Use and Risk of Asthma, Rhinoconjunctivitis, and Eczema in Adolescents: International Study of Asthma and Allergies in Childhood Phase Three," *American Journal of Respiratory and Critical Care Medicine* 183, no. 2 (2011); J. T. McBride, "The Association of Acetaminophen and Asthma Prevalence and Severity," *Pediatrics* 128, no. 6 (2011).

297 *certain genotypes appear more sensitive:* Matthew S. Perzanowski et al., "Prenatal Acetaminophen Exposure and Risk of Wheeze at Age 5 Years in an Urban Low-income Cohort," *Thorax* 65, no. 2 (2010); Seif O. Shaheen et al., "Prenatal and Infant Acetaminophen Exposure, Antioxidant Gene Polymorphisms, and Childhood Asthma," *Journal of Allergy and Clinical Immunology* 126, no. 6 (2010).

297 *Others suspect acetaminophen in autism:* K. Becker and S. Schultz, "Similarities in Features of Autism and Asthma and a Possible Link to Acetaminophen Use," *Medical Hypotheses* (2009).

297 *Perzanowski and colleagues:* J. E. Sordillo et al., "Multiple Microbial Exposures in the Home May Protect Against Asthma or Allergy in Childhood," *Clinical and Experimental Allergy* 40, no. 6 (2010).

298 *How does this finding jibe:* See chapters by Kathleen Barnes and Hurtado in Wenda Trevathan, Euclid O. Smith, and James J. McKenna, *Evolutionary Medicine* (New York: Oxford University Press, 1999). For more on human evolution, allergy, worm infection in the inner city, and asthma in animals, see chapters 5 and 9.

298 *they don't receive any calming signals:* Peter Hotez at the Sabin Vaccine Institute would dispute this. He thinks parasites cause allergy in the U.S. inner city—in this case, a worm native to dogs called *Toxocara canis*. Its eggs can be inhaled, and he's found that urban dwellers seem to have been exposed to it much more often. Still, the worm theory of allergic disease doesn't say any old parasite can protect. Only human-adapted species can tweak our immune system properly. And only chronic infections achieve the necessary immune modulation. For *Toxocara canis* exacerbating asthma, see E. Pinelli et al., "Infection with the Roundworm *Toxocara canis* Leads to Exacerbation of Experimental Allergic Airway Inflammation," *Clinical and Experimental Allergy* 38, no. 4 (2007). For a discussion by Hotez on worm infections in the U.S. inner city, and *T. canis* in particular, see P. J. Hotez, "Neglected Infections of Poverty in the United States of America," *PLoS Neglected Tropical Diseases* 2, no. 6 (2008).

298 *gene variants that predispose to both autoimmune:* Some more examples: Candelaria Vergara et al., "African Ancestry Is Associated with Risk of Asthma and High Total Serum IgE in a Population from the Caribbean Coast of Colombia," *Human Genetics* 125, nos. 5–6 (2009). Some West African groups have an enrichment of gene variants associated with autoimmune disease. See Patrizia Lulli et al., "HLA-DRB1 and -DQB1 Loci in Three West African Ethnic Groups: Genetic Relationship with Sub-Saharan African and European Populations," *Human Immunology* 70, no. 11 (2009).

298 *The surging allergy prevalence:* P. J. Cooper et al., "Asthma in Latin America: A Public Heath Challenge and Research Opportunity," *Allergy* 64, no. 1 (2009).

299 *And other parts of Africa:* Emmanuel O. D. Addo-Yobo et al., "Exercise-induced Bronchospasm and Atopy in Ghana: Two Surveys Ten Years Apart," *PLoS Medicine* 4, no. 2 (2007). Also see the most recent ISAAC surveys: Heather J. Zar et al., "The Changing Prevalence of Asthma, Allergic Rhinitis and Atopic Eczema in African Adolescents from 1995 to 2002," *Pediatric Allergy and Immunology* 18, no. 7 (2007).

299 *younger generations in former Eastern Bloc:* Triine Annus et al., "Modest Increase in Seasonal Allergic Rhinitis and Eczema Over 8 Years Among Estonian Schoolchildren," *Pediatric Allergy and Immunology* 16, no. 4 (2005).

299 *compared with chimp T cells, human T cells:* Ajit Varki, "Colloquium Paper: Uniquely Human Evolution of Sialic Acid Genetics and Biology," *Proceedings of the National Academy of Sciences* 107, Suppl. 2 (2010).

300 *Cats got asthma and colitis:* J. S. Suchodolski, "Microbes and Gastrointestinal Health of Dogs and Cats," *Journal of Animal Science* (2010).

300 *Horses could also get inflammatory bowel disease:* Darien J. Feary and Diana M. Hassel, "Enteritis and Colitis in Horses," *Veterinary Clinics of North America Equine Practice* 22, no. 2 (2006).

300 *One scientist compared the IgE levels:* Anna Ledin et al., "High Plasma IgE Levels within the Scandinavian Wolf Population, and Its Implications for Mammalian IgE Homeostasis," *Molecular Immunology* 45, no. 7 (2008).

300 *Horses even had their own:* Eman Hamza et al., "Increased IL-4 and Decreased Regulatory Cytokine Production Following Relocation of Icelandic Horses from a High to Low Endoparasite Environment," *Veterinary Immunology and Immunopathology* 133, no. 1 (2010).

300 *UC Davis scientists treated:* G. M. Halpern et al., "Diagnosis of Inhalant Allergy in a Chimpanzee Using 'in Vivo' and 'in Vitro' Tests," *Allergologia et Immunopathologia* 17, no. 5 (1989).

300 *that a female gorilla developed:* M. Hog and I. Schindera, "Neurodermatitis in Primates: A Case Description of a Female Gorilla," *Hautarzt; Zeitschrift für Dermatologie, Venerologie, und verwandte Gebiete* 40, no. 3 (1989).

301 *a captive gorilla developed:* F. Lankester et al., "Fatal Ulcerative Colitis in a Western Lowland Gorilla (*Gorilla gorilla gorilla*)," *Journal of Medical Primatology* 37, no. 6 (2008).

301 *spontaneous colitis was a significant problem:* C. A. Rubio and G. B. Hubbard, "Chronic Colitis in *Macaca fascicularis*: Similarities with Chronic Colitis in Humans," *In Vivo* 16, no. 3 (2002); C. A. Rubio and G. B. Hubbard, "Chronic Colitis in Baboons: Similarities with Chronic Colitis in Humans," *In Vivo* 15, no. 1 (2001).

301 *Mustached tamarins, small South American monkeys:* Alfonso S. Gozalo et al., "Pathology of Captive Moustached Tamarins (*Saguinus mystax*)," *Comparative Medicine* 58, no. 2 (2008).

301 *monkeys who'd lived their life in captivity:* H. Uno et al., "Colon Cancer in Aged Captive Rhesus Monkeys (*Macaca mulatta*)," *American Journal of Primatology* 44, no. 1 (1998).

301 *scientists went so far as to survey:* J. D. Wood et al., "Colitis and Colon Cancer in Cotton-top Tamarins (*Saguinus oedipus oedipus*) Living Wild in Their Natural Habitat," *Digestive Diseases and Sciences* 43, no. 7 (1998).

301 *also had a different microflora:* Gentaro Uenishi et al., "Molecular Analyses of the Intestinal Microbiota of Chimpanzees in the Wild and in Captivity," *American Journal of Primatology* 69, no. 4 (2007).

301 *individuals with colitis had:* Philip McKenna et al., "The Macaque Gut Microbiome in Health, Lentiviral Infection, and Chronic Enterocolitis," *PLoS Pathogens* 4, no. 2 (2008).

301 *Japanese macaques in one captive colony:* Michael K. Axthelm et al., "Japanese Macaque Encephalomyelitis: A Spontaneous Multiple Sclerosis–like Disease in a Nonhuman Primate," *Annals of Neurology* 70, no. 3 (2011).

302 *But in 2010, scientists from Harvard:* Joshua Kramer et al., "Alopecia in Rhesus Macaques Correlates with Immunophenotypic Alterations in Dermal Inflammatory Infiltrates Consistent with Hypersensitivity Etiology," *Journal of Medical Primatology* 39, no. 2 (2010).

303 *"Resolving these questions is of critical importance":* James E. Gern, "Barnyard Microbes and Childhood Asthma," *New England Journal of Medicine* 364, no. 8 (2011).

307 *"Darwinian medicine":* G. C. Williams and R. M. Nesse, "The Dawn of Darwinian Medicine," *Quarterly Review of Biology* 66, no. 1 (1991).

Selected Bibliography

The following titles provided background, directly influenced how I thought about this book, or served as models of style and approach.

Ashenburg, Katherine. *The Dirt on Clean: An Unsanitized History*. New York: North Point Press, 2007.

Buckman, Rob. *Human Wildlife That Lives on Us*. Baltimore: The Johns Hopkins University Press, 2003.

Clark, William R. *In Defense of Self: How the Immune System Really Works*. New York: Oxford University Press, 2008.

Crawford, Dorothy H. *Deadly Companions: How Microbes Shaped Our History*. Oxford and New York: Oxford University Press, 2007.

Holland, C., and M. W. Kennedy. *The Geohelminths: Ascaris, Trichuris, and Hookworm*. New York: Kluwer Academic Publishers, 2002.

Hoy, Suellen M. *Chasing Dirt: The American Pursuit of Cleanliness*. New York: Oxford University Press, 1995.

Jackson, Mark. *Allergy: The History of a Modern Malady*. London: Reaktion, 2006.

Kaplan, E. H. *What's Eating You?: People and Parasites*. Princeton, NJ: Princeton University Press, 2010.

Kruif, P. D., and F. Gonzalez-Crussi. *Microbe Hunters*. New York: Harcourt Brace, 2002.

Lenton, Tim, and A. J. Watson. *Revolutions That Made the Earth*. Oxford and New York: Oxford University Press, 2011.

McFall-Ngai, Margaret J., Brian Henderson, and Edward G. Ruby. *The Influence of Cooperative Bacteria on Animal Host Biology* (Advances in Molecular and Cellular Microbiology). Cambridge and New York: Cambridge University Press, 2005.

McNeill, William Hardy. *Plagues and Peoples*. New York: Anchor Books, 1989.

Melosi, Martin V. *Garbage in the Cities: Refuse, Reform, and the Environment. History of the Urban Environment*. Rev. ed. Pittsburgh: University of Pittsburgh Press, 2005.

Nakazawa, Donna Jackson. *The Autoimmune Epidemic: Bodies Gone Haywire in a World Out of Balance—and the Cutting-Edge Science That Promises Hope*. New York: Simon & Schuster, 2008.

Patterson, P. H. *Infectious Behavior: Brain-Immune Connections in Autism, Schizophrenia, and Depression*. Cambridge, MA: MIT Press, 2011.

Pollard, Tessa M. *Western Diseases: An Evolutionary Perspective* (Cambridge Studies in Biological and Evolutionary Anthropology). Cambridge and New York: Cambridge University Press, 2008.

Poulin, Robert. *Evolutionary Ecology of Parasites*. 2nd ed. Princeton, NJ: Princeton University Press, 2007.

Ridley, Matt. *The Red Queen: Sex and the Evolution of Human Nature*. New York: Perennial, 2003.

Rook, G. A. W. *The Hygiene Hypothesis and Darwinian Medicine* (Progress in Inflammation Research). Basel and Boston: Birkhäuser, 2009.

Ruebush, Mary. *Why Dirt Is Good: 5 Ways to Make Germs Your Friends*. New York: Kaplan, 2009.

Sachs, Jessica Snyder. *Good Germs, Bad Germs: Health and Survival in a Bacterial World*. New York: Hill and Wang, 2007.

Sapp, Jan. *Evolution by Association: A History of Symbiosis*. New York: Oxford University Press, 1994.

Trevathan, Wenda, Euclid O. Smith, and James J. McKenna. *Evolutionary Medicine*. New York: Oxford University Press, 1999.

Zimmer, Carl. *Parasite Rex: Inside the Bizarre World of Nature's Most Dangerous Creatures*. New York: Free Press, 2000.

Zuk, M. *Riddled with Life: Friendly Worms, Ladybug Sex, and the Parasites That Make Us Who We Are*. Orlando, Fla.: Harcourt, 2007.

Glossary

Adaptive immunity: The wing of the immune system that learns and remembers. When you receive a measles vaccine, say, you're teaching the adaptive immune system to recognize the measles virus and to remember it. Only jawed vertebrates (from fish to humans) have an adaptive immune system, which makes it a relatively recent innovation in the evolution of multicellular organisms. However, many more invertebrates than vertebrates live on this planet, so clearly an adaptive immune system isn't required to survive.

Allele: A version or variant of a gene. Consider human skin pigmentation: In the human family, we see very dark skin, very light skin, and everything in between. This great variation stems from different versions of the same pigmentation genes.

Allergy: When the immune system turns on an innocuous protein, such as tree pollen, cat dander, or peanuts, with a ferocity that's disproportionate to the threat posed. Allergic disease can manifest as food allergies, eczema, asthma, hay fever, and hives, among other symptoms. Severe anaphylactic shock, an extreme allergic reaction, can kill. In the developed world, the incidence of allergic disease increased precipitously during the late twentieth century.

Antibody: Y-shaped molecules manufactured by B cells. The forked end binds to the targeted substance, such as parasites or bacterial invaders. The back end fits into receptors on white blood cells. Many different antibody classes exist—e.g., IgA, IgE, IgG, IgM—each one calling for a slightly different type of immune response.

Asthma: A chronic inflammatory disease of the lungs characterized by reversible airway obstruction. Symptoms include wheezing, difficulty breathing, and low blood oxygen. In the long term, chronic inflammation can cause permanent thickening of the bronchial walls, restricting the passage of air in the lungs. Some cases of asthma are allergic; they're triggered by cat dander or dust mites. However, other cases aren't apparently allergic. The inflammation occurs without obvious triggers.

Autoimmunity: When the immune system turns on the body's own tissues and damages or destroys them. There are currently between eighty and one hundred identified autoimmune diseases. The incidence of many autoimmune disorders increased in the late twentieth century.

B cell: A lymphocyte of the adaptive immune system that produces antibodies. The antibodies are specific to certain substances. When you receive the measles vaccine, say, your immune system creates long-lived memory B cells specific to the measles virus. When these memory cells encounter measles proteins in the future, they spring to life and begin churning out antibodies. You fend off the invasion without illness. B cells originate in bone morrow.

Commensal: An organism that lives with, or inside, another organism without causing harm and without obviously enhancing host fitness. Many of the bacteria in our gut are considered commensal. Of course, simply occupying space without causing harm can confer benefits: it prevents worse opportunists from gaining purchase. Comes from the Latin for "sharing a table"—*com + mensa.*

Cytokine: A molecule white blood cells use to signal one another. There are many kinds. Depending on what type a lymphocyte produces, other cells will respond differently—e.g., attack or hold back. For the sake of simplicity, here they're mostly categorized as anti-inflammatory or pro-inflammatory.

Dendritic cell: A cell of the innate immune system with long armlike protrusions, or dendrites. They populate our skin and our gut, among other places. They inhabit the front line of the interface with the microbial world. They're important in activating the adaptive immune system. When they consume invaders, they chop them up and present bits and pieces of the interlopers to cells of the adaptive immune system (T and B cells). These cells then know what to seek. Importantly, dendritic cells can also tell adaptive immune cells *not* to pursue a given substance. They can induce regulatory cells primed for tolerance.

Dysregulation: The immune system exists in a state of dynamic balance. Opposing signals, pro- and anti-inflammatory, maintain the equilibrium.

Dysregulation occurs when one component of the balance weakens, unleashing the countervailing force. In disorders covered in this book, usually the anti-inflammatory signals have atrophied, giving pro-inflammatory tendencies free rein.

Endotoxin: A substance in the outer walls of Gram-negative bacteria. The innate immune system recognizes endotoxin immediately, and responds with inflammation. Some amount of exposure to endotoxin may be beneficial. Too much can cause septic shock. Low-grade immune activation by environmental bacteria may protect against allergies. Endotoxin levels are sometimes used as shorthand for bacterial load in the environment.

Genotype: Depending on what version of a given gene you have, that's your genotype. The genes of the human immune system are remarkably diverse. No individual will respond quite the same way to immune stimulus. According to genotype, some may respond with more anti-inflammatory IL-10 when exposed to environmental bacteria, say, and some with less. Consequently, one genotype may benefit from working in the cowshed, and another may suffer deleterious consequences.

Germ theory: The idea that microbes, not bad smells or miasma, cause disease. It was revolutionary in its time, and gained widespread acceptance only at the end of the nineteenth century. Germ theory is a cornerstone of modern allopathic medicine.

Helminth: A general term for parasitic worms that live inside their hosts. Three major groups infect humans, each representing a separate invasion deep in the past: cestodes (tapeworms), nematodes (roundworms), and trematodes (flukes). This book is mostly concerned with geohelminths, nematodes whose eggs require incubation in the soil to become infective.

Hookworm: A parasitic roundworm. The microscopic infective larva burrows through its victim's skin, follows the venous blood flow through the heart to the lungs, pops out, migrates over the pharynx, passes through the stomach, and latches on to the small intestine, where it grazes on tissue and mates. Fertilized eggs pass out with the host's stool. The eggs must spend a week or more in soil of the right humidity and temperature to molt into infective larvae. Two hookworm species generally infest humans, *Ancylostoma duodenale* and *Necator americanus*. Although there's considerable overlap, *Ancylostoma* tends to occur in subtropical climes and *Necator* in tropical. *Ancylostoma,* which is larger and can spread via breast milk, is considered the more pathogenic of the two. Hookworm gets its name from the hooklike shape of adults. Mouth parts are at the "sharp" end of the hook.

Host: A very general term for one organism that houses another, be it a parasite, commensal, or mutualist.

Hygiene hypothesis: A collection of observations suggesting that exposure to microbes prevents allergic disease. As a concept, it's usually traced back to a 1989 paper by the epidemiologist David Strachan. He proposed that childhood infections prevented allergy. That view has since been supplanted, however. Now scientists think abundant early-life exposure to innocuous environmental microbes protects children from developing allergic disease. Arguing that the emphasis on "hygiene" is inaccurate, some have proposed variations on the theme, including the "old friends" hypothesis, the "microbial deprivation" hypothesis, and the "disappearing microbiota" hypothesis.

Immunoglobulin: See *Antibody.*

Immunoglobulin-E (IgE): The "allergic" antibody. Levels of IgE are generally elevated in allergic diseases, such as hay fever or food allergies. IgE is also elevated during worm infection, often in the absence of allergic disease. Many think that IgE evolved to repel parasites, and that its contribution to allergic disease in modernity is an accident. In allergy, the antibody prompts increased blood flow, mucus production, and swelling of mucous membranes. In worm infection, IgE helps expel the parasite.

Inflammatory bowel disease (IBD): Unexplained inflammation of the gut. One theory holds that IBD results from a loss of tolerance to our harmless commensal microbes. There are two types: ulcerative colitis, which affects the colon; and Crohn's disease, which usually strikes the small intestine, but can afflict any portion of the gut from beginning to end. The prevalence of IBD increased dramatically during the latter twentieth century.

Innate immunity: The wing of the immune system that knows what to seek without instruction or training. Unlike the adaptive immune system, it doesn't need to learn. Cells of the innate immune system are born knowing. They possess sensors that have evolved over many millions of years to recognize conserved molecular patterns on bacteria, viruses, and parasites.

Interleukin: A class of immune-system signaling molecules. Some interleukins call for inflammation, others for calm. Some prompt an antiparasite immune response, others an antiviral or antibacterial response. One interleukin that receives a lot of attention here is interleukin-10, which is important in quashing inflammation. Another is tumor necrosis factor alpha (TNF alpha), which prompts inflammation.

Lipopolysaccharide: See *Endotoxin.*

Lymphocyte: A type of white blood cell. They congregate in lymph nodes. They learn and remember. They come in a dizzying variety, each with a slightly different function. They defend the body against invaders and they also communicate with our resident microbes.

Macrophage: A bloblike cell from the innate immune system. Unlike T cells or B cells, macrophages don't need to learn what to pursue. They automatically seek certain patterns common to intruders. Comes from the Greek for "large" + "eater."

Microbe: A microorganism, including viruses, bacteria, yeast, amoebas, and more. Usually single-celled organisms. Originates from the Greek via French, *mikros* and *bios*—"small" + "life."

Microbiota (Microbiome): Any community of microbes, but as defined in this book, usually the communities that inhabit our bodies. Scientists find microbes on every bodily surface—gut, lungs, and skin. The bulk of our native microbes live in the colon, the final loop of intestine before the exit.

Multiple sclerosis: A degenerative autoimmune disease of the central nervous system. The immune system attacks myelin, the fatty coating on neurons. As myelin degrades, the neurons cannot transmit signals. Symptoms include limb weakness, blurry vision, and, in advanced cases, difficulty breathing.

Mutualist: An organism whose relationship with another organism results in mutual benefits. However, the two organisms don't necessarily need one another to live.

Parasite: An organism that requires another organism to complete its own life cycle, but without the second organism benefitting. Parasites are usually smaller than their hosts, but not always. The cuckoo, which tricks other birds into rearing its young, is larger than its host. Some ant species enslave other ants, one colony parasitizing another. Viruses are also parasites: they hijack the host's cellular machinery to replicate. There are obligate parasites that require a host to survive, and nonobligate parasites, organisms such as *Vibrio cholerae,* the bacterium that causes cholera, which sometimes inhabit hapless humans but can also live freely. There are mutualists that can edge toward parasitism. Take the oxpecker, often seen perched on African wildebeests' backs. They eat biting insects from hard-to-reach places, helping their ungulate host, but every so often, they peck at the wildebeest itself—parasitism. At the other extreme are parasites that edge toward mutualism, like some of the helminths discussed in this book. The Greek *parasitos* means "eating at another's table."

Pathogen: An infective organism, such as the smallpox or measles virus, that causes disease. A microbe's pathogenicity is a measure of its cost to the host. Some, e.g., the common cold virus, have a relatively light impact. Others, such as *Vibrio cholerae,* which causes cholera, are heavy-handed and deadly. Yet others, such as the diarrhea-causing *Clostridium difficile* or the yeast *Candida albicans,* can inhabit us as commensals without causing harm, but then suddenly, often when the greater microbial ecosystem has been disturbed, become pathogenic.

Regulatory T cell (T-reg): These T cells are critical to restraining the immune system when necessary, which is quite often. They permit us to quash inflammation when it's no longer useful—during healing, for example. They allow us to hold fire when encountering innocuous proteins in the environment, such as cat hair. They enable us to tolerate our resident microbes. And they keep us from turning on our own tissues—autoimmune disease.

Symbiosis: One or more organisms that live in close contact with others. Scientists still debate whether symbiosis implies that they must live together, and if it includes commensals, mutualists, and parasites. In colloquial usage, it has come to mean mutualist—both parties benefit. That's generally how I use it. Some of our resident microbes, for example, help us digest otherwise indigestible plant fibers and sugars. In exchange, we give them a home that's moist, mostly anaerobic, and temperature-controlled. We also supply them with nutrients.

T cell: A type of lymphocyte from the adaptive immune system. Typically, it has receptors specific to just one substance. T cells mature in the thymus, a nut-sized organ just beneath the sternum.

T-helper-1 (Th1): Depending on what they encounter, T cells call for specific responses from the immune-system repertoire. They "help" orchestrate the counterattack, or, as the case may be, the peacekeeping. The Th1 response is usually directed at bacteria and viruses. It's cell-mediated, meaning your immune cells directly confront the invaders and destroy them. In the context of the hygiene hypothesis, stimulating Th1 early in life is thought to prevent allergies, an out-of-control Th2 response. But in the context of autoimmune diseases, an excessive Th1 response is seen as problematic.

T-helper-2 (Th2): T cells that call for a response against worms and other large parasites, such as mosquitoes, lice, and ticks. The Th2 response—runny nose, mucus production, swelling—helps repel these big-bodied invaders. Where the Th1 response is cell-mediated, the Th2 response is considered humoral, meaning that it's mediated by antibodies. Allergic diseases, like hay fever or eczema, are said to result from a dysregulated Th2 response. But in the context of chronic worm infection—what's called a modified Th2 response—allergies don't necessarily occur. A modified Th2 response is considered remarkably tolerant.

Toll-like receptor (TLR): Microbial sensors on white blood cells of the innate immune system. From the day you're born, they're capable of recognizing patterns common to microbes and parasites. Legend has it that the German researcher who, in 1985, first discovered the toll-like receptor in fruit flies exclaimed, *"Das war ja toll!"* which means, "That was weird!" Thus the name "toll." Mammals have at least eleven toll-like receptors.

Whipworm: A parasitic roundworm that inhabits the colon. The human-adapted species, *Trichuris trichiura,* threads its frontal portion into the villi, the fingerlike protrusions of the large intestine. To become infective, eggs must embryonate for between two weeks to a month in soil of the right temperature and humidity. Adults live for one year and sometimes longer. According to the CDC, 800 million people carry whipworm, mostly in the developing world.

White blood cell: The cells that defend us from invaders, and also interface with our resident microbial communities. As a group, they comprise cells of both the adaptive and innate immune systems.

Index

INDEX

pinworn as protecting from, 249
pollution thought to cause, 100
poverty and, 101
prenatal origins of, 130, 135, 136–37,
    139–40, 228, 241
preterm births and, 135
prevention vs. treatment of, 142
probiotics as cause of, 123
prospect of vaccine against, 103
rate of, 5
recent rise in, 7
and regulatory T cells, 89
stress and, 19
TB as preventing, 111
T cells and, 299
vitamin D deficiency blamed for, 115
wealth and, 8
as worm adaptation gone awry, 93–94, 95
asthmahookworm.com, 265
astrocytes, 222
Atarashi, Koji, 172
Athens, plague in, 30
Attila the Hun, 31
aurochs, 29
Australia, 8, 21, 84–85, 123, 136–37, 140,
    151, 277
Australopithecine, 22
Austria, 105, 107
autism, 214–41, 300
    as allegedly caused by MMR vaccine,
        218–19, 227
    autoimmune disease and, 220–21, 223–24,
        230–31, 236
    cerebral development in, 221–25
    fevers as improving symptoms of, 231–33
    immune dysregulation and, 223
    infections and, 228, 343
    inflammation and, 222–25, 240–41
    junk food cravings and, 239
    and problems of gut, 237
    rise in rates of, 217–18, 219, 225, 231
    schizophrenia's similarity to, 227–28
    worms and, 213, 215, 233–36, 262, 303
autism spectrum disorders, 217
autoimmune arthritis, *see* rheumatoid
    arthritis
autoimmune cells, 203
    post-birth wave of, 15
autoimmune diabetes, 6, 220
autoimmune diseases, 260, 265
    autism and, 220–21, 223–24, 230–31, 236
    causes of, 16
    cost of, 6
    and critiques of modernity, 276
    depression and, 258–59
    and Fulani's resistance to malaria, 59

GDP and, 8
genes for, 56–57, 60
gene variants and, 28, 60, 220–21, 298,
    304
H. pylori as protecting against, 161
immune dysregulation and, 223
infections and, 8
Karelia, 114–17
and lining of intestines, 136
"living environment" and, 12
outdated theory on, 203
as prevented by parasites, 2
as rare in Tsimane, 10
rates of, 6
recent rise in, 5, 224
TNF alpha as feature of, 53
in women, 6
worm as treatment for, 61, 194, 200, 244,
    285
    see also specific diseases
autoimmune myasthenia gravis, 225
autoimmune response, of mothers, and
    autism in children, 220–21, 223–24,
    227–30
autoimmunetherapies.com, 265
autoimmunity, 128

baboons, 23, 301
Bach, Jean-François, 7, 8, 228, 292
bacillus Calmette-Guérin (BCG), 61, 111–12,
    209, 253, 294
bacteria, 15, 106
    need for early-life encounters with, 108
    phyla of, 168
    ulcers caused by, 144–46, 154, 155, 156
    in water, 12
bacterial pathogens, 27
bacteriophages, 279
bacterium, 111
Bacteroides fragilis, 121, 169, 170, 171
bacteroidetes, 172, 182, 184, 239
banana hypothesis, 159
bananas, 28
Bangladesh, 256
Banjul, 82
Bankes, Isabella, 65, 66, 69
Banks, Tyra, 265
Bantu, 151, 196, 197
Barbados, 125
Barker, David J. P., 129, 131
barley, 28, 180, 183
barn dust, 117, 305
Barr, Yvonne, 202
Barreiro, Luis, 57
Barrett's esophagus, 150
Barton, Erik, 207–8, 209

367

Moises Velasquez-Manoff has covered science and the environment for *The Christian Science Monitor.* His work has also appeared in *The New York Times Magazine, Chicago Tribune,* and *The Indianapolis Star,* among other publications. He holds a master of arts, with a concentration in science writing, from the Columbia Graduate School of Journalism.